BARRY'S ADVANCED CONSTRUCTION OF BUILDINGS

BARRY'S ADVANCED CONSTRUCTION OF BUILDINGS

Fifth Edition

Stephen Emmitt
University of Bath
UK

WILEY Blackwell

This edition first published 2023
© 2023 John Wiley & Sons Ltd

Edition History
Blackwell Publishing (1e, 2005), Wiley Blackwell (2e, 2010; 3e, 2014; 4e, 2019)

Registered Offices
John Wiley & Sons, Inc., 111 River Street, Hoboken, NJ 07030, USA
John Wiley & Sons Ltd, The Atrium, Southern Gate, Chichester, West Sussex, PO19 8SQ, UK

For details of our global editorial offices, customer services, and more information about Wiley products visit us at www.wiley.com.

Wiley also publishes its books in a variety of electronic formats and by print-on-demand. Some content that appears in standard print versions of this book may not be available in other formats.

Library of Congress Cataloging-in-Publication Data

Names: Emmitt, Stephen, author.
Title: Barry's advanced construction of buildings / Stephen Emmitt.
Other titles: Advanced construction of buildings
Description: Fifth edition. | Hoboken, NJ, USA : John Wiley & Sons, Ltd.,
    [2023] | Includes index.
Identifiers: LCCN 2022058089 (print) | LCCN 2022058090 (ebook) | ISBN
    9781119734888 (paperback) | ISBN 9781119735045 (adobe pdf) | ISBN
    9781119731047 (epub)
Subjects: LCSH: Building.
Classification: LCC TH146 .E467 2023 (print) | LCC TH146 (ebook) | DDC
    690—dc23/eng/20221206
LC record available at https://lccn.loc.gov/2022058089
LC ebook record available at https://lccn.loc.gov/2022058090

Cover design: Wiley
Cover image: © sihuo0860371/Getty Images

Set in 10/12pt Minion Pro by Straive, Chennai, India
Printed and bound by CPI Group (UK) Ltd, Croydon, CR0 4YY

C9781119734888_050623

Contents

How to Navigate this Book

The design and construction of buildings is about making informed choices. The choices made will be specific to the context of the site, client requirements, building type and size, prevailing socio-economic conditions, and be underpinned by respect for our planet. Whatever their function, buildings need to offer a safe, healthy, stimulating and sustainable environment for all users.

The construction process will start with a thorough assessment of client needs and an equally thorough assessment of the building site, from which designs can be developed and options considered. Options will relate to where the building is to be positioned on the site, through to the general massing and appearance of the building. This is linked to significant decisions, such as whether to use loadbearing construction, framed construction, offsite or on site construction methods, or some form of hybrid approach. In turn, these decisions help direct choices about the materials to be used and the selection of elements such as walls and windows; and on to decisions about fixings, fittings, services provision and finishes.

To design buildings and to make choices about how they are to be constructed safely, economically and sustainably, requires knowledge of construction in the widest sense of the term. This encompasses knowledge about construction materials and technologies, construction techniques, economy and environmental credentials. This is underpinned by knowledge of building laws, regulations and guidance in the form of building standards and codes. The environmental impact of the choices made should underly all decisions, as these will have a long-term influence on the performance and sustainability of the building. To make effective decisions requires a long-term vision of the building, from inception and assembly, through the building's use and adaption, to disassembly and reuse of materials and components; what we refer to as a circular economy. The intention is to design and construct buildings that produce no waste and also generate more energy than they consume; an active building.

Design and construction are predicated on proposing and evaluating a variety of options for the given context. In all but the smallest of projects choices are made by a variety of professionals and tradespeople as they aim to satisfy the needs of the building sponsor, the client, within given parameters such as time, cost, quality and environmental impact. These parameters drive decisions and influence the choices made in a complex social and cultural context. The overall goal is to satisfy client requirements, while also creating a functional and delightful building that makes a positive impact on the planet. The criteria by which options are evaluated is related to the functional requirements of individual elements, components and materials; the parts that make up the whole. To do this effectively requires a thorough understanding of construction.

The original philosophy of Robin Barry – to address the functional requirements of building elements – is fundamental to making informed choices. *Barry's Introduction to Construction of Buildings* and *Barry's Advanced Construction of Buildings* are designed to inform readers about the underpinning construction principles relating to all buildings, regardless of size or intended use. This is achieved through the use of precedents, the description and illustration of 'typical solutions' to common construction challenges. In doing this, the intention is not to tell the reader how to solve the challenge before them, rather it is to provide an example that informs knowledge, and from which fundamental questions can be asked. Once we have developed an understanding of how buildings are constructed and how they may behave, we can then start to ask whether or not the techniques we use are appropriate for our context. We can apply our analytical skills to start to question conventional wisdom and to think about how we may go about doing things differently to respond to and anticipate changes in our climate and expectations of building users.

The Barry books are presented in two volumes, *Introduction* and *Advanced*, with the volumes designed to complement one another. The titles are used to reflect the stage at which these subjects are taught in colleges and universities in the UK. *Introduction* covers the first year, primarily dealing with loadbearing construction and domestic-scale developments. It also explores the common elements found in most buildings. The *Advanced* volume includes material usually taught in the second to third year, primarily dealing with offsite techniques framed construction and reuse of existing buildings. Combined, the two volumes take the reader through the entire life cycle of a building, from inception and construction to the building in use and eventual demolition, recycling and reuse of valuable resources.

An overview of the chapters in each volume is provided in Table 1, as an aid to navigation of the books.

Table 1   Overview of the chapters

Chapter	Introduction	Advanced
1	Introduction	Introduction
2	Site Analysis, Set-Up, Drainage and Scaffolding	Offsite Construction
3	Ground Stability and Foundations	Pile Foundations, Substructures and Basements
4	Floors	Single Storey Frames, Shells and Lightweight Coverings
5	Loadbearing Walls	Structural Timber Frames
6	Roofs	Structural Steel Frames
7	Windows	Structural Concrete Frames
8	Doors	Envelopes to Framed Buildings
9	Stairs and Ramps	Lifts and Escalators
10	Surface Finishes	Fit Out and Second Fix
11	Internal Environment and Energy Supply	Existing Buildings: Pathology, Upgrading and Demolition
12	Water Supply and Sanitation	

Chapters are designed so that they can be read from front to back or they can be dipped into as the need arises. Each chapter or section introduces the primary functional requirements and then the reader is introduced to an increasing level of detail. The illustrations and photographs are provided to enhance our understanding of the main principles. At a glance, sheets are used for each chapter to address the main what, why, how and when questions.

If readers are studying, for example, loadbearing construction, then they will need to read the *Introduction* volume and focus on specific chapters to supplement their learning in the classroom. In this situation, the reader will need to read chapters all the way through in the first instance, perhaps returning to specific issues, such as the position of the damp-proof course. Similarly, if readers are studying framed construction, the *Advanced* volume will be a valuable resource, supplemented with material on, for example, doors and windows from the *Introduction* volume. When it comes to revising for examinations in construction technology, the 'At a glance' feature will be useful in prompting one's memory, prior to revisiting key issues within the chapter. Chapters conclude with guidance on additional sources and reflective exercises. The reflective exercises aim to help readers question why and how we are constructing buildings in the way we do. These can be addressed by individual readers and also by small study groups as primers for discussion. We have set these in the context of (design) project work, so whatever the scale of the project or level of study the exercises should help readers to reflect on the most appropriate solution for a given context.

The principles and details illustrated here are intended as a guide to the construction of buildings. When readers use the books to help detail their building designs, dipping into chapters to see solutions to typical detailing problems will help with understanding. It is, however, important that we understand the principles underlying the construction of buildings – what needs to be achieved and why. Thus, the details and photographs provided give an indication of how it could be done; not how it should be done. Details should not be copied without thinking about what is really going on. This also applies to details given in guidance documents and manufacturers' information.

Readers should be asking questions such as: How is the building to be assembled, maintained and disassembled safely and efficiently? Is the detail in question entirely suitable for the task at hand? We make this point because building practices and regulations vary from region to region and country to country. For example, a building located in a wet and sheltered area of the UK may benefit from a pitched roof with a large overhang, but a similar building in a dry and exposed part of the country may benefit from a pitched roof with clipped eaves or even a flat roof. It is impossible to cover every eventuality for every reader in these books. Instead, we would urge readers to engage in critical thinking, analyse the details, and then seek out more sustainable approaches and products.

1 Introduction

In *Barry's Introduction to Construction of Buildings*, we provided an introductory chapter that set out some of the requirements and conditions relevant to all building projects, regardless of size and complexity. We continue the theme in this chapter, with some additional requirements. In this volume, the emphasis shifts from domestic to larger-scale buildings, primarily residential, commercial and industrial buildings constructed with loadbearing frames. This is supported by information on fit out and second fix, lifts and escalators, and offsite construction. Many of the principles and techniques set out in the introductory volume are, however, still appropriate to this volume. Similarly, many of the technologies described here are also used in smaller buildings. Thus, we would urge readers to consult both volumes of the *Barry* series. In this introductory chapter, we start to address some additional, yet related, issues, again with the aim of providing context to the chapters that follow.

1.1 The function and performance of buildings

Structure and fabric

It is the combined performance of the structure and building fabric, together with the integration of services, which determines the overall performance of the building during its life. In loadbearing construction, the materials forming the structural support also provide the fabric and hence the external and internal finishes. In framed structures, the fabric is independent of the structure, with the fabric applied to the loadbearing structural frame.

Loading
Buildings need to accommodate the loads and forces acting on them if they are to resist collapse. One of the most important considerations is how forces are transferred within the structure. Buildings are subject to three types of loading:

(1) *Dead loads*. Dead loads remain relatively constant throughout the life of a building, unless it is remodelled at a future date. These loads comprise the combined weight of the materials used to construct the building. Loads are transferred to the ground via the foundations. Because the weight of individual components is known, the dead load can be easily calculated.

Barry's Advanced Construction of Buildings, Fifth Edition. Stephen Emmitt.
© 2023 John Wiley & Sons Ltd. Published 2023 by John Wiley & Sons Ltd.

(2) *Live loads*. Unlike dead loads, the live loads acting on a building will vary. Live loads comprise the weight of people using the building, the weight of furniture and equipment, etc. Seasonal changes will result in (temporary) live loading from rainfall and snow. Structural design calculations assume an average maximum live load based on the use of the building (plus a safety factor). If the building use changes, then it will be necessary to check the anticipated live loading against that used at the design stage.

(3) *Wind loads*. All buildings are subject to wind loading. Maximum wind loads (gusts) are determined by considering the maximum recorded wind speed in a particular location and adding a safety factor. Wind loading is an important consideration for both permanent and temporary structures. It is also an important consideration when designing and installing temporary weather protection to protect building workers and work in progress from the elements.

When the total loading has been calculated for the proposed building, it is then possible to design the building structure (the structural frame) and the foundations. This needs to be done in conjunction with the design of the building envelope.

Structural frames

Timber, steel and reinforced concrete are the main materials used for structural frames (Photograph 1.1). In some cases, it is common to use one material only for the structural frame (e.g. timber). In other situations, it may be beneficial to use a composite frame construction (e.g. concrete and steel). Combining two or more materials is known as hybrid construction. The benefits of one material over another need to be considered against a wide variety of design and performance parameters, such as the following:

- Extent of clear span required.
- Height of the building.
- Extent of anticipated loading.
- Fire resistance and protection.
- Embodied energy and associated environmental impact.
- Ease of fixing the fabric to the frame (constructability).
- Availability of materials and labour skills.
- Extent of prefabrication desired.
- Site access (restrictions).
- Erection programme and sequence.
- Maintenance and ease of adaptability.
- Ease of disassembly and reuse of materials.
- Life cycle costs.

Dimensional stability

Stability of the building as a whole will be determined by the independent movement of different materials and components within the structure over time – a complex interaction determined by the dimensional variation of individual components when subjected to changes in moisture content, changes in temperature and not forgetting changes in loading:

Photograph 1.1   Framed building under construction.

❏ *Moisture movement.* Dimensional variation will occur in porous materials as they take up or, conversely, lose moisture through evaporation. Seasonal variations in temperature will occur in temperate climates and affect many building materials. Indoor temperature variations should also be considered.

❏ *Thermal movement.* All building materials exhibit some amount of thermal movement because of seasonal changes in temperature and (often rapid) diurnal fluctuations. Dimensional variation is usually linear. The extent of movement will be determined by the temperature range the material is subjected to, its coefficient of expansion, its size and its colour. These factors are influenced by the material's degree of exposure, and care is required to allow for adequate expansion and contraction through the use of control joints.

❏ *Loading.* Dimensional variation will occur in materials that are subjected to load. Deformation under load may be permanent; however, some materials will return to their natural state when the load is removed. Thus live and wind loads need to be considered too.

Understanding the different physical properties of materials will help in detailing the junctions between materials and with the design, positioning and size of control joints. Movement in materials can be substantial and involve large forces. If materials are restrained in such a way that they cannot move, then these forces may exceed the strength of the material and result in some form of failure. Control joints, sometimes described as 'movement joints' or 'expansion joints', are an effective way of accommodating movement and associated stresses.

Designers and builders must understand the nature of the materials and products they are specifying and building with. These include the materials' scientific properties, structural properties, characteristics when subjected to fire; interaction with other materials, anticipated durability for a given situation, life cycle cost, service life, maintenance requirements, recycling potential, environmental characteristics such as embodied energy, health and safety characteristics, and, last but not least, their aesthetic properties if they are to be seen when the building is complete. With such a long list of considerations, it is essential that designers and builders work closely with manufacturers and consult independent technical reports. A thorough understanding of materials is fundamental to ensuring feasible constructability and disassembly strategies. Consideration should be given to the service life of materials and manufactured products, since any assembly is only as durable as the shortest service life of its component parts.

Tolerances

To be able to place individual parts in juxtaposition with other parts of the assembly, a certain amount of dimensional tolerance is required. Construction involves the use of labour, either remote from the site in a factory or workshop, or on site, but always in combination. Designers must consider all those who are expected to assemble the various parts physically into a whole, including those responsible for servicing and replacing parts in the future, so that workers can carry out their tasks safely and comfortably.

With traditional construction, the craftsmen would deal with tolerances as part of their craft, applying their knowledge and skill to trim, cut, fit and adjust materials on site to create the desired effect. In contrast, where materials are manufactured under carefully controlled conditions in a factory, or workshop, and brought to site for assembly, the manufacturer, designer and contractor must be confident that the component parts will fit together, since there is no scope to make adjustments to the manufactured components. Provision for variation in materials, manufacturing and positioning is achieved by specifying allowable tolerances. Too small a tolerance and it may be impossible to move components into position on site, resulting in some form of damage; too large a tolerance will necessitate a degree of 'bodging' on site to fill the gap – for practical and economic reasons, both situations must be avoided. There are three interrelated tolerances that the designer must specify, which are related specifically to the choice of material(s):

(1) *Manufacturing tolerances.* Manufacturing tolerances limit the dimensional deviation in the manufacture of components. They may be set by a standard (e.g. ISO), by a manufacturer and/or the design team. Some manufacturers are able to manufacture to tighter tolerances than those defined in the current standards. Some designers may require a greater degree of tolerance than that normally supplied, for which there may well be a cost to cover additional tooling and quality control in the factory.

(2) *Positional tolerances.* Minimum and maximum allowable tolerances are essential for convenience and safety of assembly. However, whether the tolerances are met on site will depend upon the skills of those doing the setting out, the technology employed to erect and position components, and the quality of the supervision.

(3) *Joint tolerances.* Joint tolerances will be determined by a combination of the performance requirements of the joint solution and the aesthetic requirements of the designer. Functional requirements will be determined through the materials and technologies employed. Aesthetic requirements will be determined by building traditions, architectural fashion and the designer's own idiosyncrasies.

As a general rule, the smaller (or closer) the tolerance, the greater the manufacturing costs and the greater the time for assembly and associated costs. Help in determining the most suitable degree of tolerance can be found in the technical literature provided by trade associations and manufacturers. Once the tolerances are known and understood in relation to the overall building design, it is possible to compose the drawings and details that show the building assembly. Dimensional coordination is important to ensure that the multitude of components fit together correctly, thus ensuring smooth operations on site and the avoidance of unnecessary waste through unnecessary cutting. A modular approach may be useful, although this may not necessarily accord with a more organic design approach.

Flexibility and the open building concept

The vast majority of buildings will need to be adjusted or adapted in some way to accommodate the changing needs of the building users and owners. In domestic construction, this may entail the addition of a small extension to better accommodate a growing family, conversion of unused roof space into living accommodation or the addition of a conservatory. Change of building owner often means that the kitchen or bathroom (which may be functional and in a good state of repair) will be upgraded or replaced to suit the taste and needs of the new building owners. Thus, what was perfectly functional to one building user is not to another, necessitating the need for alterations.

In commercial buildings, a change of tenant can result in major building work, as, for example, internal partition walls are moved to suit different spatial demands. Change of retailer will also result in a complete refitting of most shop interiors. These are just a few examples of the amount of alterations and adaptations made to buildings, which, if not planned and managed in a strategic manner, will result in a considerable amount of material waste. Emphasis should be on reusing and recycling materials as they are disassembled and, if possible, the flexibility of internal space use.

Although these are primarily design considerations, the manner in which materials and components are connected can have a major influence on the ease, or otherwise, of future alterations.

Flexibility and adaptability

Designing and detailing a building to be flexible and adaptable in use presents a number of challenges, some of which may be known and foreseen at the briefing stage, but many of which cannot be predicted. Thought should be given to the manner in which internal, non-loadbearing walls are constructed and their ease of disassembly and reuse (repositioning).

Similarly, the position of services and the manner in which they are fixed to the building fabric need careful thought at the design and detailing stage. For example, a flexible house design would have a structural shell with non-loadbearing internal walls (movable partitions, folding walls, etc.), zoned underfloor space heating (allowing for flexible use of space) and carefully positioned wet and electrical service runs (in a designated service zone or service wall).

Open building

The open building concept aims to provide buildings that are relatively easy to adapt to changing needs, with minimum waste of materials and little inconvenience to building users. The main concept is based on taking the entire life cycle of a building and the different service lives of the building's individual components into account. Since an assembly of components is dependent upon the service life of its shortest-living element, it may be useful to view the building as a system of time-dependent levels. Terminology varies a little, but the use of a three-level system, primary, secondary and tertiary, is common. Described in more detail, the levels are:

- ❏   *The primary system.* Service life of approximately 50–100 years. This comprises the main building elements, such as the loadbearing walls and roof or the structural frame and floors and roof. The primary system is a long-term investment and is difficult to change without considerable cost and disruption.
- ❏   *The secondary system.* Service life of approximately 15–50 years. This comprises elements such as internal walls, floor and ceiling finishes, building services installations, doors and vertical circulation systems such as lifts and escalators. The secondary system is a medium-term investment and should be capable of replacement or adaptation through disassembly and reassembly. The shorter the service life of components, the greater the need for replacement, hence the need for easy and safe access.
- ❏   *The tertiary system.* Service life of approximately 5–15 years. This comprises elements such as fittings and furniture and equipment associated with the building use (e.g. office equipment). The tertiary system is a short-term investment and elements should be capable of being changed without any major building work.

Applying this strategy to a development of, for example, apartments, the structure and external fabric would be the primary system. The secondary system would include kitchens, bathrooms and services. The tertiary system would cover items such as the furniture and household appliances. If a discrete, modular system is used, then it is relatively easy to replace the kitchen or bathroom without major disruption and to recycle the materials. This 'plug-in' approach is certainly not a new concept but has started to become a more realistic option as the sector has started to adopt offsite production (see also Chapter 2).

Security

Security of buildings and their contents (goods and people) is a primary concern for the vast majority of building sponsors and owners. In residential developments, the primary concern is with theft of property, with emphasis on the integrity of doors and windows. In commercial developments, the concern is for the safety of the people using the building and for the security of the building's contents. The desire to keep the building users and contents safe has to be balanced with the need to allow safe evacuation in the case of a fire or an emergency. Vandalism and the fear of terrorist attacks are additional security concerns, leading to changes in the way buildings are designed and constructed. Measures may be passive, active or a combination of both.

Passive security measures
A passive approach to security is based on the concept of inherent security measures, where careful consideration at the design and detailing phase can make a major difference to the security of the building and its contents. Building layout and the positioning of, for example, doors and windows to benefit from natural surveillance need to be combined with the specification of materials and components that match the necessary functional requirements. The main structural materials and the method of construction will have a significant impact on the resistance of the structure to forced entry. For example, consideration should be given to the ease with which external cladding may be removed and/or broken through, and depending on the estimated risk, an alternative form of construction may be more appropriate. Unlawful entry through roofs and rooflights is also a potential risk. Building designers must consider the security of all building elements.

Ram raiding, the act of driving a vehicle through the external fabric of the building to create an unauthorised means of access and egress for the purposes of theft, has become a significant problem for the owners of commercial and industrial premises. Concrete and steel bollards, set in robust foundations and spaced at close centres around the perimeter of the building, are one means of providing some security against ram raiding, especially where it is inappropriate to construct a secure perimeter fence.

Active security measures
Active security measures, such as alarms and monitoring devices, may be deployed in lieu of passive measures or in addition to inherent security features. For new buildings, active measures should be considered at the design stage to ensure a good match between passive and active security. Integration of cables and mounting and installation of equipment should also be considered early in the detailed design stage. Likewise, when applying active security measures to existing buildings, care should be taken to analyse and utilise any inherent features. Some of the active measures include:

❑ Intruder alarm systems.
❑ Entrance control systems in foyers/entrance lobbies.
❑ Coded door access.
❑ CCTV monitoring.
❑ Security personnel patrols.

Health, safety and wellbeing

Various approaches have been taken to improve the health, safety and wellbeing of everyone involved in construction. These include more stringent legislation, better education and training of workers, and better management practices. Similarly, a better understanding of the sequence of construction (a combination of constructability principles and detailed method statements) has helped to identify risk hazards and to minimise or even eliminate them. This also applies to future demolition of the building, with a detailed disassembly strategy serving a similar purpose. There are four main, interrelated stages to consider. They are:

(1) *Prior to construction.* The manner in which a building is designed and detailed, i.e. the materials selected and their intended relationship to one another, will have a significant bearing on the safety of operations during construction. Extensive guidance is available via the Safety in Design (http://www.safetyindesign.org).
(2) *During construction.* Ease of constructability will have a bearing on safety during production. Offsite manufacturing offers the potential of a safer environment, primarily because the factory setting is more stable and easier to control than the constantly changing construction site. However, the way in which work is organised and the attitude of workers towards safety will have a significant bearing on accident prevention.
(3) *During use.* Routine maintenance and repair is carried out throughout the life of a building. Even relatively simple tasks such as changing a light bulb can become a potential hazard if the light fitting is difficult to access. Elements of the building with short service lives (and/or with high maintenance requirements) must be accessed safely.
(4) *Demolition and disassembly.* Attention must be given to the workers who at some time in the future will be charged with disassembling the building. Method statements and guidance on a suitable and safe demolition and disassembly strategy are required at the design stage.

1.2 New methods and products

An exciting feature of construction is the amount of innovation and change constantly taking place in the development of new materials, methods and products, many of which are used in conjunction with the more established technologies. Some of the more obvious areas of innovative solutions are associated with: changing regulations (e.g. airtightness requirements); changing technologies (e.g. new cladding systems); the trend towards greater use of offsite production (e.g. volumetric system build); advances in building services (e.g. provision of broadband); a move to the use (and reuse) of recycled materials (e.g. products manufactured from recycled material, see Photograph 1.2); and the drive for low- or zero-carbon construction, which has stimulated renewed interest in natural materials and their innovative use. Many of the changes are, however, quite subtle as manufacturers make gradual technical 'improvements' to their product portfolio. This could be as simple as gradually increasing the amount of recycled content in their products. Gradual innovations are often brought about by the use of a new production plant and automated production and/or are triggered by competition from other manufacturers, with manufacturers seeking to maintain and improve market share through technical innovation. In the vast majority of cases, this results in building products with improved performance standards and improved environmental credentials.

Combined with changing fashions in architectural design and manufacturers' constant push towards the development of new materials and products, we are faced with a very wide range of systems, components and products from which to choose. All contributors to the design and erection of buildings, from clients and architects to contractors and specialist subcontractors, will have their own attitude to new products. Some are keen to use new products and/or new techniques, while others are a little more cautious and tend to stick to what they know. Whatever one's approach, it is important to keep up to date with the

Photograph 1.2 Artificial stone made entirely from recycled rubber tyres (left of picture, rough texture) adjacent to natural stone (right of picture, smooth texture).

latest product developments and to investigate those products and methods that may well prove to be beneficial. Maintaining relationships with product manufacturers is one way of achieving this; indeed, we would urge readers to visit manufacturers and talk to them about their products. This should be balanced against independent research reports relating to specific or generic product types.

Compliance and performance monitoring

Whatever approach is taken to the use of innovative materials, components and structural systems, it is important to remember that compliance is required with the Building Regulations and appropriate Codes and Standards. And, once built and operational, it is important to monitor the performance of products in relation to the overall building performance. This applies equally to buildings constructed on site and to those produced in whole or in part in factories.

The Building Regulations and supporting guidance (*Approved Documents* in England and Wales, and in Northern Ireland; *Guidance Documents* in Scotland) are structured in such a way as to encourage the adoption of innovative approaches to the design and construction of buildings. This is done through setting performance standards, which must be achieved or bettered by the proposed construction. Acceptance of innovative proposals is in the hands of the building control body handling the application; thus, applicants must submit sufficient information on the innovative proposal to allow an accurate assessment of its performance. This is done by supplying data on testing, certification, technical approvals, CE marking and compliance with the Construction Products Directive (CPD), Eurocodes and Standards, calculations, detailed drawings and written specifications where appropriate.

Monitoring, testing and analysing the performance of new products and especially the overall performance of buildings are an important function. This has become particularly pertinent recently in our drive for carbon-neutral buildings and the use of many innovative approaches to design and construction. Although new building products and systems will have been tested by manufacturers under laboratory conditions, we can never be sure how they will perform in relation to the entire building, which will be subject to variations in local climate and patterns of use. Thus it is necessary to monitor and analyse the performance of buildings and to feed that information back to manufacturers, designers and constructors.

1.3 Product selection and specification

Both the quality and the long-term durability of a building depend upon the selection of suitable building products and the manner in which they are assembled. This applies to buildings constructed on the site and to offsite production. The majority of people contributing to the design and construction of a building are, in some way or another, involved in the specification of building products; that is making a choice as to the most appropriate material or component for a particular situation. Architects and engineers will usually specify products by brand name (a prescriptive specification) or through the establishment of performance criteria (a performance specification), which is discussed in more detail later. These choices are linked to the way in which the building is detailed and the process of construction, be it offsite or onsite. Contractors and subcontractors will be involved in the purchase and installation of the named product or products that match the specified performance requirements; that is, they will also be involved in assessing options and making a decision. Similarly, designers working for offsite manufacturers will also be involved in material and product selection; here, the emphasis will be on secure lines of supply and a transparent and ethical supply chain.

The final choice of product and the manner in which it is built into the building will have an effect on the overall quality and performance of the building. Traditionally, the factors affecting choice of building products have been the characteristics of the product (its properties, or 'fitness for purpose'), its initial cost and its availability. However, a number of other factors are beginning to influence choice, some of which are dependent on legislation, others of which are also dependent upon product safety (during construction, use and replacement/recycling and ethical resourcing) and environmental concerns as to the individual and collective impact of the materials used in the building's construction. Selection criteria will cover the following areas; the importance of one over another is dependent on the location of the product and the type of building project:

- ❑ Aesthetics.
- ❑ Availability.
- ❑ Compatibility (with other products).
- ❑ Compliance with legislation.
- ❑ Cost (whole life costs).
- ❑ Durability.
- ❑ Ease of installation (buildability).
- ❑ Environmental impact (low-carbon materials).
- ❑ Fire safety.
- ❑ Health and safety.
- ❑ Replacement and recyclability.
- ❑ Risk (associated with the product and the manufacturer).

For very small projects, it is common for contractors to select materials and products from the stock held by their local builders' merchant, choice being largely dependent upon what the merchant stocks (availability) and initial cost. For larger projects, there is a need to confirm specification decisions in a written document, the specification.

The written specification

Specifications are written documents that describe the requirements to which the service or product has to conform, such as its defined quality. It is the written specification, not the drawings, which defines and hence determines the quality of the finished work. The term specification tends to be used in the singular, which is a little misleading. In practice, the work to be carried out will be described in specifications written by the different specialists involved in the construction project. The structural engineer will write the specification for the structural elements, such as foundations and steelwork, whereas the architect will be concerned with materials and finishes. Similarly, there will be a specification for the electrical and mechanical services provision. This collection of multi-authored information is known as 'the specification.'

People from different backgrounds will use the written specification for a number of quite different tasks. It will be used during the pre-contract phase to help prepare costings and tenders. During the contract, operatives and the site managers will read the specification to check that the work is proceeding in accordance with the defined quality. Postcontract, the document will form a record of materials used and set standards, which is useful for alteration and repair work and as a source of evidence in disputes. In more recent projects this information will be held within the Building Information Models (BIMs), and will be accessible for maintenance and future repair work.

Specifying quality

Trying to define quality is a real challenge when it comes to construction, partly because of the complex nature of building activity and partly because of the number of actors who have a stake in achieving quality. The term quality tends to be used in a subjective manner and, of course, is negotiable between the project stakeholders. In terms of the written specification, quality can be defined through the quality of materials and the quality of workmanship. Designers can define the quality of materials they require through their choice of proprietary products or through the use of performance parameters and appropriate reference to standards and codes. Designers do not tell the builder how to construct the building; this is the contractor's responsibility, hence the need for method statements. The specification will set out the appropriate levels of workmanship, again by reference to codes and standards, but it is the people doing the work, and to a certain extent the quality of supervision, that determines the quality of the finished building.

Specification methods

There are a number of methods available for specifying. Some methods allow the contractor some latitude for choice and therefore an element of competition in the tendering process, while others are deliberately restrictive. The four specification methods are:

(1) *Descriptive specifying.* Where exact properties of materials and methods of installation are described in detail. Proprietary names are not used; hence, this method is not restrictive.
(2) *Reference standard specifying.* Where reference is made to established standards to which processes and products must comply, e.g. a national or international standard. This is also non-restrictive.

(3) *Proprietary specifying.* Where manufacturers' brand names are stated in the written specification. Here the contractor is restricted to using the specified product unless the specification is written in such a way to allow substitution of an equivalent. Proprietary specification is the most popular method where the designer produces the design requirements and specifies in detail the materials to be used (listing proprietary products), methods and standard of workmanship required.

(4) *Performance specifying.* Where the required outcomes are specified together with the criteria by which the chosen solution will be evaluated. This is non-restrictive and the contractor is free to use any product that meets the specified performance criteria. Performance specification is where the designer describes the material and workmanship attributes required, leaving the decision about specific products and standards of workmanship to the contractor.

The task is to select the most appropriate method for a particular situation and project context. The type of funding arrangement for the project and client preferences usually influences this decision. Typically, projects funded with public funds will have to allow for competition, so proprietary specifying is not usually possible. Projects funded from private sources may have no restrictions, unless the client has a preference or policy of using a particular approach. Obviously, the client's requirements need to be considered alongside the method best suited to clearly describe the design intent and the required quality, while also considering which method will help to get the best price for the work and, if desired, allow for innovation. In some respects, this also concerns the level of detail required for a project or particular elements of that project. Although one method is usually dominant for a project, it is not uncommon to use a mix of methods for different items in the same document.

It has been argued that performance specifications encourage innovation, although it is hard to find much evidence to support such a view. The performance approach allows, in theory at least, a degree of choice and hence competition. The advantage of one approach over another is largely a matter of circumstance and personal preference. However, it is common for performance and prescriptive specifications to be used on the same project for different elements of the building.

National Building Specification

Standard formats provide a useful template for specifiers and help to ensure a degree of consistency, as well as saving time. In the UK, the National Building Specification (NBS) and the National Engineering Specification (NES) are widely used. This commercially available suite of specification formats includes *NBS Building*, *NBS Engineering Services* and *NBS Landscape*. Available as computer software, it helps to make the writing of specifications relatively straightforward, because prompts are given to assist the writer's memory. Despite the name, the NBS is not a national specification in the sense that it must be used; many design offices use their own particular hybrid specifications that suit them and their type of work.

NBS Building is available in three different formats to suit the size of a particular project, ranging from Minor Works (small projects) to Intermediate and Standard (large projects). It is an extensive document containing a library of clauses. These clauses are selected and/or deleted by the specifier, and information is added at the appropriate prompt to suit a particular project.

With the uptake of BIM many specification decisions are tied to the product libraries held with the digital model(s) used by designers and specialist subcontractors.

Green specifications
The National Green Specification (NGS) is an independent organisation, partnered by the Building Research Establishment (BRE), to host an Internet-based resource for specifiers. It provides building product information plus work sections and clauses written in a format suitable for importing into the NBS, thus helping to promote the specification of green products.

Coordinated Project Information
Coordinated Project Information (CPI) is a system that categorises drawings and written information (specifications). CPI is used in British Standards and in the measurement of building works, the Standard Method of Measurement (SMM7). This relates directly to the classification system used in the NBS.

One of the conventions of CPI and Uniclass is the 'Common Arrangement of Work Sections' (CAWS). CAWS lists around 300 different classes of work according to the operatives who will do the work; indeed, the system was designed to assist the dissemination of information to subcontractors. This allows bills of quantities to be arranged according to CAWS. The system also makes it easy to refer items coded on drawings, in schedules and in bills of quantities back to the written specification. The main categories are shown in Table 1.1 (note there is no 'I', 'O' or 'Y'). The main sections are further divided into sub-sections.

Table 1.1   CAWS contents

A	Preliminaries
B	Complete buildings/structures/units
C	Demolition/alteration/renovation
D	Groundwork
E	In situ concrete/large precast concrete
F	Masonry
G	Structural/carcassing metal/timber
H	Cladding/covering
J	Waterproofing
K	Linings/sheathing/dry partitioning
L	Windows/doors/stairs
M	Surface finishes
N	Furniture/equipment
P	Building fabric sundries
Q	Paving/planting/fencing/site furniture
R	Disposal systems
S	Piped supply systems
T	Mechanical heating/cooling/refrigeration systems
U	Ventilation/air conditioning systems
V	Electrical supply/power/lighting systems
W	Communications/security/control systems
X	Transport systems
Z	Building fabric reference specification
	Additional rules – work to existing buildings
	Appendices

The introduction of the New Rules of Measurement (NRM) has brought a move away from CAWS to a new indexing system that aims to better reflect developments in building technologies (e.g. offsite and recycling). This numbered system contains 41 sections and no longer makes reference to CPI, as shown in Table 1.2.

Table 1.2   NRM2 contents

1	Preliminaries
2	Offsite manufactured materials, components and buildings
3	Demolitions
4	Alterations, repairs and conservation
5	Excavating and filling
6	Ground remediation and soil stabilisation
7	Piling
8	Underpinning
9	Diaphragm walls and embedded retaining walls
10	Crib walls, gabions and reinforced earth
11	In situ concrete works
12	Precast/composite concrete
13	Precast concrete
14	Masonry
15	Structural metalwork
16	Carpentry
17	Sheet roof coverings
18	Tile and slate roof and wall coverings
19	Waterproofing
20	Proprietary linings and partitions
21	Cladding and covering
22	General joinery
23	Windows, screens and lights
24	Doors, shutters and hatches
25	Stairs, walkways and balustrades
26	Metalwork
27	Glazing
28	Floor, wall, ceiling and roof finishings
29	Decoration
30	Suspended ceilings
31	Insulation, fire stopping and fire protection
32	Furniture, fittings and equipment
33	Drainage above ground
34	Drainage below ground
35	Site works
36	Fencing
37	Soft landscaping
38	Mechanical services
39	Electrical services
40	Transportation
41	Builder's work in connection with mechanical, electrical and transportation installations Appendices

Further reading

For information relating to specifications and construction information see: www.thenbs.com. See also www.greenspec.co.uk and www.bregroup.com for the Green Guide to Specification.

Reflective exercises

Your client is interested in non-conventional building materials and wishes you to specify your building project to be as 'ecologically friendly' as possible.

- ❏ What resources do you need to consult and why?
- ❏ List the potential advantages and disadvantages of changing from conventional building materials to non-conventional ones.

Offsite Construction: Chapter 2
AT A GLANCE

What? The term 'offsite' construction refers to the process of producing buildings, or parts of buildings, in factories remote from the building site. The manufacturing process is usually highly automated, resulting in prefabricated and pre-assembled components, panelised units (2D) and modular (3D, volumetric) systems. The prefabricated and pre-assembled units and modules are transported to site when required and craned into position on pre-prepared foundations or slotted into a structural frame. This is primarily a dry method of construction, although some wet trades may be employed to complete the building finishes.

Why? Offsite construction offers the potential to better control the quality of workmanship, remove the uncertainties associated with working in variable weather conditions on site, improve health, safety and wellbeing of workers, significantly reduce the amount of time spent working on the site, make financial savings through the repetitive production of units, and in many cases improve the environmental impact of buildings by reducing waste during fabrication. Offsite construction is ideally suited to buildings with a repetitive element, such as hotel bedrooms, housing and apartments, hospitals and schools. The techniques are also well suited to the pre-assembly of services, such as modular electrical cabling and services pods, elevators and staircases, and bathroom and kitchen pods. Producers of modular and system building offer a range of modules/elements that can be scaled from a small building to a very large development.

When? Using offsite fabrication changes the design and construction process. The design must be finalised before production starts and it will be necessary to work closely with the manufacturer. The design process is influenced by the possibilities and constraints of offsite production, which are specific to the technologies being used and the availability and suitability of companies that specialise in fabrication. The sequence of construction will vary depending upon the amount of prefabrication required, the type of building and site constraints. It is common to deliver the pre-assembled units when required to eliminate the need for on site storage and unnecessary handling.

How? The pre-assembled components, panels and modules will be manufactured in factories, usually to suit specific design requirements and with regard to manufacturing constraints. These pre-assembly factories manage the purchasing and handling of materials, production and delivery. Some manufacturers also offer an on-site assembly service. Off-the-shelf systems are available, in which case designers and engineers need to work within the constraints of the system. Once manufactured, the units are delivered to site, craned into position and connected to adjacent units and/or the loadbearing structural frame. Thus the construction skills, assisted with automation and robotics, are located primarily in a factory, not on the building site.

2   Offsite Construction

Construction is essentially a process of assembly, fixing and fitting of manufactured components in a precise location: the building site. The majority of components that make up buildings are factory produced (e.g. doors, windows, staircases and sanitary ware; bricks, blocks, tiles and standard sizes of timber, steel and concrete components). These are readily available from manufacturers' catalogues of standard products or may be produced to bespoke designs. Thus, what we refer to as building or construction is a little more akin to a process of assembly. It follows that moving the assembly process to a factory environment to create large 2-dimensional (2D) panels and 3-dimensional (3D) modules that can then be transported to site and craned into position is a logical development. Offsite is the term used to describe the pre-assembly of buildings and building components at a location, or locations that are remote from the building site. A wide range of terms are used in the construction sector, ranging from offsite manufacturing, fabrication and production, to industrialised building and industrialised construction, prefabrication, pre-assembled buildings and modern methods of construction (MMC), modular construction, modular building systems, volumetric construction and system build. More recently the term robotic construction has also started to be used to recognise the high degree of automation involved in offsite construction. Rapid prototyping and 3D printing (additive manufacturing) are also associated with offsite production techniques.

Offsite fabrication enables a high degree of accuracy (precision) and consistent quality of the component parts, be they panelised systems or volumetric systems (discussed below). These assemblies are then transported to the site to a precise timetable and erected in position in a clearly defined sequence. This is primarily a dry form of construction, with the majority of skilled trades being applied in a factory environment, free of the constraints of the weather. To undertake this process effectively and efficiently requires clear design decisions and planning input early in the design process. Component parts need to be accurately designed, as do the joints between, and attention must be given to fixing and positioning tolerances, as well as ease of access to the site. On a large, and usually highly repetitive, scale, offsite construction may prove to be a more cost-efficient alternative to more traditional site-based construction methods. For commercial applications, the saving in time on the site is an important economic consideration, allowing a faster return on investment and earlier occupation of the building. Improvements in accuracy, quality, environmental impact and health and safety are other important considerations when choosing offsite production.

Barry's Advanced Construction of Buildings, Fifth Edition. Stephen Emmitt.
© 2023 John Wiley & Sons Ltd. Published 2023 by John Wiley & Sons Ltd.

Producing whole buildings, or parts of buildings as an industrialised process, is a logical technological development, but by no means a recent phenomenon. The early British settlers in America took prefabricated timber houses with them in the 1620s, and records show that prefabricated buildings of timber were exported from the UK for use in other countries. With the development of cast iron, and in particular the development of prefabricated cast iron components in the 1840s and 1850s, came the development of prefabricated iron buildings, with many prefabricated houses being shipped to Africa, Australia and the Caribbean. Concrete panel systems were developed during the 1900s, steel fabrication was developed in the 1930s and aluminium fabrication followed the Second World War in response to the housing shortage. Since this time the promotion and use of offsite systems has fluctuated, coming in and out of fashion due to a variety of political, economic, social and technical reasons. More recently the shortage of skilled tradespeople, combined with material supply chain challenges and advances in digital design and manufacturing has resulted in another push to establish offsite construction. A further driver is the commitment to climate action targets and the goal of delivering net zero buildings, which has further emphasised the need to modernise the way we build to reduce material waste. Now there is a wide range of systems available in the UK that are based on lightweight framing (lightweight steel sections and timber) and also concrete systems, primarily based on loadbearing precast panels. At the time of writing this book it is estimated that approximately 10% of buildings in the UK are delivered using MMC. The main concepts relating to cut timber, lightweight metal and concrete, and the extent of offsite production associated with each technology, are discussed later.

The extent to which construction activities are moved to a factory (or workshop) setting will vary depending on the type of prefabrication and pre-assembly employed. It will also depend on the ability to generate enough demand to make it economically worthwhile. In many respects, it is the volume housebuilding sector that has the most to gain (and lose) from using offsite production. Some buildings are assembled on site from factory-produced elements, while others are delivered to site as complete 3D volumetric units and craned into position, bolted to the foundation and then 'plugged in' to the services supplies. Offsite is primarily used for the new-build market, although the techniques and methods are equally suited to the refurbishment and upgrading of buildings, physical constraints permitting. Notable examples include urban regeneration schemes and the refurbishment and upgrading of existing concrete-framed housing units.

State-of-the-art manufacturing techniques offer many potential benefits to construction clients. It also offers designers and engineers the opportunity for creativity, something that was missing in much of the earlier offsite offerings. It is entirely possible for the design team to design a building using digital technologies and BIM platforms to collaborate with offsite manufacturers. The design can then be fabricated in the factory and delivered to site and assembled, with little need for a traditional contractor or trades. The groundworks and services connections can be procured using specialists, with the remainder done by the fabricating company. The design team provides the manufacturer with a precise digital model of the building, complete with material and performance requirements for every aspect of the design, including how it is to be assembled, maintained and disassembled. Advances in robotics provides the opportunity for process automation, reliability, and guaranteed quality. Offsite fabrication also claims to have environmental benefits compared to a more traditional approaches to construction, although detailed information can be difficult to acquire from manufacturers.

2.1 Functional requirements

Offsite pre-assembled buildings are no different to those constructed on site, in that they must comply with building regulations and associated legislation relating to, for example, fire safety, thermal and acoustic insulation, and environmental footprint. Thus the functional requirements of prefabricated buildings are the same as those identified for elements of site-constructed buildings as described in Barry's Introduction to Construction of Buildings and this volume. The only exception to this is a requirement for increased strength (bracing) of the floor and wall panels of volumetric modules to resist the different loads imposed on them during loading, transportation and positioning on the site. Tolerances are required to allow for the safe positioning of units and control joints are required to allow for thermal and structural movement.

The choice to use offsite construction, or not, needs to be made early in the briefing and conceptual design stages of projects. The decision will be coloured by the individual context of the project, client wishes, site constraints and economy. Aesthetics and many other factors that do not necessarily fall under the heading of technical or functional requirements will also come into play in the decision-making process. All of these have to be offset against the appropriate functional requirements, such as thermal performance and fire safety. A number of more generic advantages and disadvantages will also need to be considered in the context of each project.

Advantages of offsite

There are a large number of reasons why offsite production may be advantageous. Some of the most consistent arguments for moving construction process to the factory are related to the age-old challenge of attaining and maintaining quality. The quality of buildings relies to a large extent on the weather at the time of construction, the availability of appropriately skilled personnel to construct the building safely, and the control of materials used in the construction of the building. All of these factors are easier to control in a factory environment compared to the building site. Cost, both the initial cost and the life cycle costs, is also a determining factor. Although the initial cost of pre-assembled units and modules may be higher than their on site equivalents, they have to be considered against the speed of assembly and the time taken to make the building habitable, and in many cases income-generating. This can be easily offset the increased cost of the building system. Life cycle costs also need to be considered, especially the ease of maintenance and replacement. For factory-based production to be economic, the number of units or modules produced must be relatively large to cover the cost of tooling in the factory. The larger the scheme and the larger the amount of repetition, the greater the economic benefit to the customer. Similarly, the greater the repeated use of a design on other sites, the more cost-effective the process of production. This has to be balanced against architectural creativity and innovative solutions. More recently the opportunity to be creative has been realised through parametric and generative design, coupled with significant advances in robotic assembly and digital fabrication.

Control of working conditions
Quality control and validation takes place in the factory, helping to ensure a consistent level of quality. This helps to reduce the likelihood of time and cost overruns associated with poor quality identified during the snagging process and the time required to correct

defects. Inclement weather usually leads to disrupted workflow and the possibility of inconsistent quality of work. Moving the majority of the work into a protected factory environment eliminates the uncertainties of the weather. With over 80% of the production process undertaken in a controlled indoor environment, the pre-assembled components remain dry during assembly. The flow of work is consistent and efficient, quality is constantly monitored and controlled, and operatives have a safe, controlled, working environment. Workers are not exposed to the uncertainties of the weather, there is better control of dust, pollutants and noise, and work can be planned to better suit the human posture resulting in, for example, less strain on the lower back.

Work on the site is much reduced. Panelised and volumetric units may be craned from the delivery vehicle directly into position on site, eliminating the need for on site storage. This operation can be achieved in most weather conditions (strong winds being an exception). There is less reliance on scaffolding and working at height, thus helping to improve the safety of workers on the construction site by reducing their exposure to risk. Because the majority of operations are conducted away from the site there is less noise, dust and disruption to neighbours of the site. There are also fewer tradesmen on site, and hence the possibility of accidents happening is much reduced. Combined, these factors result in the promotion of better health, safety and wellbeing.

Control of the quality of materials

Given the high volume of production, manufacturers are able to purchase large quantities of materials and are able to demand high-quality standards from their suppliers. Materials can be thoroughly inspected at the time of delivery to the factory and checked for compliance with the specification. All materials used are traceable as part of the ethical supply chain, providing the client with confidence that they are getting what they pay for. Such levels of material control are more difficult to achieve on buildings sites and it is difficult to prove that the contractor and subcontractors have used the materials that were specified. There is also less chance of theft of materials from the site, therefore site security can be reduced and may be required for a shorter period. Furthermore, the amount of material waste generated on the site (and sent to landfill) will be reduced, if not eliminated, with manufacturers recycling the majority of their waste within audited factory processes. The use of lean production, or lean manufacturing, techniques will also help to eliminate waste (both material and process) during assembly in the factory.

Control of environmental impact

Efficiency of the production process equates to less material waste, certainty over life cycle costs and guaranteed quality (less errors and less rework and hence less material wastage), and a traceable, ethical, supply chain. The production process offers energy saving measures and hence a reduction in carbon compared to the majority of construction activities undertaken on site. Material and process waste is minimised through the use of integrated design and manufacturing processes, resulting in very little waste material. The waste that is produced in the manufacturing and pre-assembly process can be easily recycled and not sent to landfill, as is often the case with site-based construction. Comparative savings in material waste have been estimated to be in the range of 75–90%. Life cycle management can be incorporated into offsite construction with the incorporation of asset tracking technologies (RFIs and QR codes) into all components. This allows asset tracking from design and production to operation, maintenance and recovery of components and materials at the end of its service life; usually linked to a BIM for the project. Because the supply chain is related to a manufacturer rather than individual projects it is much easier to manage the ethical supply chain, with complete traceability of materials. This allows the careful selection of material and component suppliers that meet strict ethical and environmental conditions. Many manufacturers have also provided systems that are relatively easy to deconstruct and recover at the end of the building's useful life and recycle with a minimal amount of waste, thus helping to create a closed-loop manufacturing system with a limited environmental impact. Although manufacturers claim that their systems are environmental friendly, it is not currently possible to obtain detailed information given the commercial sensitivity. Designers may need to make their own estimations of the embodied and operational carbon.

Control of time
With the majority of operations moved to the factory, the amount of time required on the site is much reduced. Groundwork for the foundations and services can be conducted in parallel to the offsite fabrication, ready to receive the pre-assembled components, panels and/or modules. For example, modules can be craned and slotted into a pre-prepared structural frame quickly and safely, resulting in a completed building in very little time. Less time on site also helps to limit disruption to the neighbours, with less noise and dust. The majority of modular systems rely on connection to pad or pile foundations, eliminating the need for strip foundations that are common in loadbearing masonry construction. Similarly, the careful grouping of services can save on pipework and connection costs (following an open building philosophy). Offsite testing and commissioning will further reduce the time spent on site. Skill on the site is required to manage the sequence of assembly, the craning and joining the pre-assembled components, panels and units together safely. In the majority of cases, scaffolding is not required, which is a considerable cost and time-saving, while also helping to improve safety. Defects can be dealt with in the factory using a zero defect approach, thus there should, in theory at least, be no problems at practical completion. Pre-assembled components that have been damaged in transit or by craning into position can be returned to the factory and replaced relatively easily and quickly.

Disadvantages of offsite

Offsite construction may not be the right choice for all clients, nor may it be appropriate for all projects and sites. Decisions to use offsite will be influenced by the site context, the type and scale of the building, and the wishes of the client, design team and contractor. Some builders and developers may also have economic and other business reasons to stick to a more site-based approach. For example, the cost of set up for the offsite manufacturing and transportation costs could override any cost and time savings for the project. Smaller house builders and contractors may prefer to use skilled labour to retain their employees, and speed of construction may not always be a primary concern. The more common reasons that may hinder the uptake and use of offsite construction include the following factors.

Physical access
Many construction sites pose physical and logistical challenges of ensuring unhindered and safe access, making the transportation and craning of large components very difficult or impossible. This is often a challenge for work to existing buildings and buildings in densely

developed areas. It may not, for example, be possible to allow safe vehicular access to the site for the delivery of large volumetric units. Many sites also have significant challenges relating to town planning conditions, thus building in conservation areas or next to a historic building may mean that offsite is not an ideal choice because of limitations of aesthetics and scale.

Choice: supply and demand

Lack of choice may be a problem for some design teams, clients and developers, as the majority of offsite manufacturers offer limited ranges due to tooling and economy of production. In recent years the ability to expand the 'basic' range of 2D panels and 3D modules has been facilitated by advances in digital manufacturing, which has started to address some of the concerns over variety and aesthetics. In a similar vein, some clients may not relish the thought of being tied into a particular manufacturer for repair work and routine maintenance. This may relate to conditions of the warranty and/or may simply be linked to the technology employed and the availability of expertise to carry out the required work. A similar concern may be expressed for future extension and adaptation of modular buildings. There is also the issue of supply and demand to consider. Manufacturers of offsite panels and modules require a large and relatively steady flow of orders to ensure a profitable and hence a sustainable business. This continues to be challenging due to fluctuating demand and inconsistencies in economic outlook.

2.2 Offsite design and production processes

Offsite construction requires a change in how we think about the design and construction process. Emphasis moves to the early design stages and what is possible, and conversely what is not, in a factory environment. The site, with the exception of ground preparation and services provision, becomes the place where prefabricated components and pre-assembled units are assembled to form a building. In the majority of cases, the construction process is simplified, with less reliance on sub-contractors overseen by a construction manager.

Design for manufacture and disassembly

The term 'design for manufacture' (DFMA) is used to describe the philosophy of designing with factory production in mind. The design is tailored for ease of manufacture, transport, assembly, and at a point in the future, disassembly and materials recovery. This concept tends to rely on the use of standardised components and methods as part of a mass customisation process. Mass customisation is central to the realisation of competitive prices and short lead times from design approval (design freeze) to site delivery. Manufacturers' standardised component parts will be contained in a CAD or BIM library to help guide designers. This is referred to as a product family library.

Using offsite places greater emphasis on the need for the design team to collaborate with manufacturers as early as possible in the design process. The design of the building will be co-created in collaboration with the manufacturer(s), using the product family library, to create the most value for the client. Information will be co-created and shared via digital models, using 3D and BIM (nD) technologies and parametric design software.

Computer numerical control (CNC) production machines and computer-aided manufacturing (CAM) will use the digital files. Once the design has been agreed and signed off by the client, manufacturing can commence in parallel to ground preparation, site drainage and foundation work. This helps to deliver much shorter construction programmes compared to on site construction methods. Less time is spent on the site. This is an important concern for businesses and property owners keen to see the generation of revenue as quickly as possible. The digital files can be used as a reference source for maintenance, recording any changes during use and for disassembly at a later date.

Work is overseen by the production manager. The factory (or factories) is responsible for the ethical sourcing, purchasing, handling and processing of materials. Supply chains are product specific, not project specific, hence they are more stable than a typical construction project, allowing greater control of quality and cost. The production manager will work closely with the design team to agree the design, oversee production and plan the delivery and erection/assembly of the building. Design and production is usually based on producing assemblies that can be transported on lorries that do not need special licences or police notification prior to transportation. In the UK, lorries can pull trailers up to 18.3 m in length, 2.9 m wide and up to 2.9 m high. Route planning is required to avoid narrow roads and bridges with height and/or weight restrictions.

The amount of work required on the site will vary, depending on the type of offsite technologies being used and the characteristics of each site. This may range from the safe coordination of the delivery and assembly of fully finished modules, through to a more complex delivery and assembly process that involves the coordination of trades to complete external and internal finishes. Cranes will be required to move the components, 2D panels and 3D volumetric units from the delivery lorry and into position.

A typical production process

Given the repetitive nature of the manufacturing process and the high levels of capital investment, it is crucial from a business perspective that customer (market) needs are clearly identified and exploited by manufacturers of offsite. Research and development activities are concerned with market trends and technical (production) factors to ensure a profitable manufacturing process. The manufacturing process involves a range of highly skilled workers and robotic production. The extent of robotic manufacturing processes will vary between manufacturers; however, most manufacturers will follow a production process similar to that described here, with rigorous quality control conducted by trained personnel at the end of each step in the production process. The main steps in a typical production process are described here for a timber- or lightweight steel-framed unit:

❑ Discussion and confirmation of the customer's technical specification (in relation to production capacity and production constraints).
❑ Planning and scheduling the manufacturing process, from ordering materials through to site delivery and hand over to the customer, is agreed prior to commencement of production.
❑ Automated pick up systems are used to coordinate production information and ensure that components are ordered from suppliers and delivered to the production line on time.

❏ Components are allocated to a specific project and supplied to the production line (approximately 3000 components may be required for an average-sized house).
❏ Main floor, ceiling panels and external wall panels are assembled (e.g. automated nailing, screwing and bolting of panels to joists).
❏ Frames are assembled in a box-shaped structure for rigidity (e.g. by automated spot-welding machines). Floor and ceiling panels are fixed to the frame, followed by the external wall panels. Fixing techniques vary but usually involve rivets, screws, nails, bolts, welds and glues. Joints between panels are filled using gaskets.
❏ Partition walls and services are installed in accordance with the specific requirements of the customer.
❏ Pre-assembled kitchen, bathroom and staircases are installed at the factory.
❏ Painting and finishes are completed (if required).
❏ Final quality control check before the modules are protected with packaging (to avoid impact damage and to protect from moisture and dust) prior to shipping.
❏ Units are loaded onto trucks by large forklifts or cranes and transported to the construction site in accordance with the customer's delivery date.
❏ Units are then craned onto pre-prepared foundations and joined together using horizontal and vertical fixings (bolts) to provide structural rigidity. Roofing units are delivered at the same time as the modules, craned into position and fixed.
❏ Interior finishing work (if needed) is completed.
❏ Final quality control check before the completed building is handed over to the client.

Selecting a manufacturer

Before investing in offsite construction, potential purchasers (specifiers) should:

❏ visit the factory to see how the units are assembled, the quality control methods in place and the degree of flexibility available in the construction of the units (physical layout and choice of materials).
❏ check the experience and financial stability of the manufacturer, ask for and take up references, check independent reports (if available); do not rely solely on the promotional material produced by the manufacturer.
❏ look for independent approvals. Check that the modular building system has been accredited by the British Board of Agrément (BBA); International Organization for Standardization (ISO) approval should apply to the whole process; functional performance has been independently tested and endorsed (e.g. for quality, fire, acoustic insulation, thermal insulation and air leakage, and structural stability).
❏ speak to fellow architects, engineers and contractors to gain feedback on their experience with a particular manufacturer. What went well? What could have been done better?
❏ if applicable, investigate how the modular system will interface with traditional construction techniques and/or existing buildings.
❏ visit some of the built schemes. How are they weathering externally and standing up to use internally? What do the clients and users think, based on their use of the buildings?
❏ as with all other decisions about building components and products, try to consider at least three manufacturers and compare them to see who offers the best overall value for the given context of the site and client.

Logistics and transportation

Manufacturers will provide detailed transportation plans and delivery information to ensure the pre-assembled components and units can be delivered to site when required. It is not efficient to deliver components to site and store them, nor may it be practical. The pre-assembled components and units should be scheduled to be delivered to a 'just-in-time' schedule. This will involve route planning (height and width restrictions) and a detailed lifting and positioning plan in relation to the proposed reach capacity of the crane(s) to be used. This will also involve coordination with the master programme for the project.

2.3 Pre-assembly

A wide range of terms and definitions exist in relation to offsite methods. The most commonly known terms are used in this book.

On site pre-assembly

There are some situations where pre-assembly is undertaken at, or immediately adjacent to, the construction site. The term 'on site' prefabrication or pre-assembly is used to describe this activity. Sufficient space is required around the proposed building to allow for safe and efficient pre-assembly activities to take place. This is sometimes done in the open, but more commonly is carried out under the protection of a temporary structure, the field factory. Examples may include the casting of concrete components for positioning in the building and the pre-assembly of timber components, such as roof trusses, for subsequent craning into position. This technique is not common, but it may be useful where access to the site for large construction components and the vehicles on which they are transported is not possible, or the shear size of the project lends itself to a field factory approach.

Sub-assemblies and components

Sub-assemblies and components cover a wide range of familiar sub-assemblies and components. These would include familiar sub-assemblies and components such as doors, windows, staircases and balconies manufactured in factories (see *Barry's Introduction to Construction of Buildings*). More recent innovations include prefabricated reinforcement cages and mats, and precast foundations.

Prefabricated reinforcement cages and mats

The fixing of reinforcement bars (sometimes referred to as 'rebar') on the site is a labour-intensive activity. Many of the positions in which the reinforcement is laid and tied require stress of the workers' posture. Fixing reinforcement also poses a risk to health and safety of the workforce. Moving this activity to the factory allows the reinforcement cages and mats to be made in conditions where the working conditions can be controlled and hence the workers' posture and health and safety can be addressed. The resultant cages and mats can be delivered to site and quickly positioned by crane, considerably reducing the amount of time spent on site by labourers. Reinforcement mats can simply be rolled out, fixed into position and the concrete poured onto them.

Prefabricated foundations

The use of driven piled foundations removes the need for mass excavation of the site: this can save time and limits the amount of ground works. The development of brownfield sites is an example where it may be necessary to limit the amount of disturbance to the ground because of ground contamination. The use of driven piles and bored displacement piles removes the need for soil excavation and disposal that would be needed for traditional foundations. Prefabricated foundations also reduce the problems associated with working around wet concrete foundations. As soon as the piles are driven into the required position, the pile caps can be positioned and prefabricated foundation beams craned into position. Prefabricated units can then be delivered and positioned on the foundations. If the modular units are sufficiently strong in their construction, the foundation beams can be omitted; thus, the modular units sit on and span between, the pile caps.

Panelised pre-assembly (2D)

Panelised systems are also referred to as 2D systems or non-volumetric pre-assembly. The term 'flatpack' is also used. Panels are used for the construction of walls, floors and roofs, in conjunction with a structural frame. Panels are produced in two forms, either as uninsulated panels (termed 'open') or thermally insulated panels (termed 'closed'). Panels are relatively easy to manoeuvre and fix on the site using cranes and site personnel. Once the units are fitted together, the corner finishing pieces are attached and the units sealed. Effective fitting of prefabricated units relies on the use of sealants between individual panels to create an airtight seal.

Prefabricated panels range from simple unfinished wall, floor and roof panels to fully finished panels and cassettes.

❏ Open wall and roof panels with skin on one side only. Uninsulated floor panels comprising exposed joists or beams with floor decking only to the upper face.

❏ Thermally insulated open and closed wall, roof and floor panels finished on both sides but without surface finishes.

❏ Thermally insulated closed wall and roof panels finished on one side (internal or external face, depending on the design). Floor panels will be finished on the upper or lower side.

❏ Thermally insulated closed wall, roof and floor panels finished on both sides and including the integration of mechanical and electrical services. Wall panels also include doors and windows. Roof panels also include roof windows. These panelised systems are sometimes referred to as 'enhanced' panels because of the level of finishing.

The word 'cassette' is used to describe a floor or wall or roof panel. Cassettes are usually closed systems with built-in lifting straps. They are fully insulated and include service runs. Timber roof cassettes allow clear spans of up to 12 m by incorporating engineering joists. Metal cassettes also allow long clear spans, the structural support provided by lattice cross-section joists built into the cassettes. Panels and cassettes are designed and produced to be loadbearing so that they can be joined on site to create usable space. They can also be used in conjunction with a structural frame.

Volumetric (or modular) pre-assembly (3D)

Volumetric and modular pre-assembly are terms that tend to be used interchangeably to describe the process of making large parts of buildings, or entire buildings, in a factory. Volumetric is used to describe the enclosure of space, hence the term 3D modules. Fully complete modules will include wiring and plumbing, surface decoration, fixtures and fittings. In some cases, they may also include furniture. Modular bathroom and toilet pods and plant rooms are well-suited to modular pre-assembly. Volumetric units include bedrooms, kitchens, bathrooms and toilet pods . The size of the units is limited by transportation, with $8 \times 3.6\,m$ being a typical size. Completed modules are transported to site and positioned by crane on a pre-prepared foundation or slotted into a structural frame. Rarely is there any need for scaffolding or on site storage facilities, allowing for rapid assembly and completion of the project on site. The typical range includes the following, ranging from the most simple to the most complete modules:

- ❑ Modules with surface skin to inside or outside face only and no thermal insulation (the insulation and finishes are added on site).
- ❑ Thermally insulated modules with skin to both sides but no surface finishes.
- ❑ Thermally insulated modules with the inside or outside face finished.
- ❑ Modules with finishes to both sides, complete with thermal insulation and integration of services, windows and doors.

Modular building is one means of helping to achieve efficiency, reduce wastage of materials and deliver improved quality of the finished product. Some specialist commercial applications, such as chains of hotels, supermarkets and fast food outlets, have exploited factors such as time and repetition of a particular style (associated with brand image) particularly well to make prefabrication and modularisation work for their business needs. For commercial applications, the slight increase in initial build cost can be offset against savings in time and longer-term savings in the repetition of units. Supermarkets, hospitals, schools, airports, hotel chains and volume house builders have successfully used modular construction techniques.

Economic advantages are achieved with projects that have long runs of identical modules. For smaller developers, the speed of construction may not be their primary concern; however, other factors such as more consistent quality and improved working conditions may be determining factors.

Logistics is a major consideration when constructing buildings using large volumetric units. Once assembly has been completed in the factory the modules need to be labelled and temporarily stored, often in a protected environment to prevent damage from the weather, until they are required on the site. Each unit is labelled so that it is clear where they belong in the assembly sequence for the building, and stored so that they can be delivered in the correct order. To reduce demands on storage, 'just-in-time' manufacturing processes may be used. This means that the units are completed just before they are required on site, which reduces demands on storage, although any unexpected delays at the factory or during transportation will result in delays on the site. Consideration must be given to the sequence of lifting the modules into or onto the structural frame or building, and also to safely manoeuvring them into their final position. Clear access must be maintained to ensure the modules are not damaged during the lifting and positioning process. If scaffolding is being used, it must be designed to allow adequate space for access.

Hybrid systems (2D and 3D)

Combining panelised (2D) and volumetric (3D) systems is known as a hybrid system. An example would be a development that uses a flatpack design along with volumetric modular bathroom pods. Combining several different techniques can help to resolve design challenges that the use of one technology (system) cannot achieve in isolation. Coordination is required to ensure compatibility between the systems, both in terms of their constructability and their long-term durability.

Modular building services

Building services is one area of construction that can benefit in a major way from prefabrication and pre-assembly. A considerable part of building services is repetitive work. Many components can be pre-assembled and grouped together, for example horizontal pipework, vertical risers, complex wiring systems, pre-wired and assembled electrical installations (light fittings, switches, heating units, etc.).

It is beneficial to do as much as possible of the assembly of services offsite in clean and controlled environments. While it may not be possible to prefabricate long runs of cables and pipes that have to be fed around the building, it is possible to assemble the fixtures, fittings and plant. This has led to the development of innovative jointing and fitting systems to ease assembly and future replacement, repair and disassembly work. It is also becoming common to break services down into units that can be delivered as discrete modules. Plant rooms with boilers, air handling units, power terminals and connecting cables and pipework can be made up offsite in structural frames, tested, then transported to site and lifted into place, where they are subsequently commissioned for service (Figure 2.1). Photograph 2.1 shows an example of a standard steel frame structure fitted out with prefabricated bathroom and toilet pods as part of a hotel development.

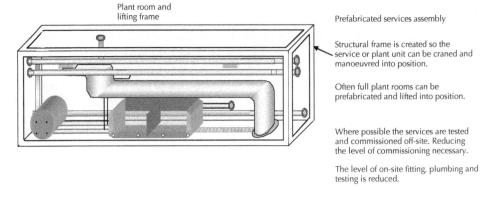

Plant room and lifting frame

Prefabricated services assembly

Structural frame is created so the service or plant unit can be craned and manoeuvred into position.

Often full plant rooms can be prefabricated and lifted into position.

Where possible the services are tested and commissioned off-site. Reducing the level of commissioning necessary.

The level of on-site fitting, plumbing and testing is reduced.

Figure 2.1   Modular building services.

Standard panels and units are stored ready for delivery.

Panels can be assembled around a frame or can form a self-supporting structure.

Photograph 2.1   Framed construction with prefabricated bathroom and service pods.

2.4 Joints and joining

Whatever offsite system is chosen, the quality and performance of the completed building will depend upon the way in which the unit is attached to the foundations and how individual modules and panels are joined together. Structural integrity of the whole building relies on the way in which the joints are designed and how the pre-assembled components and units are positioned and fixed to one another on the site. Acoustic performance needs careful consideration to avoid direct and indirect sound transition through structural frames and/or fixings. Acoustic insulation is usually achieved with the use of double-layer walls and floor that are designed to be acoustically isolated, hence reducing unwanted sound transmission. Similarly, the fire resistance of the completed building will be affected by the integrity of joints and associated fire and smoke stopping in cavities and at junctions. Most modules are designed with fire resistance of 30–60 minutes, which can be increased if required.

Tolerances

Tolerances are crucial to the final quality and durability of the building. The three interrelated tolerances to consider are manufacturing, positioning and joint tolerances:

(1) *Manufacturing tolerances.* Factory production methods are capable of producing components, panels and modules to very precise dimensions. These are dimensionally consistent, regardless of how much product is required. Manufacturers will provide full details for their product families.
(2) *Positional tolerances.* Maximum and minimum allowable tolerances are essential for safe and convenient positioning and assembly. The specified tolerance will depend upon the size of units being manoeuvred and the technologies employed to position the units. It is important to ensure that all units are positioned without damaging them or their neighbouring units. Consideration needs to be given to the reach of cranes.
(3) *Joint tolerances.* These will be determined by the materials used in the construction of the units, which determine the extent of thermal and structural movement; the size of units and their juxtaposition with other units. The design of the joint will impact on performance requirements such as air infiltration, sound attenuation and thermal properties. The sealing of the joint after assembly will be instrumental to the long-term durability of the building.

Site work and connections

The setting out and construction of the foundations must be carried out accurately since there is no (or very little) room for error. This applies to modules that are placed and bolted directly onto the foundations and also to the erection of the structural framework to house a number of modules. Accuracy also applies to the positioning of utilities connections such as electricity, gas, mains water and waste drainage. Modules are usually bolted to a ground beam, directly to the foundation, or directly to the structural frame, after which the service connections are made and subsequently tested. Manufacturers provide detailed information relating to their particular system; however, it is common for structural connections to be made in both the horizontal and vertical planes to ensure structural integrity. Most systems rely on bolted connections and/or self-tapping screws. Metal modular systems usually incorporate welded connecting plates to the modules to allow them to be bolted to the structural frame and/or adjacent module. Timber modular systems are usually constructed on a steel subframe, to which metal shoes are welded to allow the module to be bolted to the foundation connection. Some system manufacturers also include cover plates to provide protection to the joint between modules (which are usually sealed with expanding foam prior to the cover plate being fixed). Other systems rely on sealing with a flexible weatherproof sealant, or a bespoke cover to suit the design of the building.

2.5 Prefabricated housing

Offsite production methods have long been used in housing. In the UK, the most common systems are based on precut timber, lightweight metal and to a lesser extent precast concrete. Hybrid systems utilise the benefits of two or more materials, for example, a steel subframe onto which a framed timber module is assembled. The range of designs and cost ranges from high-end bespoke individual housing to larger housing developments and apartment buildings. More recently the mass housebuilding companies have moved towards offsite modular systems, either investing in their own production facilities or partnering with manufacturers and suppliers of modular systems.

Pre-cut timber systems

Timber was the first material to be used for prefabrication, being readily available and easy to work in the factory with large machines and on site with hand-held tools. Timber also has the advantage of being easy to work (repair) if damaged in transit or during construction. A variety of systems that encompass varying degrees of factory production are available from manufacturers located in the UK, Europe and North America. These range from the fully built factory house, delivered to site and craned onto suitable foundations, to the 'kit-of-parts' which are assembled on site by hand and are popular with self-build (DIY) and self-help schemes.

The advantages of timber systems relate to the thermal properties of timber, the fact that there is no waste (all 'waste' can be recycled), the ease of handling, working and fixing, and the relative ease of repair and replace. The disadvantages tend to relate to the uniqueness of the manufacturer's system and the associated difficulties of extending or altering the loadbearing structure. Given that timber is relatively lightweight, care needs to be taken with the detailing and fixing to prevent the unwanted transmission of sound to adjoining properties. Attention to detail is also required due to the potential reduction in thermal insulation around timber studs. Further information on timber and structural timber frames can be found in Chapter 5 of this book.

The Segal self-build method

The architect Walter Segal developed a simple method of construction based on timber frame construction and modern materials, specifically for self-help community building projects in the UK. The Segal method is based on a modular grid system that uses standard sizes of building materials as supplied by builders' merchants. The timber frame is built off simple pad foundations, which are dug at existing ground levels to avoid the need for expensive site levelling. Once the structural frame has been erected the roof can be constructed, walls completed and services installed. The lightweight and simple design allows both men and women to build their home (individually or as part of a cooperative group) using simple tools and with limited knowledge or experience of building. This dry construction method eliminates what Segal called the 'tyranny of wet trades' (plastering, bricklaying, etc.) and forms a lightweight, adaptable, ecologically sound building that is designed to suit the requirements of the users (and also the builders). Considerable cost savings are possible due to savings on labour and, to a lesser extent, materials due to the simplicity of the design.

Volumetric timber-framed units

Volumetric production of timber units has been greatly assisted by developments in IT (especially building information modelling), allowing the production of a large variety of standard house types and providing the means to computer-generated bespoke designs. As a general guide, the timber-framed houses built in a factory use 20–30% more material in the framing than those framed on the site. This is to ensure safe lifting and transportation. The additional cost of the material is offset against time and labour savings. The majority of factories will glue and nail or screw the components together to form a solid, volumetric, assembly. The main principles used are those outlined in Chapter 5. The main difference is that it is easier to control quality in the factory and the whole building assembly can be kept dry during manufacture, transportation and positioning, thus significantly reducing concerns about the moisture content of the timber. Photograph 2.2 shows a panel of a timber-framed house being assembled under factory conditions. These panels are then joined together to form a modular unit.

The Yorkon assembly line demonstrates the cleanliness and efficiency of off-site construction. Lifting gear, flat clean floors and the controlled factory environment provide much improved working conditions compared with that of a construction site.

Volumetric assembly–fully fitted apartments being assembled in the factory

Roof assembly

The externals walls for the modular apartments are lifted into position

Photograph 2.2 Factory production of a timber-framed house. Insulation and windows installed (a); exterior timber cladding being fixed (b).

Metal systems

Lightweight steel is the material most used for metal-framed units. Advantages and disadvantages are not so different to those for cut-timber systems. The advantages of metal-framed volumetric units include the possibility of zero waste in manufacturing, the capacity for long clear spans, ease of handling and fixing, and easy to repair and replace. The disadvantages tend to relate to the uniqueness of the manufacturer's system and the associated difficulties of extending, altering and replacing the metal components. Given that steel is relatively lightweight, care needs to be taken with the detailing and fixing to prevent the unwanted transmission of sound to adjoining properties. Attention to detail is also required due to the potential reduction in thermal insulation around the metal frame. Further information on steel and structural steel frames can be found in Chapter 6 of this book.

Steel-framed housing

A typical steel-framed modular housing development would comprise a 75 mm deep galvanised steel frame with insulation and vapour barrier sheathing similar to a timber framed module. Wall frames are delivered with integral bracing, which are easily placed into position. Floor joists are usually 150 mm deep 'Z'-sections that are attached to the wall panels. The roof structure is assembled at ground level and lifted into place in one piece. The windows and doors can be screwed to the steel frame before the external finish, for example, brickwork, is constructed.

Modular steel framing

Volumetric house construction comprises steel-framed modular units that may be joined together to form semi-detached and terraced units. The pitched roof is also prefabricated and craned into position. Average construction times are between 6 and 8 weeks for a house, with the steel frame taking around 3 days to erect. The steel frame construction comprises cold-formed lightweight steel stud sections, commonly 75 mm deep galvanised steel framing members, which are sheathed in insulation on the external face and finished with fire-resisting board on the inside face of the wall (thus creating a panel construction). The wall frames include integral steel diagonal cross-bracing members and are designed to be easy to manoeuvre and fix on site. Floor construction typically comprises 150 mm deep steel joists fixed to a Z-section element attached to the wall panels. Windows are installed on site and the external cladding (usually brickwork) is built on site once the frame is complete. Advantages include quick construction, dimensional accuracy, and long-life and long-span capabilities (thus allowing for future adaptability). The modules and materials are also relatively easy to recover and recycle at the end of their service life.

Concrete systems

Concrete panels have been in use since the early 1900s in the UK. Concrete is cast in large moulds and the reinforced units transported to site before being craned into position. Early pioneers would cast the concrete on site, but with concerns over quality control and efficiency, the casting of units has now moved to a few specialist factories where quality can be carefully controlled and the casting process made cost-effective. Some of these units may be made from standard mould shapes and are effectively available off the shelf; others are designed and cast to a special order. The completed units are then delivered to site to suit the contractor's programme and are lifted into position using a crane.

Considerable investment is required in making the moulds, and the units are heavy for transporting and positioning. More recent developments have been in the use of lightweight reinforced concrete units; however, there is still a large amount of work required on site to finish the concrete units, and this can add considerable time to the site phase. Reinforced concrete frames (rather than structural panels) can be used to provide the structural support for modules or pods, which are craned into position. It is, however, more usual for a steel frame to be used.

The advantages of concrete are largely associated with the material's inherent properties of high thermal mass, good sound and fire resistance and its loadbearing capacity. Disadvantages are associated with the material as it is heavy and awkward to manoeuvre, and it is difficult to make changes during and after construction. For example, drilling holes into the panels for services produces waste (if the holes are not pre-formed in the factory). Care is also needed to avoid thermal bypasses at junctions, and a high degree of finishing may be required on site. Further information on concrete and structural concrete frames can be found in Chapter 7 of this book.

Concrete and steel hybrid systems

Modular building is suited to small, inner city and urban sites. Several hybrid systems are available that use a combination of materials. Typical examples are modules constructed of steel and concrete, bolted together and clad in brickwork or stone to suit the streetscape and provide variety to different projects. Closed panel timber systems, with windows, doors and insulation can be combined with roof and floor cassettes and beam and block floors. There is plenty of variety available. U values of $0.15\,\text{W/m}^2\,\text{K}$ are easily achieved for external walls and roofs (or ceilings) with thermally insulated cassettes.

2.6 Additive manufacturing (3D printing)

The term 'rapid prototyping' is used to describe techniques that allow the fabrication of a building component or assembly using digital files and 3D printing. This technique allows manufacturers and design teams to quickly and cheaply produce a scale model of a prototype prior to full-scale production. The rapid development of 3D printing technologies also makes it possible to produce full-scale building components from additive manufacturing techniques. Components and buildings are printed directly from digital data files that are contained in 3D models. A technique similar to inkjet printing on paper. The technique relies on 'printers' producing artifacts by adding one layer of material to another to create the whole; hence the term additive manufacturing. Other terms such as direct digital manufacturing and freeform construction are also used. Material is sprayed from a nozzle to 'print' buildings and building components, ranging from simple to relatively complex shapes. Typically, this is carried out in factories due to the large size of the printers, but it is possible to use very large printers on the construction site with the print heads travelling along scaffold to print materials in consecutive layers. Materials used range from concrete and cementitious materials, clay, mud, plastics, polymers and nylon to metals. These comprise organic, non-organic and composite materials such as fibre-reinforced polymer (FRP), along with reclaimed and recycled content materials, all with varying environmental footprints. The main characteristic of the sprayed materials is a low viscosity so that they can be pumped through the printer head. Along with many other innovative approaches to construction these techniques are promoted as environmental friendly given the low level of construction waste generated in the construction process. Some of the materials used have a better carbon footprint than others.

There are four main approaches to additive manufacturing as given below:

(1) *Concrete printing.* The materials used are cement-based mortars and gypsum using a wet process. The material is sprayed through a nozzle to build up layers of material into walls and other features. Reinforcement is required to provide tensile strength to the artifact. There are limitations with this approach because of the slow curing process, although this has not stopped manufacturers from printing concrete houses using recycled construction materials comprising glass fibre, cement and additives. Concrete printing may be used to produce components without using formwork, which is known as freeform construction. There is a saving in materials and time as there is no need to construct a formwork or mould.

(2) *Contour crafting (CC).* Contour crafting is a rapid prototype or 3D printing process to fabricate large components and full-scale structures. These are generated from digital data. Extrusion nozzles are housed on robotic arms on a gantry and controlled by computer software to quickly produce large structures, such as houses, by printing a quick-setting material.

(3) *D-shape printing.* The D-shape printing process involves a large 3D printer that prints objects in layers. The printer mixes sand with an inorganic liquid (seawater) and a magnesium-based binder to produce objects with a stone-like appearance.

(4) *Direct metal laser sintering.* This process relies on lasers precisely heating layers of powdered metal material until it forms into a solid mass. This process can produce components with an extremely fine level of detail and accuracy.

Subtractive manufacturing

Some manufacturing techniques rely on removing material, such as routing machines and laser cutters. This is known as subtractive manufacturing and the two most common techniques are:

❑ *CNC router and CNC milling machines.* A CNC router is a computer-controlled machine that can be used for cutting hard materials such as timber, steel, plastics and composites. The cutting is controlled from a 3D digital file. A CNC milling machine works in a similar way, using rotary cutters to remove material from the surface of a hard material. Both machines enable a high level of detail to be achieved.

❑ *Laser cutting.* This technique involves firing a laser to melt, burn or vaporise material to cut or engrave a design from a digital computer file. Materials that can be cut and engraved range from wood sheets to acrylics and plastics.

Further reading

For additional information on modern methods of construction (MMC) see; www.designingbuildings.co.uk; and for steel systems see Steel Construction Info: www.steelconstruction.info.

Reflective exercises

This reflective exercise relates to your current design project, regardless of its location, size and intended use.

❏   What aspects of the building could be fabricated offsite. How would this influence your design and why would you decide to use offsite construction for the project?
❏   What benefits does offsite fabrication offer (i) the client, (ii) the design team and (iii) the builder? Justify your thinking.
❏   Are there any challenges with using offsite fabrication for your design project?
❏   How could offsite fabrication impact the environmental impact of your building project?
❏   What is the likely impact on embodied carbon and operation carbon?

Pile Foundations, Substructures and Basements: Chapter 3 AT A GLANCE

What? Pile foundations are columns that are driven or cast into the ground to provide a supporting framework for the building to rest on. Steel and concrete columns, known as pile foundations, can be inserted (driven) or bored into the ground, transferring the building loads to loadbearing strata. Hardwood timber may also be used or piles, especially for temporary low-rise buildings. 'Substructures' is a term used to describe a variety of work that takes place below ground level to form supporting structures and basements.

Why? It is common for framed buildings to bear onto piled foundations, especially when the loadbearing strata is a considerable depth below the surface of the ground. Substructures and basements provide additional space within the building footprint. In situations where it is not possible to build high, for example, in semi-urban or rural areas, it may be possible to provide additional space under the building, subject to ground conditions and town planning restrictions.

When? The piles are placed or driven into the ground early in the construction process. This ensures any local ground disturbance caused by the piling does not affect other works, for example, site drainage. On sites that have previously had buildings on them, it is common practice to identify and remove existing foundations or to avoid them when placing the piles. Substructure works and the construction of basements are usually one of the first tasks to undertake to ensure a logical flow of work on the site. Adding a basement to an existing building is likely to have an impact on adjacent buildings and foundations.

How? There are a variety of piling techniques available, the choice of one over another largely dictated by ground conditions, the building design and appropriate physical access for machinery to conduct the piling. Basements tend to be constructed of reinforced concrete or dense concrete blocks, which is then tanked to prevent water and damp from entering the building.

3 Pile Foundations, Substructures and Basements

It is common for framed buildings to bear onto piled foundations. Steel and concrete columns, known as pile foundations, can be inserted (driven) or bored into the ground, transferring the building loads to loadbearing strata. The loadbearing strata may be a considerable depth below the surface of the ground. When building on previously developed sites, it is common practice to identify and remove existing foundations or to avoid them when placing the piles. Alternatively, some or all of the existing foundations may be reused to avoid unnecessary ground disturbance and help to reduce the environmental impact (and cost) of the new development. For temporary structures, it is possible to use timber piles that can be left in the ground to decay naturally or steel screw piles that can be removed (unscrewed) when the building is taken down. In this chapter, the emphasis is on piled foundations, substructures and basements. The main functional requirements of foundations are described in *Barry's Introduction to Construction of Buildings*.

3.1 Pile foundations

The word 'pile' is used to describe columns, usually of reinforced concrete, driven into or cast in the ground to carry foundation loads to deep underlying firm stratum, or to transmit loads to the subsoil by the friction of their surfaces in contact with the subsoil (see Figure 3.1). The main function of a pile is to transmit loads to lower levels of ground by a combination of friction along their sides and end bearing at the pile point or base. Piles that transfer loads mainly by friction to clays and silts are termed friction piles, and those that mainly transfer loads by end bearing to compact gravel, hard clay or rock are termed end-bearing piles. Four or more piles may be used to support columns of framed structures. The columns are connected to a reinforced concrete pile cap connected to the pile, as illustrated in Figure 3.2. Piles may be classified by their effect on the subsoil as displacement piles or non-displacement piles. Displacement piles are driven, forced or cut (by an auger) into the ground to displace subsoil. The strata are penetrated. No soil is removed during the operation. Solid concrete or steel piles and piles formed inside tubes, which are driven into the ground and which are closed at their lower end by a shoe or plug, which may either be left in place or extruded to form an enlarged toe, are all forms of displacement pile. Non-displacement piles are formed by boring or other methods of excavation that do not substantially displace subsoil. Sometimes, the borehole is lined with a casing or tube that is either left in place or extracted as the hole is filled.

Barry's Advanced Construction of Buildings, Fifth Edition. Stephen Emmitt.
© 2023 John Wiley & Sons Ltd. Published 2023 by John Wiley & Sons Ltd.

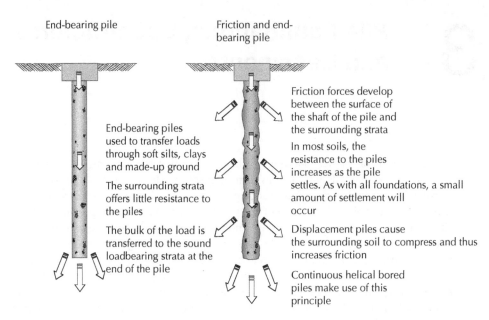

Figure 3.1   End-bearing and friction piles.

Driven piles are those formed by driving a precast pile and those made by casting concrete in a hole formed by driving. Bored piles are those formed by casting concrete in a hole previously bored or drilled in the subsoil.

Driven piles – concrete

Square, polygonal or round section reinforced concrete piles are cast in moulds in the manufacturer's yard and are cured to develop maximum strength. The placing of the reinforcement and the mixing, placing, compaction and curing of the concrete can be accurately controlled to produce piles of uniform strength and cross-section. The precast piles are often square section with chamfered edges, as illustrated in Figure 3.3. The head of the pile is reinforced with helical binding wire; this helps prevent damage that would otherwise be caused by driving the pile into the ground. Once the pile is in place, the concrete at the top of the pile is removed to expose the main reinforcement. The helical reinforcement can be removed once the main reinforcement bars, which will be tied into the pile cap, are exposed.

To assist driving, the foot of the pile may be finished with a cast iron shoe, as illustrated in Figure 3.4. Figure 3.5 is an illustration of a typical pile. The piles are lifted into position and driven into the ground by means of a mechanically operated drop hammer attached to a mobile piling rig. To increase the length of the pile to the required depth, additional segments are added and the pile is driven in until a predetermined 'set' is reached. The word 'set' is used to describe the distance that a pile is driven into the ground by the force of the hammer falling a measurable distance. From the weight of the hammer and the distance it falls, the resistance of the ground can be calculated, and the bearing capacity of the pile calculated.

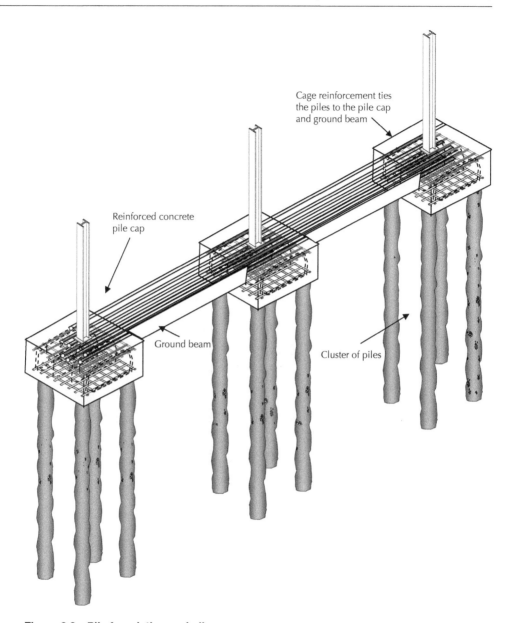

Cage reinforcement ties
the piles to the pile cap
and ground beam

Reinforced concrete
pile cap

Ground beam

Cluster of piles

Figure 3.2   Pile foundation and pile caps.

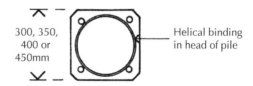

300, 350,
400 or
450mm

Helical binding
in head of pile

Figure 3.3   Head of precast concrete pile.

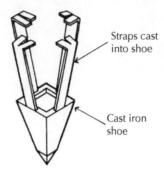

Figure 3.4   Shoe of precast concrete pile.

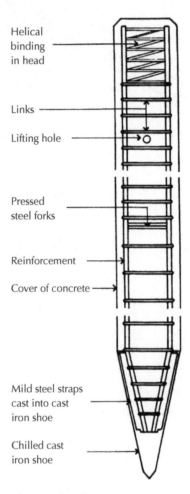

Figure 3.5   Precast reinforced concrete pile.

To connect the top of the precast pile to the reinforced concrete foundation, the top 300 mm of the length of the pile is broken to expose reinforcement to which the reinforcement of the foundation is connected.

Precast-driven piles are not generally used on sites in built-up areas. Difficulties may be experienced when attempting to move large precast piles through narrow streets, although using smaller sections may overcome this problem. The logistics of moving precast and prefabricated objects should always be considered when selecting construction methods. The noise, vibration and general disturbance caused by driving (hammering) piles into the ground can be a nuisance. Where vibration is excessive, or buildings and structures are sensitive to vibration, damage may be caused to adjacent buildings, structures and services. Driven piles are used as end-bearing piles in weak subsoils, where they are driven into a firm underlying stratum. Driven piles give little strength in bearing due to friction of their sides in contact with the soil, particularly when the surrounding the soil is clay. This is due to the fact that the operation of driving moulds the clay around the pile and so reduces frictional resistance between the pile and the surrounding clay. In coarse-grained cohesionless soils, where the piles do not reach a firm stratum, driven piles act as friction-bearing piles due to the action of pile driving, which compacts the coarse particles around the sides of the pile and so increases frictional resistance and in compacting the soil increases its strength. This type of piled foundation is sometimes described as a floating foundation, as is a cast-in-place piled foundation, as bearing is mainly by friction and in effect, the piles are floating in the subsoil rather than bearing on firm soil.

Driven tubular steel piles

Tubular steel piles are similar in principle to precast concrete piles. Typically the piles are 6 m long, although shorter 3 m segmental piles can be used in areas where access and headroom are restricted. The piles are particularly suitable for driving in difficult or uncertain ground conditions up to 50 m deep. Hard driving conditions caused by fill, obstructions and boulders can also be dealt with. The piles are capable of taking large axial loads. Tubular steel piles can also accommodate horizontal loads resulting from bending moments and horizontal reactions, and can resist vertical tension loads that are a result of uplift and heave reactions.

Normally the piles are driven with an open end so that soil fills and plugs the void; in exceptional circumstances, the end of the pile can be closed by welding and end plate. The void at the top of the pile is filled with concrete. Reinforcement can be positioned in the pile, allowing it to be tied into the pile caps reinforcement cage. The piles are top driven using rigs with hydraulic hammers. The site should be firm, dry and level ready to receive the piling rigs, which can weigh up to 35 tonnes.

Driven cast-in-place piles

Driven cast-in-place piles are of two types: the first has a permanent steel or concrete casing and the second uses a temporary casing. The purpose of driving and maintaining a permanent casing is to consolidate the subsoil around the pile casing by the action of driving.

The lining is left in place to protect the concrete cast inside the lining against weak strata of subsoil that might otherwise fall into the pile excavation. Permanent casings also protect the green concrete (concrete which has not set) of the pile against static or running water that may erode the concrete. The lining also protects the concrete against contamination. Figure 3.6 shows a driven cast-in-place pile with a permanent reinforced concrete casing. Precast reinforced concrete shells are threaded on a steel mandrel. Metal bands and bitumen seal the joints between shells. The mandrel and shells are lifted on to the piling rig and then driven into the ground. At the required depth, the mandrel is removed, a reinforcing cage is lowered into the shells and the pile completed by casting concrete inside the shells. This type of pile is used principally in soils of poor bearing capacity and in saturated soils where the concrete shells protect the green concrete cast inside them from static or running water.

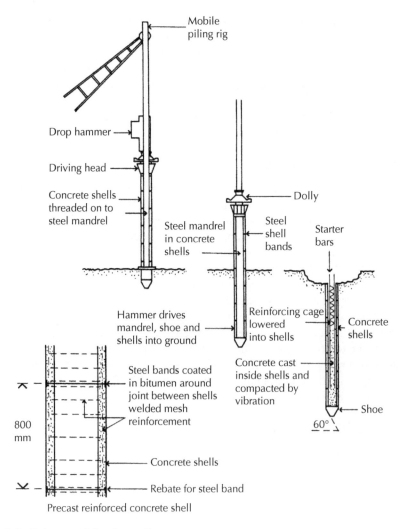

Figure 3.6   Driven cast-in-place pile.

A driven cast-in-place pile without permanent casing is illustrated in Figure 3.7. The base of a steel lining tube, supported on a piling rig, is filled with ballast. A drop hammer rams the ballast and the tube into the ground. At the required depth, the tube is restrained and the ballast is hammered in to form an enlarged toe. Concrete is placed by hammering it inside a lining tube; the tube is gradually withdrawn. The effect of driving the tube and the ballast into the ground is to compact the soil around the pile, and the subsequent hammering of the concrete consolidates it into pockets (voids) and weak strata. The enlarged toe provides additional bearing area at the base of the pile. This type of pile acts mainly as a friction pile.

Another type of driven cast-in-place pile without permanent casing is formed by driving a lining tube with a cast iron shoe into the ground with a piling hammer operating from a piling rig, as illustrated in Figure 3.8. Concrete is placed and consolidated by the hammer as the lining tube is withdrawn. The particular application of this type of pile is for piles formed through a substratum so compact as to be incapable of being taken out by drilling. The purpose of the cast iron shoe, which is left in the ground, is to penetrate the compact stratum through which the pile is formed.

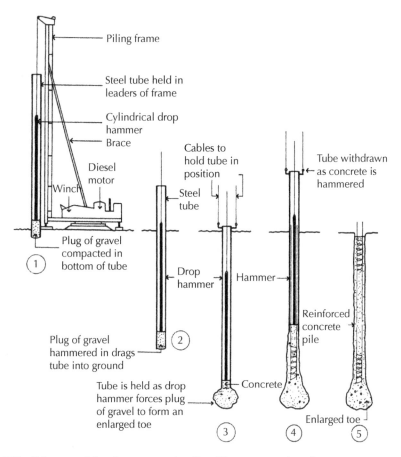

Figure 3.7   Driven cast-in-place concrete pile with permanent casing.

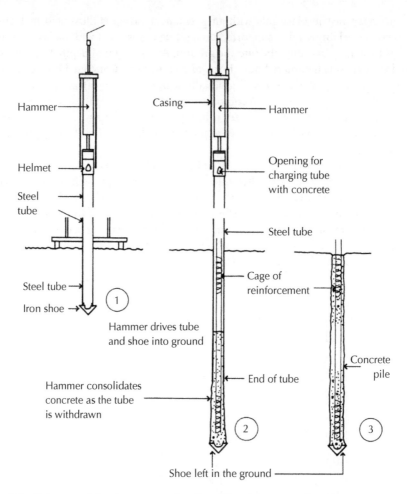

Figure 3.8   Driven cast-in-place concrete pile without permanent casing.

Jacked piles

Figure 3.9 illustrates a system of jacked piles that are designed for use in cramped working conditions, as for example, where an existing wall is to be underpinned and headroom is restricted by floors and in situations where the vibration caused by pile driving might damage existing buildings. Where the wall to be underpinned has a sound concrete base, a small area below the foundation is excavated. This provides sufficient space for small beams, the pile jack and the first section of pile to be inserted.

First the pile cap with the steel shoe is inserted and driven into the ground. The jack then retracts and the next section is inserted between the driven pile section and the jack. The pile sections are then repeatedly inserted between the jack and then driven into the ground (Figure 3.9). The precast concrete sections are jacked into the ground, as illustrated. Some

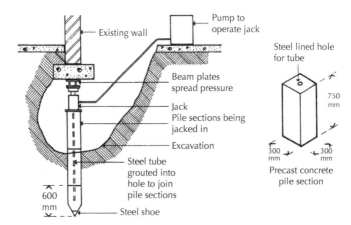

Figure 3.9   Precast concrete jacked pile.

systems use hollow precast concrete pile sections. Where hollow sections are used, reinforcement can be inserted into the void and concreted in position; alternatively, lengths of steel tube are often inserted and grouted in position, making a strong connection between all of the sections. Once the jack is removed, a concrete cap is cast on top of the pile and up to the underside of the concrete base.

When the wall to be underpinned has a poor base and the wall or structure above might be disturbed by either the area excavated for underpinning or the jacking, then an alternative process must be used. One option is to insert pairs of piles on each side of the wall. Steel or reinforced concrete beams (often called needles) are then inserted through the wall above the foundation but below ground level. The needles will be used to support the wall and transfer the loads to the piles on either side of the original foundation. When piles are formed on both sides of the wall, they are jacked in against temporary units loaded with kentledge. As there is no building foundation to jack against, a temporary loaded structure (kentledge) must be used so that the jack can drive piles from the structure into the ground. Once the piles are jacked into position, the jack and kentledge are removed. The piles and needles can then be tied together using a reinforced concrete pile cap. Figure 3.10 shows some underpinning arrangements.

Bored piles

Auger bored piles
A hole is bored or drilled by means of earth drills (mechanically operated augers), which withdraw soil from the hole into which the pile is to be cast. Occasionally, it is necessary to lower or drive in steel lining tubes as the soil is taken out, to maintain the sides of the drilled hole. As the pile is cast, the lining tubes are gradually withdrawn. The mechanical rigs used to install the piles come in a range of sizes, from small units weighing just a few tonnes to large rigs exceeding 20 m and weighing in excess of 50 tonnes.

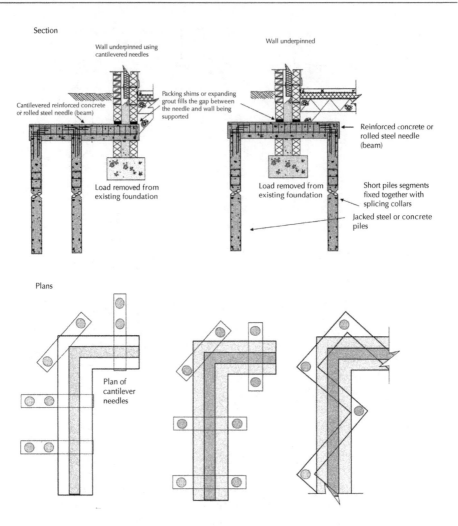

Section

Wall underpinned using cantilevered needles

Wall underpinned

Cantilevered reinforced concrete or rolled steel needle (beam)

Packing shims or expanding grout fills the gap between the needle and wall being supported

Reinforced concrete or rolled steel needle (beam)

Load removed from existing foundation

Load removed from existing foundation

Short piles segments fixed together with splicing collars

Jacked steel or concrete piles

Plans

Plan of cantilever needles

Figure 3.10 Various underpinning arrangements.

Although not common nowadays, some boring can be fixed to tripods rather than the typical tracked rigs (Figure 3.11). Advantages of such equipment are that the piling rigs are light and easily manipulated. Because all of the arisings are brought to the surface, a precise analysis of the subsoil strata is obtained from the soil withdrawn. A disadvantage of piles cast in the ground is that it is not possible to check that the concrete is adequately compacted and whether there is adequate cover of concrete around the reinforcement.

Figure 3.11 illustrates the drilling and casting of a bored cast-in-place pile. Soil is withdrawn from within the lining tubes with a cylindrical clay cutter that is dropped into the hole, which bites into and holds the cohesive soil. The cutter is then withdrawn and the soil knocked out of it. Coarse-grained soil is withdrawn by dropping a shell cutter (or bucket) into the hole. Soil, which is retained on the upward hinged flap, is emptied when the cutter is withdrawn. The operation of boring the hole is more rapid than might be supposed, and a pile can

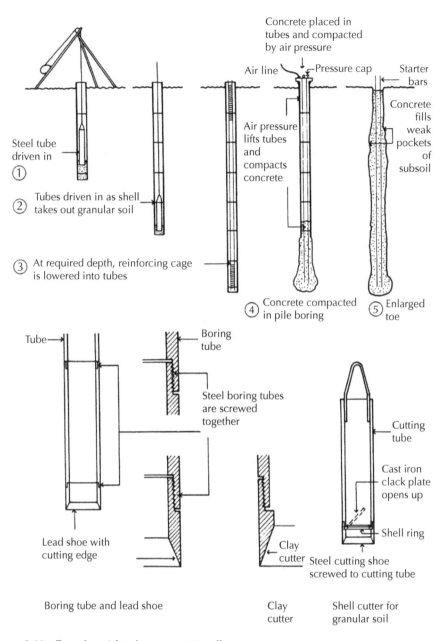

Figure 3.11   Bored cast-in-place concrete pile.

be bored and cast in a matter of hours. Concrete is cast under pressure through a steel helmet, which is screwed to the top of the lining tubes. The application of air pressure immediately compacts the concrete and simultaneously lifts the helmet and lining tubes as the concrete is compacted. As the lining tubes are withdrawn, protruding sections are unscrewed and the helmet refixed until the pile is completed. As the concrete is cast under pressure, it extends beyond the circumference of the original drilling to fill and compact weak strata and pockets in the subsoil. Because of the irregular shape of the surface of the finished pile, it acts mainly as a friction pile to form what is sometimes called a floating foundation. As the pile continues to settle into the soil, the friction forces surrounding the pile increase.

Large-diameter bored pile

Figure 3.12a and Photograph 3.1a illustrate the formation and casting of a large-diameter bored pile formed in cohesive soils. Figure 3.12b shows the same operation performed in non-cohesive soils with a coring barrel. A tracked crane supports hydraulic rams and a diesel engine, which operates a kelly bar and rotary bucket drill. The diesel engine rotates the kelly bar and bucket. In the bottom of the bucket are angled blades that rotate, excavating the strata and filling the bucket with soil. The hydraulic rams force the bucket into the ground. The filled bucket is raised and emptied and drilling proceeds. In non-cohesive soils, the excavation is lined with steel lining tubes. To provide increased end bearing, the drill can be belled out to twice the diameter of the pile (Figure 3.12c). The augers and

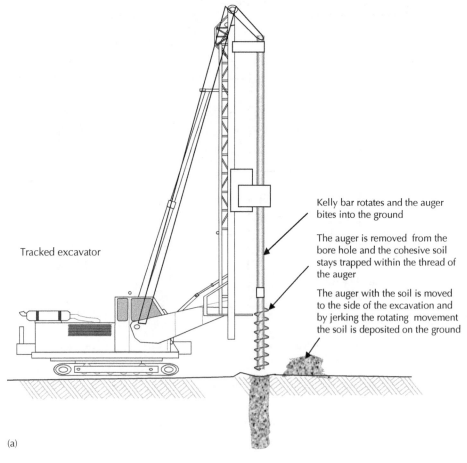

Tracked excavator

Kelly bar rotates and the auger bites into the ground

The auger is removed from the bore hole and the cohesive soil stays trapped within the thread of the auger

The auger with the soil is moved to the side of the excavation and by jerking the rotating movement the soil is deposited on the ground

(a)

Figure 3.12   (a) Bored pile with augur: cohesive soils. (b) Bored pile with core barrel: non-cohesive soils. (c) Forming a large toe with belling tool.

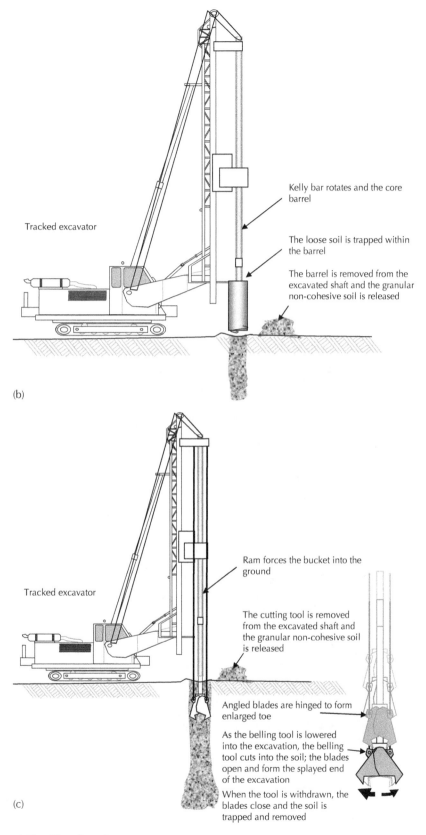

Kelly bar rotates and the core barrel

The loose soil is trapped within the barrel

The barrel is removed from the excavated shaft and the granular non-cohesive soil is released

Tracked excavator

(b)

Ram forces the bucket into the ground

The cutting tool is removed from the excavated shaft and the granular non-cohesive soil is released

Tracked excavator

Angled blades are hinged to form enlarged toe

As the belling tool is lowered into the excavation, the belling tool cuts into the soil; the blades open and form the splayed end of the excavation

When the tool is withdrawn, the blades close and the soil is trapped and removed

(c)

Figure 3.12 *(Continued)*

core barrels for cutting through rock, cohesive soils and non-cohesive soils are shown in Photographs 3.1a–e.

Rotary drilling equipment is commonly used for piles to be cast in cohesive soils. A tractor-based rig supports a diesel engine and crane jib. A cable run from the motor up the jib supports a large, square drilling rod or kelly bar that passes through a turntable, which

Photograph 3.1 (a) Piling rig with auger and kelly bar. (b) Bored cast-in-place pile with steel sleeve excavation support. (c) Core barrel with trap for excavating granular soils. (d) Core barrel with bullet-cutting teeth for cutting through rock. (e) Auger for cohesive soils. (f) Reinforcement cage for pile.

Photograph 3.1   *(Continued)*

rotates the bar to which is attached a drilling auger. The weight of the rotating kelly bar causes the augur to drill into the soil. The augur is withdrawn from time to time to clear it of excavated soil. Where the subsoil is reasonably compact, the reinforced concrete pile is cast in the pile hole and consolidated around the reinforcing cage. In granular subsoil, the excavation may be lined with steel lining tubes that are withdrawn as the pile is cast in place. This type of pile is often used on urban sites where a number of piles are to be cast, because it will cause the least vibration to disturb adjacent buildings and create the least noise disturbance. See the series of Photograph 3.1a–f for illustration of the plant and equipment associated with bored cast-in-place piles.

Continuous flight auger piles
Continuous flight auger (CFA) piles are formed using hollow stem auger boring techniques. The auger has a hollow central tube surrounded by a continuous thread. The helical cutting edge is continuous along the full length of the auger. The hollow tube that runs down the centre of the shaft is used for pumping concrete into the hole as the cutting device is withdrawn (Figure 3.13). As the CFA rig cuts into the ground, arisings are brought to the surface. The arisings allow the soil to be inspected at regular intervals, giving an indication of the strata below the surface. Once the rig has produced a bore to a calculated depth, through known strata, the auger is steadily withdrawn as concrete is pumped under pressure through the hollow stem. The concrete simultaneously fills the void left by the auger as it is extracted. The concrete reinforcement cage is pushed into the bore after the pile has been concreted. Spacers are fixed to the side of the reinforcement so that the cage is positioned centrally in the pile and adequate concrete cover is maintained around the reinforcement. CFA piles are suitable for a range of ground conditions, are relatively quiet, and cause less noise and vibration when compared to hammer-driven piles; they can accommodate large working loads and are quick to install.

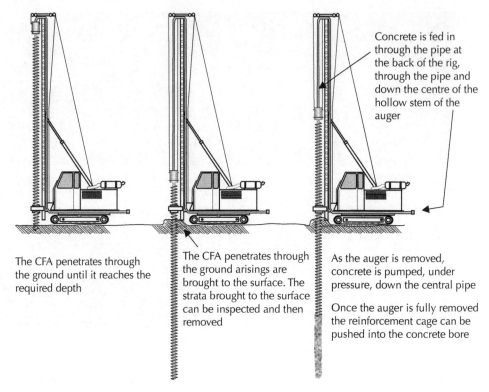

Concrete is fed in through the pipe at the back of the rig, through the pipe and down the centre of the hollow stem of the auger

The CFA penetrates through the ground until it reaches the required depth

The CFA penetrates through the ground arisings are brought to the surface. The strata brought to the surface can be inspected and then removed

As the auger is removed, concrete is pumped, under pressure, down the central pipe

Once the auger is fully removed the reinforcement cage can be pushed into the concrete bore

Figure 3.13   CFA piles.

Bored displacement piles

Continuous helical displacement piles

Continuous helical displacement (CHD) piles are becoming a popular alternative to CFA piles. The displacement piles have the advantage that they do not produce arisings; this is particularly useful on contaminated sites. In most ground conditions, the CHD piles have enhanced load-carrying capacity, compared with a CFA pile of similar dimensions. Due to the compaction of the soil, friction between the pile and the strata is increased. The increased strength gained in some ground conditions enables the pile length to be shortened, resulting in shorter installation times and more economical foundations.

The piles are formed using a multi-flight, bullet-ended shaft, which is driven by a high torque rotary head (Figure 3.14). This enables the ground to be penetrated without bringing any material to the surface (Figure 3.15). Some slight heaving of the surface may occur as the ground is compressed; however, this is normally negligible.

The pile is drilled to the calculated or proven depth; the shaft is then reversed and extracted while concrete is simultaneously pumped under pressure into the helical void that remains. Once the auger is totally extracted, reinforcement can be pushed into the concrete as a single bar or cage

Testing piles

Piles are often tested using dynamic load, sonic integrity or static load methods. These three methods are described briefly here.

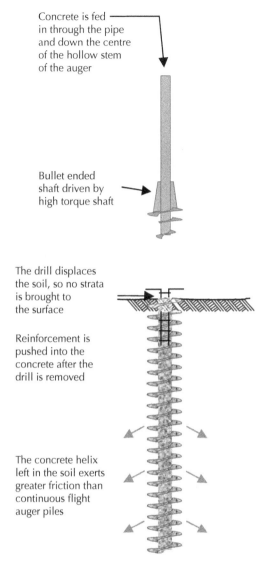

Concrete is fed
in through the pipe
and down the centre
of the hollow stem
of the auger

Bullet ended
shaft driven by
high torque shaft

The drill displaces
the soil, so no strata
is brought to
the surface

Reinforcement is
pushed into the
concrete after the
drill is removed

The concrete helix
left in the soil exerts
greater friction than
continuous flight
auger piles

Figure 3.14   Components of CHD piles.

Dynamic load methods

Dynamic load methods of testing are suitable for most types of pile, but are more frequently used on precast concrete or tubular steel piles. The test is commonly used on small piling works where the cost of static load testing cannot be justified. The test determines the loadbearing capacity of the pile, skin friction and end bearing. Other characteristics such as hammer energy transfer, pile integrity, pile stresses, driving and load-displacement behaviour can also be determined.

To dynamically test a pile, it is struck by a hammer, using the piling rig. Two strain trans-ducers and accelerometers (measures speed and acceleration) are firmly attached to the face of the pile near to the head of the pile. As the pile is struck, the equipment measures the force and acceleration of the pile. The information is relayed to the monitoring equipment. Once analysed, the data provides models of shaft friction distribution, bearing capacity and load-settlement behaviour.

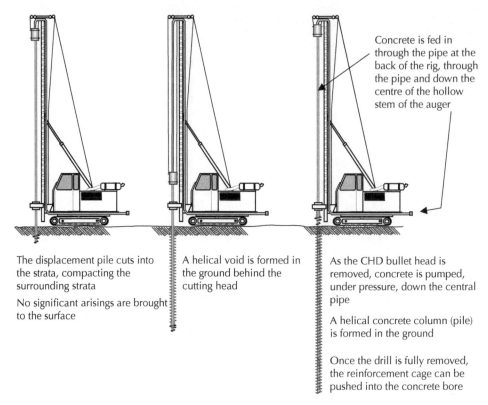

Concrete is fed in
through the pipe at the
back of the rig, through
the pipe and down the
centre of the hollow
stem of the auger

The displacement pile cuts into
the strata, compacting the
surrounding strata

No significant arisings are brought
to the surface

A helical void is formed in
the ground behind the
cutting head

As the CHD bullet head is
removed, concrete is pumped,
under pressure, down the central
pipe

A helical concrete column (pile)
is formed in the ground

Once the drill is fully removed,
the reinforcement cage can be
pushed into the concrete bore

Figure 3.15   CHD piles.

Sonic integrity testing

Sonic integrity testing is normally used on CFA, CHD or other piles foundations formed using in situ concrete. The integrity method is fast and reliable. A large number of piles can be tested in a single visit. The pile determines the reliability, morphology (form and composition) and quality of construction of the pile.

Before the pile can be tested, it must be sufficiently cured, free of latency and trimmed back to sound concrete. It is preferable to carry out the test at the final cut-off level of the pile. A small hand-held hammer is used to strike the pile. A series of low-strain acoustic shock waves are sent through the piles. As the waves pass down the pile, the sound waves rebound where changes in impedance occur. The rebound (echo) is then recorded by a small accelerometer (instrument for measuring speed and acceleration), which is held against the pile head. The response is monitored and stored, and a graphical representation produced for immediate inspection.

Static load testing

Static load testing is used to determine the displacement characteristics of a pile. All piles are suited to static load testing. Static load testing frames are assembled specifically for the test. Major piling contractors assemble frames capable of accommodating loads up to 4000 kN.

A known load has to be applied in the form of kentledge (loaded test frame) or tension pile reaction. Load can also be applied by fixing a frame to piles already installed in the ground. Other piles can then be tested against the frame load. Once an adequate reaction has been provided (load), the test is carried out using a hydraulic jack and calibrated digital load cell. Time, load, temperature and displacement data are recorded.

Pile caps and spacing of piles

Piles may be used to support pad, strip or raft foundations. Commonly a group of piles is used to support a column or pier base. The load from the column or pier is transmitted to the piles through a reinforced concrete pile cap, which is cast over the piles. To provide structural continuity, the reinforcement of the piles is linked to the reinforcement of the pile caps through starter bars protruding from the top of the cast-in-place piles or through reinforcement exposed by breaking off the top concrete from precast piles. The exposed reinforcement of the top of the piles is wired to the reinforcement of the pile caps. Similarly, starter bars cast in and linked to the reinforcement of the pile caps protrude from the top of the pile caps for linking to the reinforcement of columns. Figure 3.16 illustrates typical arrangements of pile caps. The spacing of piles should be wide enough to allow for the necessary number of piles to be driven or bored to the required depth of penetration without damage to adjacent construction or to other piles in the group. Piles are generally formed in comparatively close groups for economy in the size of the pile caps to which they are connected. Photographs 3.2a–c show the stages of

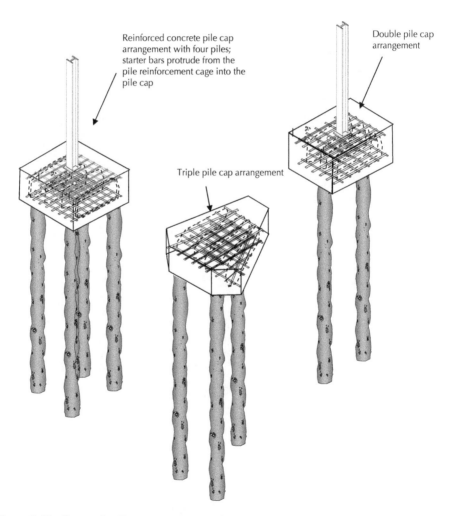

Reinforced concrete pile cap arrangement with four piles; starter bars protrude from the pile reinforcement cage into the pile cap

Double pile cap arrangement

Triple pile cap arrangement

Figure 3.16   Concrete pile cap arrangements.

Photograph 3.2   (a) Top of concrete pile broken away to expose reinforcement. (b) Triple concrete pile arrangement ready to receive pile cap. (c) Pile cap reinforcement and formwork.

the pile cap construction. As a general rule, the spacing, centre to centre of friction piles, should be not less than the perimeter of the pile, and the spacing of end-bearing piles not less than twice the least width of the pile.

Geothermal piles

Geothermal piles, also known as 'energy piles' and 'heat exchanger piles', are pile foundations that are designed to work in conjunction with a closed-loop ground source (geothermal) heat pump. These piles use the thermal mass of the ground to store unwanted heat energy in the summer (for cooling) and to provide heat energy in the winter. As a source of renewable energy, these piles can help to reduce carbon emissions.

In simple terms, a geothermal pile has one or more polyethylene pipes incorporated into the pile during construction. These are subsequently filled with a heat exchange fluid, usually water. The temperature of the pile is likely to be somewhere around 12 °C, with the heat pump producing temperatures in the range of 25–40 °C, making this system ideal for low-temperature heating and cooling systems. Fibre optic sensors can be used to establish the structural integrity of the pile, especially when the piles are large and contain considerable amounts of reinforcing. The use of geothermal piles is a relatively new innovation, and research is ongoing to investigate the structural and thermal performance of geothermal piles.

3.2 Ground stabilisation

There are a number of different methods that can be used for improving the general ground condition of a site. In many cases, improving the ground reduces the cost of foundations. Where the ground has been improved by compaction and consolidated, traditional foundation methods may be used rather than an expensive system of piles. Some sites which have been built up or are unstable may need to be improved, just to provide a sound hard standing so that heavy plant can operate on the site safely. Other sites require more permanent improvement, ensuring that the new building or structure and access to and around the structure remain stable.

Dynamic compaction

Dynamic compaction and consolidation using tamping systems can enhance the ground conditions up to considerable depths. The ground is consolidated by repeatedly dropping specially designed tampers into the ground. The ground is tamped on a grid with the objective of densifying deep layers. Different shaped tamper heads are available with a variety of weights, depending on the degree of consolidation and compaction specified. Two systems are commonly used. The first uses a flat-bottomed tamper; the alternative, more modern method uses cone-shaped tampers. Flat-bottomed tampers can be slow and tend to create more noise and vibration than the cone system of tamping. When lifting and lowering the weight, time is needed for the tamper to stabilise. Modern methods tend to use vertical guiders (or leaders) to control the fall and rise of the tamper.

Ground conditions suitable for dynamic compaction include natural granular soils, made-up ground and former landfill sites. The technique can also be used as part of a more significant earthworks operation, where the ground is built up in layers, compacted and consolidated. Where fill is built up in layers, the fill may take the form of unmodified material (as previously excavated) or soils that are modified or stabilised using additives, such as quicklime and pulverised fuel ash (PFA) cement.

To achieve the desired effect, several passes may be required. Careful monitoring and testing is required; grid levels may need to be taken before and after each pass. Trial drops should be taken to determine the optimum treatment regime, monitor the imprint and depths, and measure pore water pressures, as necessary. To determine the allowable bearing capacity, accurate measurements are taken of the penetration achieved by application of a known load from a known height. Analysis of the levels can be used to calculate the amount of void closure and the degree of densification. Using dynamic compaction, bearing capacities of $50-150 \, kN/m^2$ can be achieved, with greater bearing capacities achievable depending on the ground conditions.

Vibro compaction

Vibro compaction (also known as vibro displacement or vibro replacement) uses large vibrating mandrels (vibrating shafts or rods) to penetrate, displace and compact the soil. When the mandrel is removed from the ground, the subsequent void is filled with stone. The mandrel is then forced back through the stone, further displacing and compacting the ground and stone. The method that produces stone columns in the ground compacts the surrounding strata, enhancing the ground-bearing capacity and limiting settlement. Typical applications include support of foundations, slabs, hard standings, pavements, tanks or embankments. Soft soils, man-made and other strata can be reinforced to achieve improved specification requirements, while slopes can be treated to limit the risk of slip failure. The allowable ground-bearing capacities for low- to medium-rise buildings and industrial developments are in the region of $100-200 \, kN/m^2$. Beneath tanks, embankments or slopes of $100 \, kN/m^2$ can be achieved. Ground conditions may allow heavier loads to be supported.

The high-powered mandrel penetrates the ground, in a vertical plane to the designed depth. When the mandrel is extracted, the resulting bore is filled with suitable aggregate or stabilised solids; these are compacted in layers of $200-600 \, mm$ increments. The shape of the mandrel (poker) tip is designed to ensure high compaction of the stone or stabilised soil column and surrounding strata. The taper on the mandrel causes increased densification of the strata.

Vibro displacement and compaction can be used in granular and cohesive soils. In granular soils, the ground-bearing capacity is improved by the introduction of columns of compacted stone or stabilised soil, and the compaction and densification of the granular soil that surrounds each column.

Cohesive soils are not compacted in the same way that granular soils are. In clays, the stone columns help to share and distribute the loads. The columns of stone carry the loads down the pile and distribute them through the strata; however, they are not end bearing. Where the density and consistency of the ground varies, the installation of stone columns helps to stabilise the ground, enhances loadbearing performance and makes the conditions more uniform, thus limiting differential settlement. The benefits of vibro compaction include:

❑ Buildings can be supported on conventional foundations (normally reinforced and shallow foundations).
❑ Work can commence immediately following the vibro displacement. Foundations can be installed straightaway.
❑ The soil is displaced. No soil is produced.
❑ Contaminants remain in the ground – reducing disposal and remediation fees.
❑ Economical, when compared with piling or deep excavation works.
❑ Can be used to regenerate brownfield sites.
❑ Can use reclaimed aggregates and soils.

Vibro flotation

Vibro flotation uses a similar process to vibro compaction, except that the vibrating poker has high-pressure water jets at the tip of the poker. The water jets help to achieve the initial penetration into the ground. The advantage of this system is that the water jets help the vibrator penetrate hard layers of ground. A major disadvantage is that the system is messy and imprecise, thus rarely used.

Pressure grouting

In permeable soils or soils where it is known that small cavities may be within the ground, pressure grouting may be used to fill the voids. Holes are drilled into the ground using mechanically driven augers. As the auger is withdrawn, cement slurry is forced down a central tube into the bore under pressure. Pressures of up to $70\,000\,N/mm^2$ can be exerted by the grout on the surrounding soil. The slurry contains cementous additives, such as PFA, microsilica, chemical grout, cement or a mixture. PFA is cheap and often used as a bulk filler to improve the bearing capacity of the ground. As the grout enters the void, it forces itself into the voids, cavities and fissures in the soils and rock. In weak soils, it will displace and compact the ground as it fills the voids. As the voids are filled, the ground becomes stiffer, more stable and water resistant. Pressure grouting can also be used around basements and coffer dams to reduce the hydrostatic pressure on the structure. To create a water-resistant barrier, the bores and subsequent columns of grout are placed at close centres. PFA is generally used as the ground modification and stabilisation material, whereas the expensive chemical mixes (resin or epoxy mixes) or those containing microsilica are used to fill small voids and improve the ground's water resistance.

Soil modification and recycling

Shortage of land in cities and towns, combined with a desire to bring redundant land back into use, has necessitated the need to improve ground conditions. Specialist companies have developed plant that is capable of modifying and stabilising the ground relatively quickly. Mobile plant can cut into the ground and break it up, grade and crush the material and then combine it with additives before laying it to provide a more suitable surface. The additives used in soil stabilisation increase the strength of the soil, providing more workable materials, which can be better compacted to maximise bearing capacity and minimise settlement. Soil stabilisation may be used to treat and neutralise certain contaminants or encapsulate the contaminants, removing the need for expensive removal and disposal.

3.3 Substructures and basements

Over the centuries, many buildings have been designed and constructed with substructures and basements. Traditionally, the basement was used for storage of goods and coal that were not prone to damp or damage from occasional flooding. Over time, as property values have increased and drainage has improved, many of these basements have been turned into living accommodation. The basements were usually built using engineering bricks, which provided a degree of resistance to groundwater penetration. More recently it has become common to construct basements from reinforced concrete and/or dense blockwork, which is then tanked to prevent water from entering the building. Given suitable ground conditions, it is usually relatively economic to construct basements to buildings. This is especially true of high-rise buildings, which require substantial substructure works to accommodate the dynamic forces on the structure. Deep excavations naturally lend themselves to several levels of accommodation below ground. These spaces will need to be lit using artificial light and ventilated with mechanical ventilation systems.

The foundation substructure of multi-storey buildings is often constructed below natural or artificial ground level. In towns and cities, the ground for some metres below ground level has often been filled, over the centuries, to an artificial level. Filled ground is generally of poor and variable bearing capacity and is not good material on which to place foundations. It is generally necessary to remove the artificial ground and construct a substructure or basement of one or more floors below ground. Similarly, where there is a top layer of natural ground of poor and variable bearing capacity, it is often removed and a substructure formed. Where there are appreciable differences of level on a building site, a part or the whole of the building may be below ground level as a substructure.

The natural or artificial ground around the substructure is usually permeable to water and may retain water to a level above that of the lower basement level or floor of a substructure. Groundwater in soil around a substructure will impose pressure on both the walls and the floor of a substructure. This is known as hydrostatic pressure and is sufficiently strong to penetrate small cracks and joints. The cracks in the construction may be due to construction joints that are not watertight, shrinkage or movement of dense concrete walls and floors, and even dense, solidly built brick walls may allow water to penetrate. To limit the penetration of groundwater under pressure, it is usual practice to build in waterstops across construction and movement joints in concrete walls and floors, and to line walls and concrete floors with a layer of impermeable material in the form of a waterproof lining like a tank, hence the term 'tanking to basements'. Typical basement tanking methods are illustrated in Figures 3.17–3.19.

Another approach is to accept that there will be some penetration of groundwater through the external concrete wall, but not to allow this water into the usable part of the basement

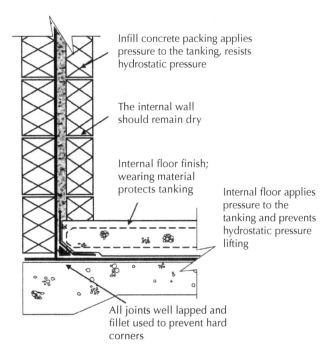

Infill concrete packing applies pressure to the tanking, resists hydrostatic pressure

The internal wall should remain dry

Internal floor finish; wearing material protects tanking

Internal floor applies pressure to the tanking and prevents hydrostatic pressure lifting

All joints well lapped and fillet used to prevent hard corners

Figure 3.17   Type A: internal basement tanking system.

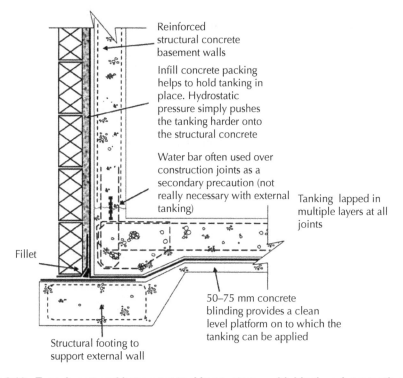

Reinforced structural concrete basement walls

Infill concrete packing helps to hold tanking in place. Hydrostatic pressure simply pushes the tanking harder onto the structural concrete

Water bar often used over construction joints as a secondary precaution (not really necessary with external tanking)

Tanking lapped in multiple layers at all joints

Fillet

50–75 mm concrete blinding provides a clean level platform on to which the tanking can be applied

Structural footing to support external wall

Figure 3.18   Type A: external basement tanking system – with blockwork protection.

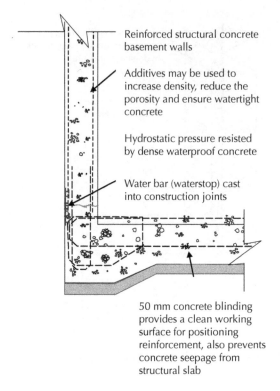

Reinforced structural concrete
basement walls

Additives may be used to
increase density, reduce the
porosity and ensure watertight
concrete

Hydrostatic pressure resisted
by dense waterproof concrete

Water bar (waterstop) cast
into construction joints

50 mm concrete blinding
provides a clean working
surface for positioning
reinforcement, also prevents
concrete seepage from
structural slab

Figure 3.19   Type B: waterproof concrete basement.

(Figure 3.20). Cavity walls are constructed with an external wall that retains the soil (the structural wall), a clear cavity that is drained at the bottom and an internal wall that provides a dry surface. Water that manages to penetrate the external structure runs down the external face of the cavity and is guided through channels to a sump where it is pumped out of the building. The external structural wall can be constructed of dense reinforced concrete with waterstops, or can be formed using contiguous, secant or steel piles (Figure 3.20).

The design of a basement is dependent on use, site conditions, construction conditions and waterproofing system. Table 3.1 provides a brief summary of the types of construction that are suitable for different basements.

Waterstops to concrete walls and floors
Dense concrete, which is practically impermeable to water, would by itself effectively exclude groundwater (Figure 3.19). However, in some situations, it is difficult to prevent movement and the formation of cracks caused by shrinkage, structural, thermal and moisture movement. As concrete dries out and sets, it shrinks, and this inevitable drying shrinkage causes cracks, particularly at construction joints, through which groundwater will penetrate.

Waterstops
As a barrier to the penetration of water through construction joints and movement joints in concrete floors and walls underground, it is usual practice to either cast 4 (PVC) waterstops against and across joints or to cast rubber waterstops into the thickness of concrete. The first

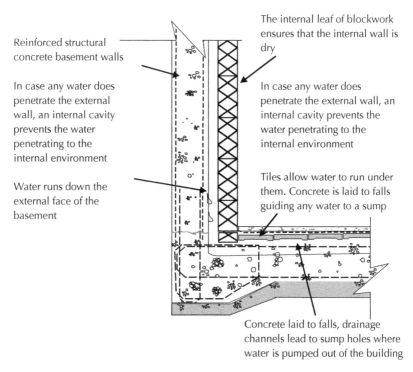

Reinforced structural concrete basement walls

In case any water does penetrate the external wall, an internal cavity prevents the water penetrating to the internal environment

Water runs down the external face of the basement

The internal leaf of blockwork ensures that the internal wall is dry

In case any water does penetrate the external wall, an internal cavity prevents the water penetrating to the internal environment

Tiles allow water to run under them. Concrete is laid to falls guiding any water to a sump

Concrete laid to falls, drainage channels lead to sump holes where water is pumped out of the building

Figure 3.20 Type C: traditional drained cavity basement.

method is generally used where water pressure is low, and the alternative, second method where water pressure is high. The first method is the most economical as it merely involves fixing the PVC stops to the formwork. Movement joints are formed right across and up the whole height of large buildings and filled with an elastic material that can accommodate the movement due to structural, thermal and moisture changes. These movement joints are formed at intervals of not more than 30 m.

Movement joints are formed in the main to accommodate thermal movement due to expansion and contraction of long lengths of solid structure and at angles and intersections right across the width and up the whole height of buildings, including floors and roofs. In effect, movement joints create separate structures each side of the joint. In framed structures, movement joints are usually formed between a pair of columns and pairs of associated beams.

PVC waterstops, illustrated in Figure 3.21, are fixed to the inside face of the timber formwork to the outside face of walls and to the concrete base under reinforced concrete floors so that the projecting dumbbells are cast into the concrete floors and walls. The large dumbbell in the centre of the waterstops for movement joints is designed to accommodate the larger movement likely at these joints. Provided the concrete is solidly consolidated up to the stops, this system will effectively act as a waterstop. At the right-angled joints of waterstops, preformed cross-over sections of stops are heat welded to the ends of straight lengths of stop. Rubber waterstops are cast into the thickness of concrete

Table 3.1 Type and level of protection required to suit use of basement

Grade of construction	Use of basement	Level of performance and conditions required	Type of basement construction	Comment
Grade 1. Basic utility rooms	Plant rooms, car park	Some seepage and some damp patches may occur (tolerable), >65% relative humidity, 15–32 °C temperature	Reinforced concrete to BS 8110 Type B structure	Check that the groundwater does not contain chemicals. Some chemicals may degrade or have other deleterious effects on the structure and internal finishes.
Grade 2. Improved utility	Workshops and plant rooms, retail storage areas	No water penetration or seepage, but moisture vapour is tolerable, 35–5% relative humidity, <15 °C storage, up to 42 °C for plant rooms	Type A internal tanking system Type B to BS 8007 Watertight concrete	Good supervision of all stages of construction is necessary to ensure watertight construction. Membranes should be applied in multiple layers and lapped joints.
Grade 3. Habitable	Ventilated residential and working areas, e.g. offices, restaurants and leisure centres	Dry environment, tightly controlled, 40–60% relative humidity, 18–29 °C temperature, depending on use	Type A internal tanking system Type B to BS 9007 Monolithic concrete structure Type C to BS 8110 Drained cavity system	Good supervision of all stages of construction is necessary to ensure watertight construction. Membranes should be applied in multiple layers and lapped joints.
Grade 4. Special	Archives, paper stores and computer rooms	Controlled environment that is totally dry, 35–50% relative humidity, 13–22 °C temperature	Type A Internal tanking system Type B to BS 8007 Monolithic concrete structure, combined with vapour proof membrane Type C Drained cavity system. Ventilated wall cavity and vapour barrier to inner skin and floor protection	Good supervision of all stages of construction is necessary to ensure watertight construction. Membranes should be applied in multiple layers and lapped joints.

Brief description of basement types.

Type A – Tanking membrane. Waterproof membrane, formed out of mastic asphalt, bitumen, rubber/bitumen compound, bonded sheet membranes, Bituthene, or polymer-modified bitumen, is either applied externally, internally or sandwiched between the basement walls. Bentonite clay and bentonite day sheets are also becoming common.
Type B – Monolithic concrete. The structure itself provides the necessary waterproofing. The structure is formed with dense reinforced concrete, water bar is often used at construction joints, and additives may be introduced to the concrete to make it denser and more water resistant.
Type C – Drained cavity. It is anticipated that water will penetrate the external structure. An internal skin of blockwork or concrete (with drainage former) is used to form a cavity. Water that enters the structure is drained off and pumped out of the building.
Source: Adapted from BS 8102:1990, BSI.

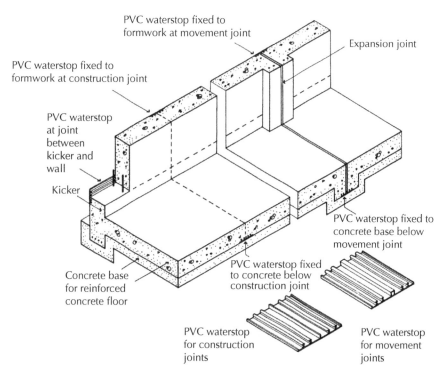

PVC waterstop fixed to
formwork at movement joint

Expansion joint

PVC waterstop fixed to
formwork at construction joint

PVC waterstop
at joint
between
kicker and
wall

Kicker

PVC waterstop fixed to
concrete base below
movement joint

Concrete base
for reinforced
concrete floor

PVC waterstop fixed
to concrete below
construction joint

PVC waterstop
for construction
joints

PVC waterstop
for movement
joints

Figure 3.21 PVC waterstops.

walls and floors, as illustrated in Figure 3.22. Plain web stops are cast in at construction
joints and centre bulb stops at expansion joints. These stops must be firmly fixed in place
and supported with timber edging to one side of the stop, so that concrete can be placed
and compacted around the other half of the stop without moving it out of place. At the
junction of the joints, hot vulcanising joins the stops. Hot vulcanising is where a hot iron
heats the PVC, and as the PVC melts the two ends merge together.

For waterstops to be effective, concrete must be placed and firmly compacted up to the
stops, and the stops must be secured in place to avoid them being displaced during placing
and compacting of concrete. Waterstops will be effective in preventing penetration of water
through joints, provided they are solidly cast up to or inside sound concrete and there is no
gross contraction at construction joints or movement at expansion joints.

Tanking

The term 'tanking' is used to describe a continuous waterproof lining to the walls and floors
of substructures, to act as a tank to exclude water.

Mastic asphalt

The traditional material for tanking is mastic asphalt, which is applied and spread hot in three
coats to a thickness of 20 mm for vertical work and 30 mm for horizontal work. Joints between
each layer of asphalt in each coat should be staggered at least 75 mm for vertical work and

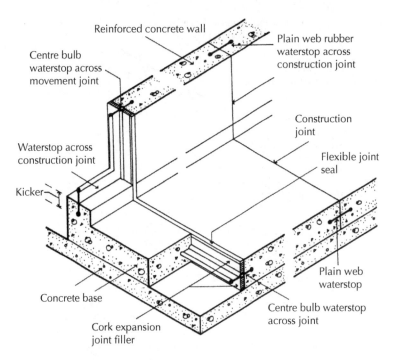

Figure 3.22   Rubber waterstops.

150 mm for horizontal work with the joints in succeeding coats. Angles are reinforced with a two-coat fillet of asphalt. Asphalt tanking should be applied to the outside face of structural walls and under structural floors so that the walls and floors provide resistance against water pressure on the asphalt and the asphalt keeps water from the structure. Figure 3.23 is an illustration of asphalt tanking applied externally to the reinforced concrete walls and floor of a substructure or basement. The horizontal asphalt is spread in three coats on the concrete base and over pile caps and extended 150 mm outside of the junction of the horizontal and vertical asphalt and the angle fillet. The horizontal asphalt is then covered with a protective screed of cement and sand 50 mm thick. The reinforced concrete floor should be cast on the protective screed as soon as possible, to act as a loading coat against water pressure under the asphalt below.

When the reinforced concrete walls have been cast in place and have dried, the vertical asphalt is spread in three coats and fused to the projection of the horizontal asphalt with an angle fillet. A half brick protective skin of brickwork is then built, leaving a 40 mm gap between the wall and the asphalt. The gap is filled solidly with mortar, course by course, as the wall is built. The half brick wall provides protection against damage from backfilling and the mortar-filled gap ensures that the asphalt is firmly sandwiched up to the structural wall. In Figure 3.23, the asphalt tanking is continued under a paved forecourt. Where vertical asphalt is carried up on the outside of external walls, it should be carried up at least 150 mm above ground to join a damp-proof course (dpc).

Figure 3.24 is an illustration of mastic asphalt tanking to a concrete floor and loadbearing brick wall to a substructure. The protective screed to the horizontal asphalt and protecting outer wall and mortar-filled gap to the vertical asphalt serve the same functions as they do for a concrete substructure. As a key for the vertical asphalt, the horizontal joints in the external face of the loadbearing wall should be lightly raked out and well brushed when the mortar has hardened sufficiently.

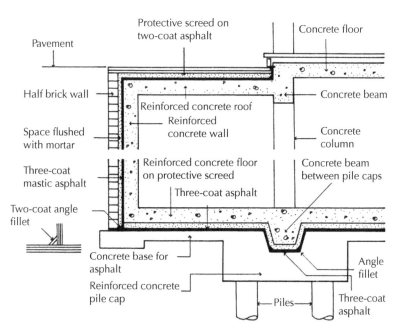

Figure 3.23   Mastic asphalt tanking applied externally to the reinforced concrete walls and floor of a substructure or basement.

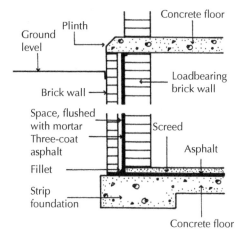

Figure 3.24   Mastic asphalt tanking applied to a concrete floor and loadbearing brick wall to a substructure.

Where the walls of substructures are on site boundaries and it is not possible to excavate to provide adequate working space to apply asphalt externally, a system of internal tanking may be used. The concrete base and structural walls are built, the horizontal asphalt is spread on the concrete base and a 50 mm protective screed spread over the asphalt. Asphalt is then spread up the inside of the structural walls and joined to the angle fillet reinforcement at the junction of horizontal and vertical asphalt. A loading and protective wall, usually of brick, is then built with a 40 mm mortar-filled gap up to the internal vertical asphalt. The internal protective and loading wall, which has to be sufficiently thick to resist the pressure of water on the asphalt, is usually one brick thick. A concrete loading slab is then cast on the protective screed to act against water pressure (also called hydrostatic pressure) on the horizontal asphalt. An internal asphalt lining is rarely used for new buildings, because of the additional floor and wall construction necessary to resist water pressure on the asphalt. Internal asphalt is sometimes used where a substructure to an existing building is to be waterproofed.

Service pipes for water, gas and electricity and drain connections that are run through the walls of a substructure that is lined with asphalt tanking, are run through a sleeve that provides a watertight seal to the perforation of the asphalt tanking and allows for some movement between the service pipe or drain and the sleeve. The sleeve is coated with asphalt that is joined to the vertical asphalt with a collar of asphalt, which runs around the sleeve, as illustrated in Figure 3.25.

Asphalt that is sandwiched in floors as tanking has adequate compressive strength to sustain the loads normal to buildings. The disadvantages of asphalt are that asphalting is a comparatively expensive labour-intensive operation and that asphalt is a brittle material that will readily crack and let in water if there is differential settlement or appreciable movements of the substructure. In general, the use of asphalt tanking is limited to substructures with a length or width of not more than about 7.5 m to minimise the possibility of settlement or movement cracks fracturing the asphalt.

Bituminous membranes
As an alternative to asphalt, bituminous membranes are commonly used for waterproofing and tanking to substructures. The membrane is supplied as a sheet of polythene or polyester film, or sheet bonded to a self-adhesive rubber/bitumen compound or a polymer-modified bitumen. The heavier grades of these membranes are reinforced with a meshed fabric sandwiched in the self-adhesive bitumen. The membrane is supplied in rolls about 1 m wide and 12–18 m long, with the self-adhesive surface protected with a release paper backing.

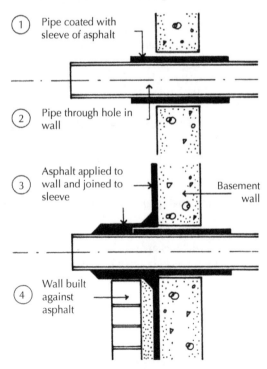

Figure 3.25   Four stages in forming an asphalt collar around pipe.

The particular advantage of these membranes is that their flexibility can accommodate small shrinkage, structural, thermal and moisture movements without damage to the membrane. Used in conjunction with waterstops to concrete substructures, these membranes may be used as tanking (see Figures 3.17–3.26).

The surface to which the membrane is applied by adhesion of the bitumen coating must be dry, clean and free from any visible projections that would puncture the membrane. The membrane is applied to a dry, clean float finished screed for floors and to level concrete wall surfaces on which all projecting nibs from formwork have been removed and cavities filled. The vertical surface to which the membrane is to be applied is first primed. The rolls of sheet are laid out, the paper backing removed and the membrane laid with the adhesive bitumen face down or against walls and spread out and firmly pressed on to the surface with a roller. Joints between long edges of the membrane are overlapped 75 mm and end joints 150 mm, and the overlap joints are firmly rolled in to compact the join. Laps on vertical wall surfaces are overlapped so that the sheet above overlaps the sheet below. At construction and movement joints, the membrane is spread over the joint with a PVC or rubber waterstop cast against or in the concrete.

Bitumen membranes are formed outside structural walls and under structural floors, as illustrated in Figure 3.26, with an overlap and fillet at the junction of vertical and horizontal membranes. To protect the vertical membrane from damage by backfilling, a protective concrete or half brick skin should be built up to the membrane. At angles and edges, a system of purposely cut and shaped cloaks and gussets of the membrane material is used

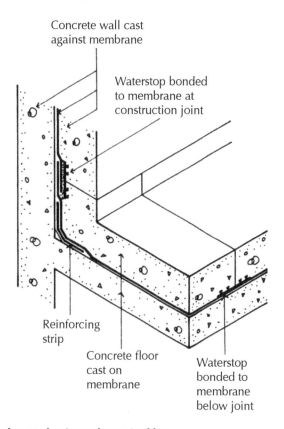

Concrete wall cast against membrane

Waterstop bonded to membrane at construction joint

Reinforcing strip

Concrete floor cast on membrane

Waterstop bonded to membrane below joint

Figure 3.26   Bituminous sheet membrane tanking.

over which the membrane is lapped, as illustrated in Figure 3.27. To be effective as a seal to the vulnerable angle joints, these overlapping cloaks and gussets must be carefully shaped and applied. The effectiveness of these membranes as waterproof tanking depends on dry, clean surfaces free from protrusions or cavities, and careful workmanship in spreading and lapping the sheets, cloaks and gussets.

Cavity tanking
Cavity drain structures make allowances for the small amount of water that may pass through the external wall. The basement is constructed with two walls forming a void between the external and internal leaf, and a cavity is formed in the walls around the

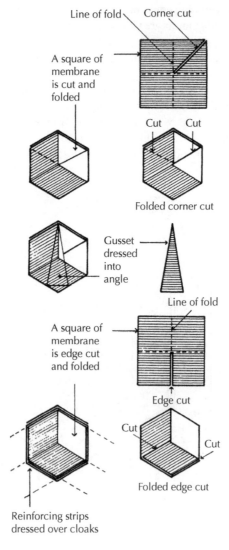

Figure 3.27   Internal angle cloaks to bituminous membrane.

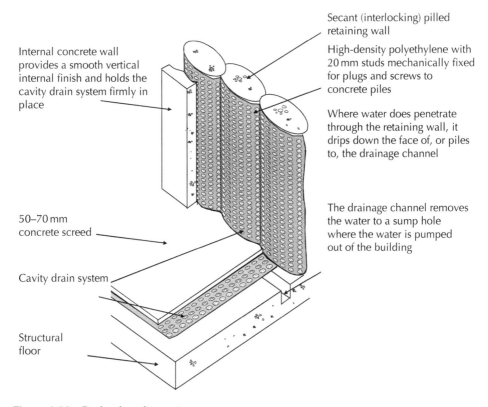

Secant (interlocking) pilled retaining wall

High-density polyethylene with 20 mm studs mechanically fixed for plugs and screws to concrete piles

Internal concrete wall provides a smooth vertical internal finish and holds the cavity drain system firmly in place

Where water does penetrate through the retaining wall, it drips down the face of, or piles to, the drainage channel

50–70 mm concrete screed

The drainage channel removes the water to a sump hole where the water is pumped out of the building

Cavity drain system

Structural floor

Figure 3.28   Drained cavity system.

basement and below the floor (Figure 3.28). Traditionally, cavities were formed with floor tiles, which created a void, and two separate leafs of masonry walling. Nowadays, rolls of studded 1 mm thick polyethylene are used, with raised studs (domes), which stand between 5 and 20 mm from the surface, depending on the manufacturer and purpose. The sheets of polyethylene, which come in rolls 20 m long by 1.4 m wide, are applied to the structural walls and floors to form a very small but effective void. The sheets are fixed to the wall using an effective plug and screw system, and then permanently held and protected by a concrete screed floor and in situ concrete wall. Any water that penetrates through the external wall or floor is guided to drainage channels, where it is then pumped out of the building.

If the water flow is expected to be relatively high, a sealed cavity system is used. All of the water that penetrates the wall flows, under gravity, to a sump pit where it is then pumped away. In existing buildings, where the walls of a basement are damp and not subject to water flow, a ventilated cavity system may be used. The cavity helps to keep the damp surface away from the internal wall and wall finishes. Moisture on the internal face of the wall will evaporate and the ventilation system guides the moist air out of the structure (Photograph 3.3 shows a ventilated and sealed drain cavity applied to a wall).

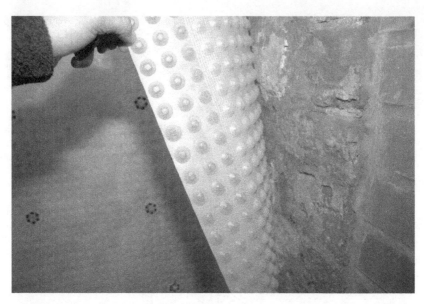

Photograph 3.3   Polypropylene ventilated cavity drain sheet applied to brick wall to control damp. The sealed cavity drain allows water to flow to a sump where it is pumped away.

Basement walls

There is often insufficient space to provide for the basement excavation and adequate batter so that the basement walls can be constructed out of the ground. It is becoming more common for deep basements to be constructed using diaphragm walls, with the aid of large clam grabs and bentonite slurry, and auger-bored piled walls using secant, interlocking and contiguous piles.

Contiguous piles

With the development of CFA piling rigs, bored piles are becoming more common as a form of permanent wall and foundation. Contiguous piles (Figures 3.29 and 3.30a) are formed by drilling bored piles at close centres. Piles vary in diameter from 300 to 2400 mm, although piles greater than 1200 mm are rarely used. A small gap is left between each pile; typically, this ranges between 20 and 150 mm. The size of the gap depends on the soil strength. The gaps may be filled to provide a more water-resistant structural concrete-facing wall. The use of CFA rigs to form the piles limits the depth of pile to 30–55 m, depending on the type of rig. In practice, walls are usually constructed to a maximum of 25 m, although some piles may be driven much deeper to provide vertical load capacity. A ring beam is cast along the top of the piles linking all of the piles together; this provides extra rigidity and strength and helps to distribute any loads placed on top of the piles. Ground anchors can also be used to help resist the overturning forces caused by the surrounding strata and hydrostatic pressure. Grout is normally forced through the anchor to tie it securely into the ground. Where the area surrounding the retaining structure accommodates roads, structures or other property, grout may be forced into the ground to produce a positive pressure on the ground, which counteracts the possibility of settlement caused by the excavation of the basement and movement of the retaining wall. The excavation of the basement may cause

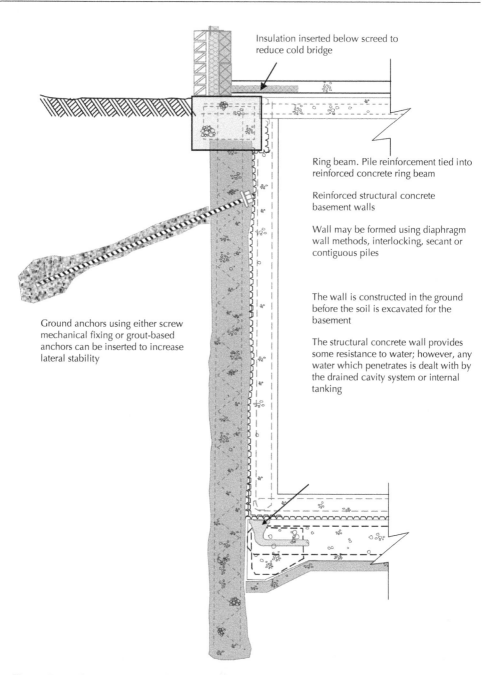

Insulation inserted below screed to reduce cold bridge

Ring beam. Pile reinforcement tied into reinforced concrete ring beam

Reinforced structural concrete basement walls

Wall may be formed using diaphragm wall methods, interlocking, secant or contiguous piles

The wall is constructed in the ground before the soil is excavated for the basement

The structural concrete wall provides some resistance to water; however, any water which penetrates is dealt with by the drained cavity system or internal tanking

Ground anchors using either screw mechanical fixing or grout-based anchors can be inserted to increase lateral stability

Figure 3.29   Secant, interlocking and contiguous piles. Section through an anchored-bored pile which, subject to spacing, can be used to form secant, interlocking and contiguous retaining walls.

(a) Contiguous piled wall – gaps left between each pile

Spacing 50–100 mm

Diameter 300–2400 mm

(b) Hard/soft (interlocking) piles – soft primary piles bored and cast first, hard secondary piles secant (cut into) soft piles, also called interlocking piles. Typical strength of female piles 1–3 N/mm^2

Soft piles cement and bentonite or cement, bentonite and sand

1st phase of piles

2nd phase (male piles)

(c) Hard/firm piles. Female piles which are intersected by the male piles have characteristic strength of 10–20 N/mm^2. Retarder is also used in female piles

Minimum overlap 25 mm

Female piles   Male piles

Diameter 600–1200 mm

(d) Hard/hard piles. Both male and female piles are reinforced and cast with full strength concrete. Reinforcement in the female piles is positioned so that it is not damaged by the auger when forming the male pile. Male piles can also be cased in steel liners for extra strength

(e) Hard/hard piles with I section steel columns. Rolled steel I section columns can be introduced to add extra strength

Figure 3.30   Contiguous and secant piles.

settlement due to the vibration, which causes the surrounding ground to compact. Also, as the surrounding ground applies its load on the retaining wall, some slight movement will occur. Vibration and movement of the retaining wall will result in settlement of the surrounding ground; the pressure exerted by pressurised grout can be used to remove the potential of settlement.

Secant piles

Secant piles consist of overlapping and interlocking piles (Figures 3.30b–e). Female (primary) piles are bored using CFA rigs and cast first. The secondary male piles are then drilled, secanting (cutting into) into the female pile. The system is often used in the construction of deep basement walls. Secant walls are often considered to be a more economical alternative to diaphragm walls.

Depending on the type of secant pile construction, either one or both piles are reinforced to resist the lateral loads. When the secant wall is in place, the excavated face can be covered with a layer of structural concrete. The concrete can be either sprayed or cast against the wall, providing a fair-faced concrete finish. A reinforced concrete ring beam connects all of the piles together, improving the structural stability of the wall. The beam will also help to distribute any loads placed on top of the wall. The depth of the wall is usually limited to 25 m; however, it is possible to construct secant walls to a depth of 55 m. Multi-storey basements are usually stabilised by ground anchors and cross-bracing to resist lateral forces. Secant piles can be constructed as hard/soft, hard/firm or hard/hard walls.

Hard/soft secant piles

For hard/soft piles, the female piles, which are cast first, are constructed of a soft pile mix; this usually takes the form of a cement and bentonite mix or cement, bentonite and sand mix. The mix has a weak characteristic strength of $1-3\,N/mm^2$. The female piles are used as a water-retaining structure rather than a loadbearing column. Soft piles can retain up to 8 m head of groundwater. The unreinforced soft pile is not commonly used as a permanent wall material. As the bentonite and cement mix dries, it will shrink and crack, losing its water-resisting properties. Some soft piles have been designed to retain their water-resisting properties for the life of the structure; often this necessitates the mix remaining hydrated throughout the life of the building.

Hard/firm secant piles

In the hard/firm pile arrangement, the female pile has a characteristic strength of $10-20\,N/mm^2$; during the wall's construction, the strength of the pile is held low by adding a retarding agent to the concrete mix (Figure 3.30c). Female piles are usually designed to hit their target strength within 56 days rather than the more typical 28 days. Obviously, such practice ensures that the construction of the male pile is easier as the auger has to exert less force when secanting the female pile. Piles usually overlap a minimum of 25 mm.

Hard/hard secant piles

With hard/hard secant piles, both male and female piles are cast with full-strength concrete and both are fully reinforced. The female piles are cast first and a high-torque-cutting casing is used to drill through the female pile. The reinforcement in the female pile is positioned so that the rig does not cut through it when boring the male pile. The depth of overlap is usually about 25 mm; considerable care is required to ensure that this is maintained along the full length of the pile. I-section beams can also be added to the pile to further increase the lateral strength of the wall (Figure 3.30e).

Diaphragm walls

Diaphragm walls are formed by excavating a segmented trench to form a continuous wall (Figure 3.31). While the deep trench is being excavated, it is filled with bentonite slurry (supporting fluid). The slurry fills the trench and exerts pressure on the sides of the excavation, thus preventing the excavation walls collapsing. The excavation is carried out using a clamshell grab, which digs the material out through the bentonite slurry. The width of the wall is determined by the width of the grab. The diaphragm wall is cast in the ground. Using a tremie tube of about 200 mm diameter, the concrete is fed into the trench and placed in position. As the concrete settles at the base of the excavation, the bentonite slurry is displaced, drawn out of the excavation and cleaned for reuse. The disposal of bentonite at the end of the operation is expensive, as the mixture is treated as a contaminated material. The diaphragm wall is constructed in alternate panels, usually around 5 m long. The excavators are usually guided by shallow concrete beams that are cast so that the beam faces form the desired position of the wall. Diaphragm walls have been constructed to depths of 120 m; however, there are some practical difficulties when attempting to splice and link the reinforcement cages over such depths.

The joints between each section can be cast using steel tubes or interlocking junctions to reduce the ingress of water through the joints (Figures 3.31b and c). The hydrofraise machine is used to cut an interlocking surface into the previously cast segment of wall (Figure 3.31d). Interlocking precast concrete sections can also be used. Once the precast concrete sections are in place, grout is used to fill any remaining joints.

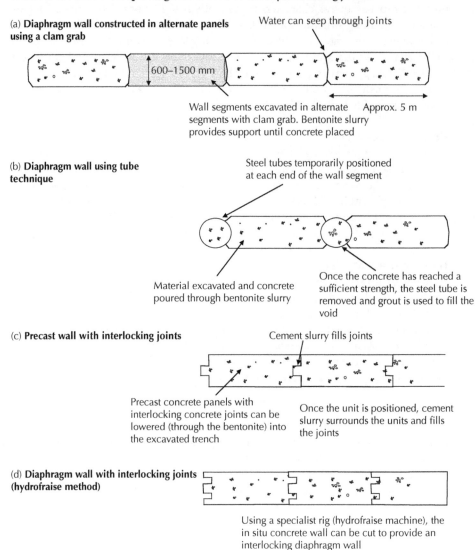

(a) **Diaphragm wall constructed in alternate panels using a clam grab**

Water can seep through joints

600–1500 mm

Wall segments excavated in alternate segments with clam grab. Bentonite slurry provides support until concrete placed

Approx. 5 m

(b) **Diaphragm wall using tube technique**

Steel tubes temporarily positioned at each end of the wall segment

Material excavated and concrete poured through bentonite slurry

Once the concrete has reached a sufficient strength, the steel tube is removed and grout is used to fill the void

(c) **Precast wall with interlocking joints**

Cement slurry fills joints

Precast concrete panels with interlocking concrete joints can be lowered (through the bentonite) into the excavated trench

Once the unit is positioned, cement slurry surrounds the units and fills the joints

(d) **Diaphragm wall with interlocking joints (hydrofraise method)**

Using a specialist rig (hydrofraise machine), the in situ concrete wall can be cut to provide an interlocking diaphragm wall

Figure 3.31   Diaphragm walls, plan sections.

Adding basements to existing buildings

In urban areas, it may be difficult or not possible to extend an existing property above ground due to space restrictions and/or town planning policy. Thus the only way to provide additional space is to go down into the ground. This is, of course, not without a number of challenges relating to, for example, ground conditions and groundwater pressure; the position of existing services; the condition and position of existing foundations, including those of adjoining/neighbouring properties; legal conditions relating to the property; town planning policy; access for conducting the work safely; and cost of the work. This is a specialist area in which some architects, building surveyors and contractors have expert knowledge and expertise. The primary concern will be related to the practicalities of conducting the building work without damaging neighbouring properties or causing nuisance; for example, by the transfer of sound from the new basement to adjoining properties.

Further reading

See www.structuralguide.com and www.designingbuildings.co.uk.

Reflective exercises

Your client owns a large terraced Victorian property located in a conservation area of a city. The planning department have rejected plans to extend at the rear of the property and therefore your client is considering extending under the existing building to provide additional living space. There is an existing, unused, cellar that runs under half of the house, but it has limited head height and no ventilation, and is damp most of the year.

❑ What do you need to consider to ensure the proposed changes comply with building legislation?
❑ What are the implications of digging down to create one or two storeys of accommodation?
❑ Your client is a musician and requires some of the space as a music studio. How do you prevent noise transmission to the neighbouring properties?

Single-storey Frames, Shells and Lightweight Coverings: Chapter 4 AT A GLANCE

What? Many industrial and commercial enterprises require a relatively cheap, single storey, clear-span structure to house their activities. Single-storey frames, shells and their associated lightweight coverings are a familiar feature of industrial and commercial zones. Although exceptions do exist, these buildings tend to be functional, practical and somewhat lacking in aesthetic appeal. Concrete shells and timber glulam buildings offer greater creative freedom compared to steel and concrete frames and tend to be used for buildings that are in more prominent locations and are used by the general public, for example, for recreational purposes.

Why? Framed buildings are an economical and quick form of construction suited to a variety of industrial and commercial uses. The majority of the materials used for the structural frames can be recovered and reused or recycled at the end of their service life. Walling and roofing materials tend to be selected on (low) cost and usually have a short design life compared to the structural frame. This means that the building envelope may be replaced several times over the life of the building.

When? Some single-storey units may be assembled from prefabricated, off-the-shelf manufacturers' systems. Others will be designed to meet specific functional and performance requirements, relying primarily on the assembly of prefabricated components.

How? The structural frame will be erected on foundations, to which lightweight cladding and insulation will be added to form the walls and roof. Internal finishes, services and internal walls and floors will be added once the fabric is watertight. The techniques of construction vary, depending on the type of structure, be it, for example, a steel framed building or a timber glue laminated frame.

4 Single-storey Frames, Shells and Lightweight Coverings

This chapter describes the construction of single-storey buildings such as sheds, warehouses, factories, lightweight mast and fabric structures, and other buildings, generally built on one floor and constructed with a structural frame of steel, reinforced (prefabricated) concrete or timber, supporting lightweight roof and wall coverings (see also Chapters 5–8). A large proportion of the buildings in this category are constructed to serve a very specific purpose for a relatively short period of time, after which the market and hence the required performance of the building will have changed. It is not uncommon for sheds and warehouses to have a specified design life of between 15 and 30 years. After this time, the building will be deconstructed and materials recovered, reused and recycled. Alternatively (and less likely), considerable works of repair and renewal are required to maintain minimum standards of comfort and appearance. As a consequence, the materials used are selected primarily for economy of initial cost, tend to have limited durability and are often prone to damage in use.

In traditional building forms, one material could serve several functional requirements; for example, a solid loadbearing brick wall provides strength, stability, exclusion of wind and rain, resistance to fire and to a small extent thermal and sound insulation. In contrast, the materials used in the construction of lightweight structures are, in the main, selected to perform specific functions. For example, steel sheeting is used as a weather envelope and to support imposed loads, layers of insulation for thermal and sound resistance, thin plastic sheets for daylight, and a slender frame to support the envelope and imposed loads. The inclusion of one material for a specific purpose is likely to have a significant impact on the performance of adjacent materials; thus, the designer needs to look at the performance of individual materials and the performance of the whole assembly.

4.1 Lattice truss, beam, portal frame and flat roof structures

To reduce the volume of roof space that has to be heated and also to reduce the visual impact of the roof area, it is common practice to construct single-storey buildings with low-pitch roof frames, either as portal frames or as lattice beam or rafter frames (Figure 4.1). The pitch may be as low as 2.5°. Alternatively, flat roof structures may be used.

Barry's Advanced Construction of Buildings, Fifth Edition. Stephen Emmitt.
© 2023 John Wiley & Sons Ltd. Published 2023 by John Wiley & Sons Ltd.

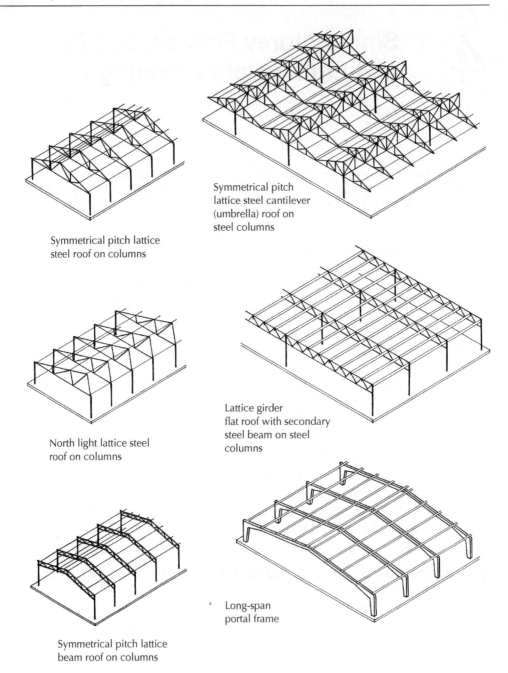

Symmetrical pitch
lattice steel cantilever
(umbrella) roof on
steel columns

Symmetrical pitch lattice
steel roof on columns

North light lattice steel
roof on columns

Lattice girder
flat roof with secondary
steel beam on steel
columns

Symmetrical pitch lattice
beam roof on columns

Long-span
portal frame

Figure 4.1   Typical lattice and portal frame construction.

Functional requirements

The primary functional requirements of single-storey framed structures are:

- ❏ Strength and stability.
- ❏ Durability and freedom from maintenance.
- ❏ Fire safety.

Strength and stability
The strength of a structural frame depends on the strength of the material used in the fabrication of the members of the frame and also on the stability of the frame, which is dictated both by the way in which the members are connected and on the bracing across and between the frames. Steel is most used in framed structures because of its good compressive and tensile strength, and good strength-to-weight ratio. Hot-rolled steel and cold-formed strip steel provide a wide range of sections suited to the economical fabrication of structural frames. These sections are also relatively easy to recover and reuse at the end of the building's life. Concrete has good compressive strength but poor tensile strength and so it is used as reinforced concrete in structural frames to benefit from the tensile strength of the steel and the compressive strength of the concrete. The concrete also provides protection against corrosion and damage by fire to the steel. Timber is often used in the fabrication of roof frames because it has adequate tensile and compressive strength to support comparatively light loads. Timber tends to be used instead of steel to form lightweight roof frames because of its ease of handling and fixing.

Durability and freedom from maintenance
On exposure to air and moisture, unprotected steel corrodes to form an oxide coating, such as rust, which is permeable to moisture and thus encourages progressive corrosion, which may in time adversely affect the strength of the material. To inhibit rust, steel is painted, coated with zinc or encased in concrete. Painted surfaces will require periodic repainting. Any cutting and drilling operations will damage zinc or painted coatings. Reinforced concrete is highly durable and the surface will need little maintenance other than periodic cleaning. Seasoned, stress-graded timber treated against fungal and insect attack should require little maintenance during its useful life, other than periodic staining or painting.

Fire safety
All loadbearing structures (including roofs) should be designed so that they do not fail prematurely during a fire. Providing the structure with the necessary fire resistance helps to reduce the risk posed by falling debris to building users, pedestrians and fire fighters.

Elements of the structure that give support or stability to another element of the building must have no less fire resistance than the other supporting elements. Similarly, if a roof provides stability and support to columns, then the roof must have at least the same fire resistance as the columns. All roofs should have sufficient fire resistance to resist exposure from the underside of the roof, remaining sound for a minimum of 30 minutes. The same provision also applies to roofs that form part of a fire escape. Where the roof performs the function of a floor, the minimum period of fire resistance is dependent on the purpose of the building and the height of the building (Table 4.1). If the building is constructed with a basement, this will also have an impact on the required fire resistance.

Table 4.1   Typical fire resistance periods for roofs that form floors

Minimum period of fire resistance (minutes)			
Upper storey height (height in metres above ground)			
Purpose of building	Not more than 5m	Not more than 18m	Not more than 30m	More than 30m
---	---	---	---	---
Residential				
Flats and maisonettes	30	60	90	129
Office				
Not sprinklered	30	60	90	Not permitted
Sprinklered	30	30	60	120
Shop and commercial				
Not sprinklered	60	60	90	Not permitted
Sprinklered	30	60	60	120
Assembly and recreation				
Not sprinklered	60	60	90	Not permitted
Sprinklered	30	60	60	120
Industrial				
Not sprinklered	60	90	120	Not permitted
Sprinklered	30	60	60	120

Lattice truss roof construction

'Lattice' is a term used to describe an open grid of slender members fixed across or between each other, usually in a regular pattern of cross-diagonals or as a rectilinear grid. 'Truss' is used to define the action of a triangular roof framework, where the spread under load of sloping rafters is resisted by the horizontal tie member that is secured to the feet of the rafters (which trusses or ties them against spreading) (Figures 4.2 and 4.3, Photographs 4.1–4.3).

Symmetrical pitch steel lattice truss construction
The single-bay frame illustrated in Figure 4.4 is a relatively economic structure. The small section, mild steel members of the truss can be cut and drilled with simple tools, assembled with bolted connections and speedily erected without the need for heavy lifting equipment. Similarly, the structure can be readily dismantled and reused or recycled when no longer required.

The small section, steel angle members of the truss are bolted to columns and purlins, and sheeting rails are bolted to cleats to support roof and side wall sheeting. These frames can be fabricated offsite and quickly erected on comparatively slender mild steel I-section columns fixed to concrete pad foundations. The bolted, fixed base connection of the foot of the columns to the concrete foundation provides sufficient strength and stability against wind pressure on the side walls and roof. Wind bracing provides stability against wind pressure on the end walls and gable ends of the roof.

The depth of the roof frames at mid-span provides adequate strength in supporting dead and imposed loads, as well as rigidity to minimise deflection under load. For maximum structural efficiency, the pitch of the rafters of the frames should be not less than 17° to the horizontal. This large volume of roof space cannot be used for anything other than housing services such as lighting and heating, and where the activity enclosed by the building needs

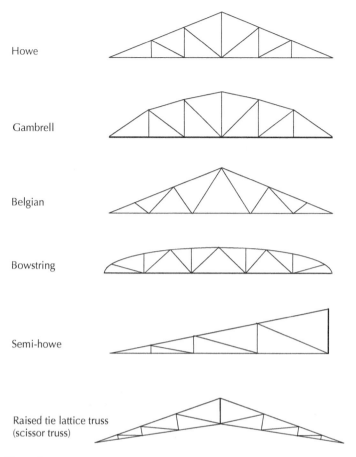

Howe

Gambrell

Belgian

Bowstring

Semi-howe

Raised tie lattice truss
(scissor truss)

Figure 4.2 Truss types.

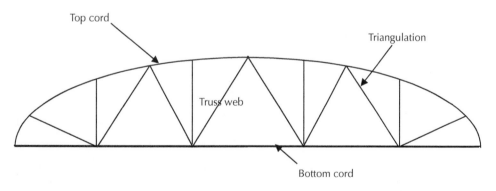

Top cord

Triangulation

Truss web

Bottom cord

Figure 4.3 Truss components.

Photograph 4.1   Bowstring truss.

Photograph 4.2   Large-span bowstring trusses spanning.

Photograph 4.3   (a) Bowstring truss across airport. (b) Bowstring truss against external wall place alongside.

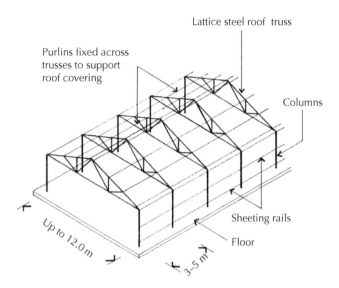

Figure 4.4   Single-bay symmetrical pitch lattice steel roof on steel columns.

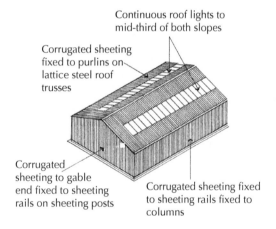

Figure 4.5   Single-bay symmetrical pitch lattice steel roof on columns with corrugated sheeting.

to be heated, it makes for an uneconomical solution. Trusses are usually spaced between 3 and 5 m apart (for economy in the use of small section purlins and sheeting rails) and are often limited to spans of approximately 12 m. Larger trusses can be fabricated to provide large clear spans.

Rooflights are commonly used to provide reasonable penetration of daylight to the interior of the building, as illustrated in Figure 4.5. The thin sheets of profiled steel sheets used to clad the walls have poor resistance to accidental damage and vandalism. As an alternative to steel columns and cladding, loadbearing brick walls may be used for single-bay buildings to provide support for the roof frames. The masonry walls provide better durability to accidental damage and vandalism. The roof frames are positioned on brick piers, which provide additional stiffness to the wall and transfer the loads of the roof to the foundations, as shown in Figure 4.6. As an alternative, a low brick upstand wall may be constructed to a height of around 1500 mm as protection against accidental impact damage, with wall sheeting above.

North light steel lattice truss construction

The north light roof has an asymmetrical profile with the south-facing slope pitched at 17° or more to the horizontal and the north-facing slope at around 60°, as illustrated in Figures 4.7 and 4.8. To limit the volume of unusable roof space (that has to be heated), most north light roofs are limited to spans of up to about 10 m.

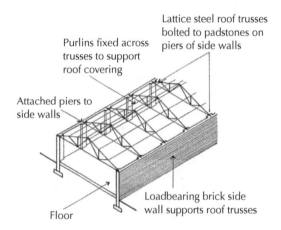

Figure 4.6   Single-bay symmetrical pitch lattice steel roof on brick side walls.

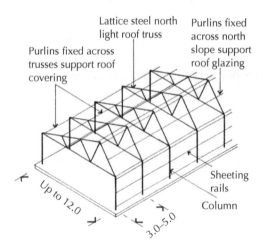

Figure 4.7   Single-bay north light lattice steel roof trusses on steel columns.

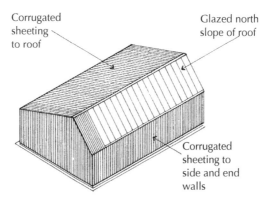

Figure 4.8   Single-bay lattice steel north light roof on columns with corrugated sheeting.

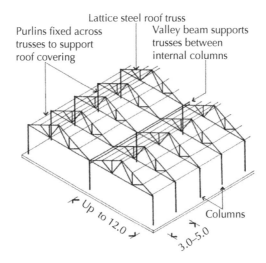

Figure 4.9   Two-bay symmetrical pitch lattice steel roof and columns with valley beam.

Multi-bay lattice steel roof truss construction

To cover large areas, it is common practice to use two or more bays of symmetrical pitch roofs to both limit the volume of roof space and the length of the members of the trusses.

To avoid the use of closely spaced internal columns (which may obstruct the working floor area) to support roof trusses, a valley beam is used. The valley beam supports the roof trusses between the internal columns, as illustrated in Figure 4.9. The greater the clear span between internal columns, the greater the depth of the valley beam and the greater the volume of unused roof space, as illustrated in Figure 4.10.

Cantilever (umbrella) multi-bay lattice steel truss roof

Figure 4.11 shows a cantilever (or umbrella) roof with lattice steel girders constructed inside the depth of each bay of trusses at mid-span. The lattice girder supports half of each truss with each half cantilevered each side of the truss.

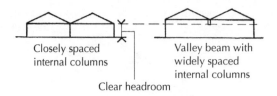

Figure 4.10   Increased volume of unused roof space with widely spaced internal columns.

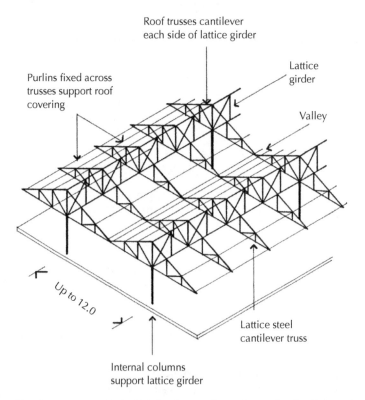

Figure 4.11   Two-bay symmetrical pitch lattice steel cantilever (umbrella) roof.

Lattice steel truss construction

Lattice steel trusses are often fabricated from one standard steel angle section with two angles positioned back to back for the rafters and main tie, and a similar angle for the internal struts and ties, as illustrated in Figure 4.12. The usual method of joining the members of a steel truss is with steel gusset plates, which are cut to shape to contain the required number of bolts at each connection. The flat gusset plates are fixed between the two angle sections of the rafters and main tie and to the intermediate ties and struts. Bearing plates

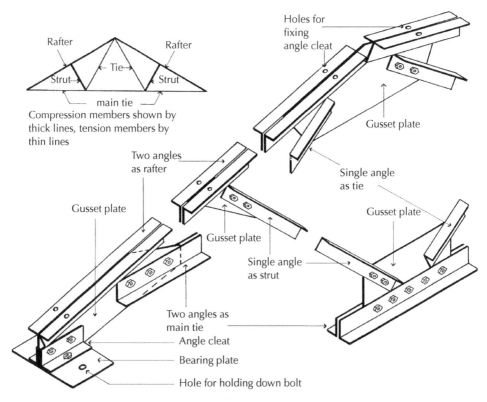

Figure 4.12   Lattice steel truss construction.

fixed to the foot of each truss provide a fixing to the columns. The members of the truss are bolted together through the gusset plates.

Standard I-section steel columns are used to support the roof trusses. A steel base plate is welded or fixed with bolted connections, with gusset plates and angle cleats, to the base of the columns. The column base plate is levelled with steel packing plates and then grouted in position with non-shrinkable cement. The base plate rests on the concrete pad foundation, to which it is rigidly fixed with four holding-down bolts, cast or set into the foundation, as illustrated in Figure 4.13. The rigid fixing of the columns to the foundation bases provides stability to the columns, which act as vertical cantilevers in resisting lateral wind pressure on the side walls and the roof of the building. A cap is welded or fixed with bolted connections to the top of each column and the bearing plates of truss ends are bolted to the cap plate (Figure 4.14).

Lattice trusses can be fabricated from tubular steel sections that are cut, mitred and welded together, as illustrated in Figure 4.15. Because of the labour involved in the cutting and welding of the members, they tend to be more expensive than a similar-sized angle section truss; however, they have greater structural efficiency and are visually more attractive. The truss illustrated in Figure 4.15 has a raised tie, which affords some increase in working height below the raised part of the tie.

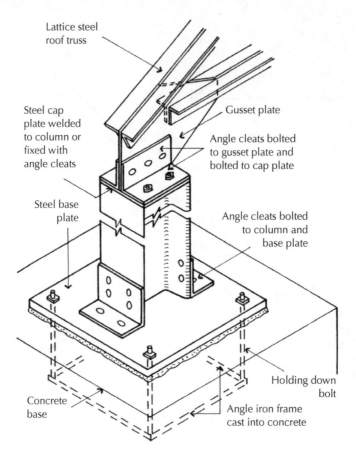

Lattice steel
roof truss

Steel cap
plate welded
to column or
fixed with
angle cleats

Gusset plate

Angle cleats bolted
to gusset plate and
bolted to cap plate

Steel base
plate

Angle cleats bolted
to column and
base plate

Holding down
bolt

Concrete
base

Angle iron frame
cast into concrete

Figure 4.13   Cap and base of steel column support for lattice steel truss (3D view).

Steel lattice beam roof construction

The two structural forms best suited to the use of deep profiled steel roof sheeting are lattice beam and portal frame. The simplest form of lattice beam roof is a single-bay symmetrical pitch roof constructed as a cranked lattice beam or rafter.

Symmetrical pitch lattice steel beam roof construction

The uniform depth lattice beam is cranked to form a symmetrical pitch roof with slopes of between 5° and 10°, as illustrated in Figure 4.16. Because of the low pitch of the roof, there is little unused roof space and this form of construction is preferred to lattice truss construction where the space is to be heated. The beams are fabricated from tubular and hollow rectangular section steel, which is cut and welded together with bolted site connections at mid-span to facilitate the transportation of half-lengths to the site. The top and bottom chords of the beams are usually of hollow rectangular section for ease of fabrication. End plates, welded to the lattice beams, are bolted to the flanges of I-section columns. Service pipes and small ducts may be run through the lattice frames, and larger ducts suspended below the beams inside the roof space.

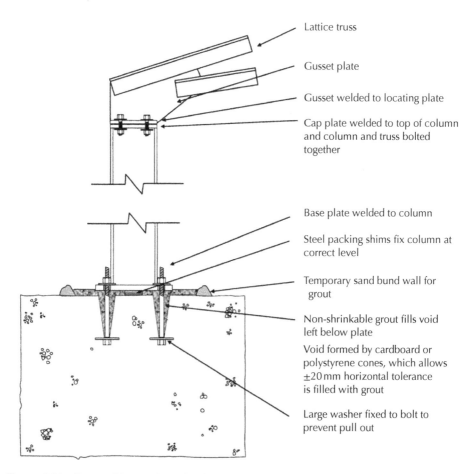

Lattice truss

Gusset plate

Gusset welded to locating plate

Cap plate welded to top of column
and column and truss bolted
together

Base plate welded to column

Steel packing shims fix column at
correct level

Temporary sand bund wall for
  grout

Non-shrinkable grout fills void
left below plate

Void formed by cardboard or
polystyrene cones, which allows
±20 mm horizontal tolerance
is filled with grout

Large washer fixed to bolt to
prevent pull out

Figure 4.14   Cap and base of steel column support for lattice steel truss.

Multi-bay symmetrical pitch lattice steel beam roof construction

For multi-bay symmetrical pitch lattice beam roofs, it is usual to fabricate a form of valley beam roof, as illustrated in Figure 4.17. The valley beam is designed to be the same depth as the beams to prevent any increase in the unwanted volume of roof space. To provide the maximum free floor space, a form of butterfly roof with deep valley beams is used, as illustrated in Figure 4.18. The deeper the valley beams, the greater the spacing between internal columns and the greater the unused roof space.

Steel portal frames

Rigid portal frames are an economic alternative to lattice truss and lattice beam roofs, especially for single-bay buildings. To be effective, a pitched roof portal frame should have as low a pitch as practical to minimise spread at the knee of the portal frame (spread increases with the pitch of the rafters of a portal frame). The knee is the rigid connection of the rafter to the post of the portal. Portal frames with a span of up to 15 m are defined as short span, frames with a span of between 16 and 35 m as medium span, and frames with a span of

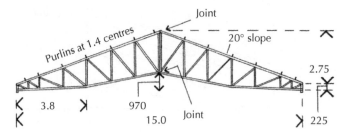

Raised tie tubular steel lattice truss

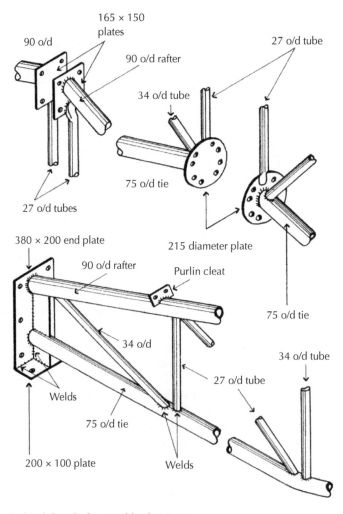

Figure 4.15 Raised tie tubular steel lattice truss.

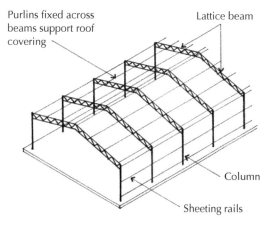

Purlins fixed across beams support roof covering

Lattice beam

Column

Sheeting rails

Figure 4.16   Single-bay symmetrical pitch lattice beam and column frame.

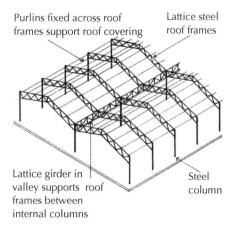

Purlins fixed across roof frames support roof covering

Lattice steel roof frames

Lattice girder in valley supports roof frames between internal columns

Steel column

Figure 4.17   Two-bay lattice beam roof on steel columns.

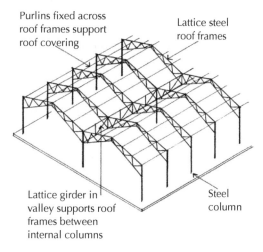

Purlins fixed across roof frames support roof covering

Lattice steel roof frames

Lattice girder in valley supports roof frames between internal columns

Steel column

Figure 4.18   Two-bay lattice steel butterfly roof.

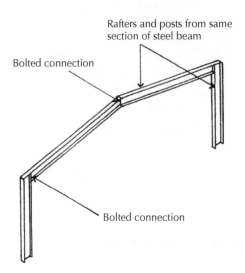

Figure 4.19   Short-span steel portal frame.

36–60 m as long span. Because of the considerable clear spans afforded by the portal frame, there is little advantage in using multi-bay steel portal systems, where the long-span frame would be sufficient. For short- and medium-span frames, the apex or ridge, where the rafters join, is usually made as an on site, rigid bolted connection for convenience in transporting half portal frames to the site. Long-span portal frames may have a pin joint connection at the ridge to allow some flexure between the rafters of the frame, which are pin jointed to the foundation bases.

For economy, short- and medium-span steel portal frames are often fabricated from one mild steel I-section for both rafters and posts, with the rafters welded to the posts without any increase in depth at the knee, as shown in Figure 4.19. Short-span portal frames may be fabricated offsite as one frame, transported to site and craned into position. Larger-span portals are assembled on site with bolted connections of the rafters at the ridge with high-strength friction grip (hsfg) bolts (see Chapter 5) (Figure 4.20).

Many medium- and long-span steel portal frames have the connection of the rafters to the posts haunched at the knee to make the connection deeper and hence stiffer, as illustrated in Figure 4.21. The haunched connection of the rafters to the posts can be fabricated either by welding a cut I-section to the underside of the rafter (as illustrated in Figure 4.21) or by cutting and bending the bottom flange of the rafter and welding in a steel gusset plate.

In long-span steel portal frames, the posts and lowest length of rafters, towards the knee, may often be fabricated from cut and welded I-sections so that the post-section and part of the rafter is wider at the knee than at the base and ridge of the rafter (Figures 4.22 and 4.23).

The junction of the rafters at the ridge is often stiffened by welding cut I-sections to the underside of the rafters at the bolted site connection, as shown in Figure 4.24.

Steel portal frames may be fixed or pinned to bases to foundations. For short-span portal frames, where there is relatively little spread at the knee or haunch, a fixed base is often used. The steel plate, which is welded through gusset plates to the post of the portal frame,

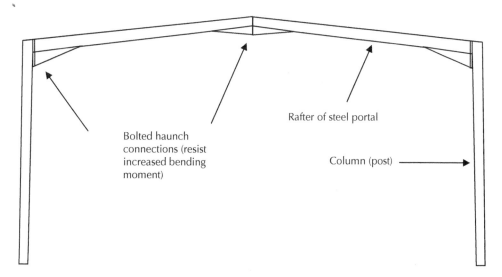

Figure 4.20   Long-span steel portal frame.

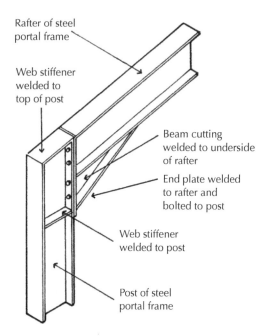

Figure 4.21   Haunch to steel portal frame.

is set level on a bed of cement grout on the concrete pad foundation and is secured by four holding-down bolts, which are set or cast into the concrete foundation (illustrated in Figure 4.25). A pinned base is made by positioning the portal base plate on a small steel packing piece onto a separate base plate, which bears on the concrete foundation. Two anchor bolts, either cast or set into the concrete pad foundation, act as holding-down bolts

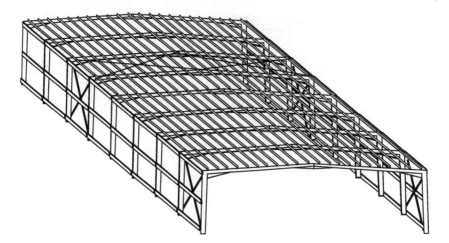

Figure 4.22   Portal bays simply added together to increase the length of the building.

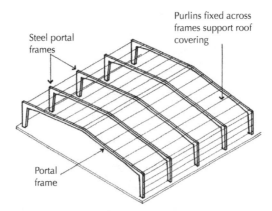

Figure 4.23   Long-span steel portal frames.

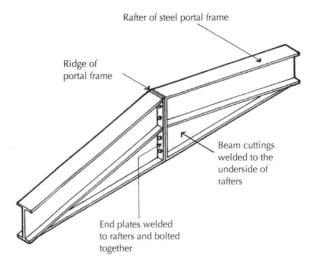

Figure 4.24   Stiffening at ridge of steel portal frame.

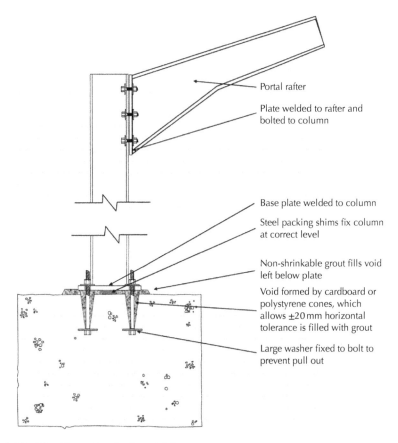

Portal rafter

Plate welded to rafter and bolted to column

Base plate welded to column

Steel packing shims fix column at correct level

Non-shrinkable grout fills void left below plate

Void formed by cardboard or polystyrene cones, which allows ±20 mm horizontal tolerance is filled with grout

Large washer fixed to bolt to prevent pull out

Figure 4.25   Fixed base to steel portal frame.

to the foot of the portal frame, as illustrated in Figure 4.26. The packing between the plates allows some flexure of the portal post independent of the foundation.

Short-span portal frames are usually spaced between 3 and 5 m apart and medium-span frames at between 4 and 8 m apart, to suit the use of angle or cold-formed purlins and sheeting rails. Long-span portals are usually spaced at between 8 and 12 m apart to economise on the number of comparatively expensive frames. Channel, Zed, I-section or lattice purlins and sheeting rails support roof sheeting or decking and wall cladding. With flat and low-pitch portal frames, it is difficult to achieve a watertight system of roof glazing; therefore, a system of monitor lights is sometimes used. These lights are formed by welded, cranked I-section steel purlins fixed across the portal frames (Figure 4.27). The monitor lights finish short of the eaves to avoid any unnecessary complications and can be constructed to provide natural and controlled ventilation to the interior.

Wind bracing
The side wall columns (stanchions) and their fixed bases that support the roof frames are designed to act as vertical cantilevers to carry the loads in bending and shear that act on them from horizontal wind pressure on the roofing and cladding. The rigid knee joint

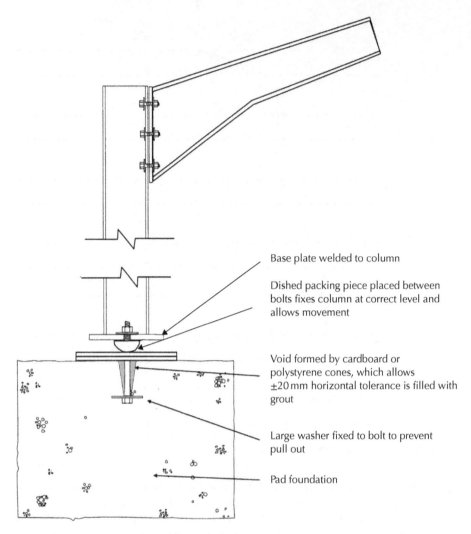

Base plate welded to column

Dished packing piece placed between bolts fixes column at correct level and allows movement

Void formed by cardboard or polystyrene cones, which allows ±20 mm horizontal tolerance is filled with grout

Large washer fixed to bolt to prevent pull out

Pad foundation

Figure 4.26   Pinned base to steel portal frame.

between rafter and post will carry the loads from horizontal wind pressure on roof and side wall cladding. Where internal columns are comparatively widely spaced, it is usually necessary to use a system of eaves bracing to assist in the distribution of horizontal loads. The system of eaves bracing shown in Figure 4.28 consists of steel sections fixed between the tie or bottom chord of roof frames and columns. To transfer the loads from wind pressure on the gable ends, a system of horizontal gable girders is formed at tie or bottom chord level.

Structural bracing and wind bracing
Additional bracing is used to assist in setting out the building, to stabilise the roof frames, square up the ends of the building and offer additional resistance to the wind. The rafter bracing between the end frames, illustrated in Figure 4.28, helps to stabilise the rafters of

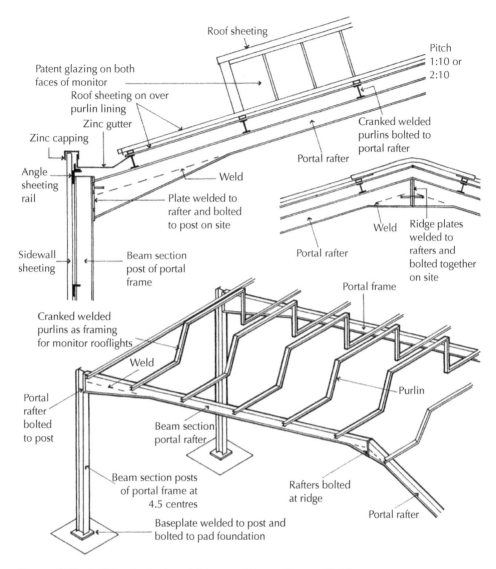

Figure 4.27   Solid web steel portal frame with monitor rooflights.

the roof frames. Longitudinal ties between roof frames stabilise the frames against probable uplift due to wind pressure. The vertical bracing in adjacent wall frames at gable end corners hold the building square and serve as bracing against wind pressure on the gable ends of the building.

Purlins and sheeting rails
Purlins are fixed across rafters and sheeting rails across the columns to provide support and fixing for roof and wall cladding and insulation (Figures 4.29 and 4.30). The spacing and size of the purlins and the sheeting rails are determined by the type of roof and wall

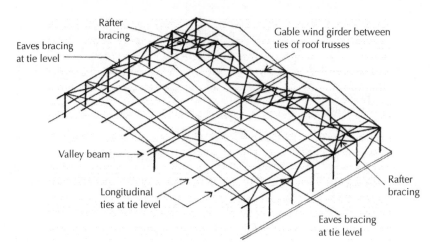

Figure 4.28   Wind bracing to steel truss roof on steel columns.

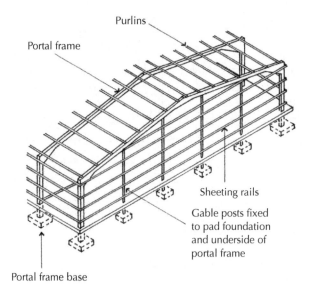

Figure 4.29   Gable end framing.

cladding used. As a general rule, the deeper the profile of the sheeting, the greater its safe span and the further apart the purlins and sheeting rails may be fixed.

Mild steel angles and purlin rails are sometimes used, but these tend to have been replaced by a range of standard sections, purlins and rails in galvanised, cold-formed steel strip. The sections most used are Zed and Sigma (Figure 4·31), with more complex sections with stiffening ribs also produced. These thin-section purlins and rails help to facilitate direct fixing of the sheeting by self-tapping screws.

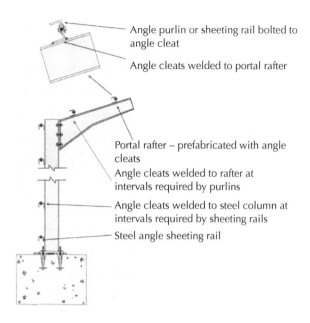

Angle purlin or sheeting rail bolted to
angle cleat

Angle cleats welded to portal rafter

Portal rafter – prefabricated with angle
cleats

Angle cleats welded to rafter at
intervals required by purlins

Angle cleats welded to steel column at
intervals required by sheeting rails

Steel angle sheeting rail

Figure 4.30   Connection of purlin to truss and sheeting rails to columns.

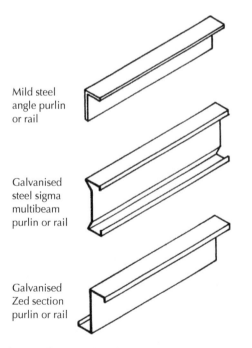

Mild steel
angle purlin
or rail

Galvanised
steel sigma
multibeam
purlin or rail

Galvanised
Zed section
purlin or rail

Figure 4.31   Steel section purlin and sheeting rails to support sheet metal and composite sheeting.

Purlins and sheeting rails are fixed to structural supports with cleats, washer plates and sleeves, as illustrated in Figure 4.32. Anti-sag bars are fixed between cold-formed purlins to stop them twisting during the fixing of roof sheeting and to provide lateral restraint to the bottom flange against uplift due to wind pressure. The purlins also derive a large degree of stiffness from the sheeting. Anti-sag bars and apex ties are made from galvanised steel rod that is either hooked or bolted between purlins, as illustrated in Figure 4.33. The apex ties provide continuity over the ridge. For the system to be effective, there must also be some form of stiffening brace or strut at the eaves.

The secret fixing for standing seam roof sheeting for low-pitch roofs does not provide lateral restraint for cold-formed purlins; thus, it is necessary to use a system of braces between purlins. The braces are manufactured from galvanised steel sections and bolted between purlins with purpose-made apex braces, as illustrated in Figure 4.34.

To support the wall sheeting (cladding), sheeting rails are fixed across, or between, the steel columns and/or vertical frame members (Figures 4.31 and 4.32). Zed or Sigma section rails are bolted to cleats and then bolted to the structural frame. A system of side rail struts is fixed between rails to provide strength and stability against the weight of the sheeting.

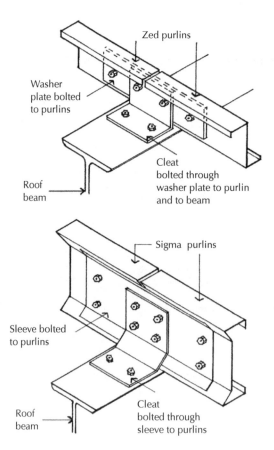

Figure 4.32   Washer plates and sleeves for continuity over supports.

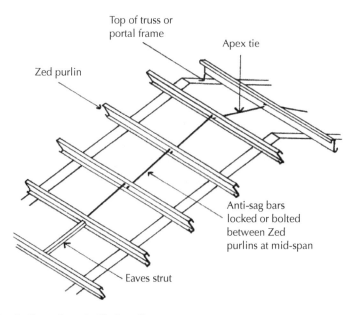

Figure 4.33   Anti-sag bars to Zed purlins.

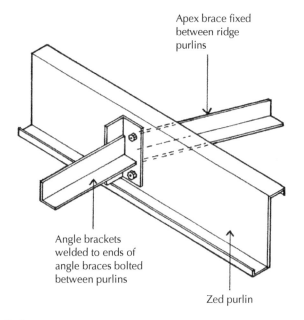

Figure 4.34   Purlin braces.

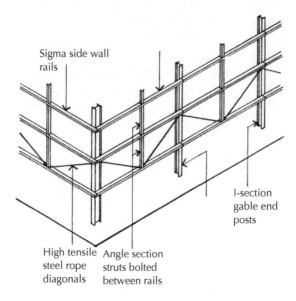

Figure 4.35   Struts and ties to side wall rails.

The side rails are fabricated from lengths of galvanised mild steel angle, with a fixing plate welded to each end, thus enabling the rails to be bolted to the sheeting rails. A system of tie wires is also used to provide additional restraint, as shown in Figure 4.35.

Timber provides an alternative material for short- and medium-span purlins between structural frame members. The ease of cutting and simplicity of fixing make treated timber a convenient and economic alternative to steel.

Precast-reinforced concrete portal frames

Following the end of the Second World War (1945), there was a shortage of steel, which led to the widespread use of reinforced concrete portal frames for single-storey structures, such as agricultural sheds, storage and factory buildings. A limited range of standard frames is cast in standard moulds under factory conditions. The comparatively small spans, limited sizes and bulky nature of the frames resulted in this method being used much less than steel. The advantages of concrete are its good fire resistance and relative freedom from maintenance.

For convenience of casting, transportation and erection on site, precast-reinforced concrete portal frames are usually cast in two or more sections, which are bolted together on site at the point of contraflexure in the rafters and/or at the junction of post and rafter (Figure 4.36). The portal frames are typically spaced between 4.5 and 6 m apart to support precast-reinforced concrete purlin and sheeting rails. Alternatively, timber or cold-formed steel Zed purlins and sheeting rails may be used. The bases of the concrete portals are placed in mortices cast in concrete foundations and grouted in position. Alternatively, base plates can be used in the same way that they are used in steel portal frames. The base plate is welded to the reinforcement and cast into the foot of the concrete frame at the same time as the rest of the precast frame. The clear span for standard single-bay structures may be up to 24 m, as shown in Figure 4.36.

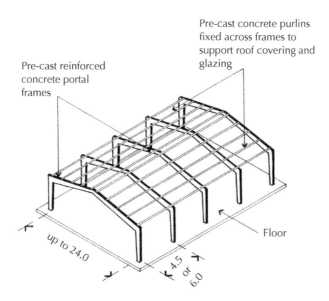

Pre-cast concrete purlins
fixed across frames to
support roof covering and
glazing

Pre-cast reinforced
concrete portal
frames

Floor

up to 24.0

4.5 or 6.0

Figure 4.36   Single-bay symmetrical pitch portal frames.

Figure 4.37 is an illustration of a two-bay symmetrical pitch concrete portal frame. In this example, the rafter is bolted to the post at the point of contraflexure. The internal posts are shaped to accommodate a precast-reinforced concrete valley gutter, which is bolted to the rafters and laid to a fall. The concrete purlins are fixed by loops protruding from their ends, which fit over studs cast in the rafters, as shown in Figure 4.38.

Timber portal frames

In the middle of the twentieth century, the technique of gluing timber laminae improved dramatically with the development of powerful, waterproof, synthetic resin adhesives. Later improvements in the technique of selecting wood of uniform properties and gluing laminations together under stringent quality control led to the development of factories capable of producing laminated timber sections suitable for use in buildings, in lieu of steel and reinforced concrete for all but the more heavily loaded structural elements.

Glue laminated timber (Glulam)

'Glulam' is the generic name that has been adopted for the product of a system of making members such as beams and roof frames from laminae of natural wood glued together to form longer lengths and shapes than is possible with natural wood by itself. Glulam is defined as a structural member made from four or more separate laminations of timber arranged with the grain parallel to the longitudinal axis of the member: the individual pieces being assembled with their grain approximately parallel and glued together to form a member, which functions as a single structural unit. The advantage of glulam is that both straight and curved sections can be built up from short, thin sections of timber glued together in long sections, up to 50 m, without appreciable loss of the beneficial properties of the natural wood from which they were cut.

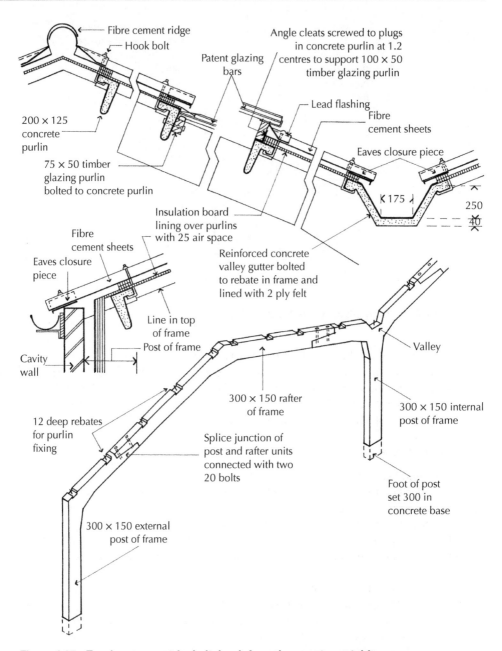

Figure 4.37 Two-bay symmetrical pitch reinforced concrete portal frame.

A range of standard glulam straight roof and floor beams are produced in a variety of sizes up to 20 m long and 4.94 m deep. These beams can be cut, holed and notched in the same way as the timber from which they were made. A wide range of purpose-made portal frames, flat pitched and cambered roof beams and arched glulam structures are practical where the curved forms and natural colour and grain can be displayed and where medium to wide clear spans are required.

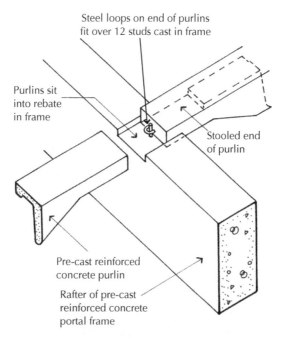

Steel loops on end of purlins
fit over 12 studs cast in frame

Purlins sit
into rebate
in frame

Stooled end
of purlin

Pre-cast reinforced
concrete purlin

Rafter of pre-cast
reinforced concrete
portal frame

Figure 4.38   Connection of concrete purlins to concrete portal frame.

Glulam structural members

Because of the labour costs involved in the fabrication of glulam members, glulam cannot compete with any of the basic steel frames in initial cost. However, glulam comes into its own in one-off, purpose-designed, medium-span buildings, where the durability of glulam and the appearance of natural wood are an intrinsic part of the building design, for example, in sports halls, assembly halls and swimming pools (Photographs 4.4–4.6). The advantages of timber in this form as a structural material are its low self-weight, minimal maintenance requirements to preserve and maintain its strength, and that it does not suffer from corrosion. Such properties are particularly important where there are levels of high humidity, as in swimming pools.

Timber laminae are mostly cut from European white wood, imported from Scandinavia. The knots in this wood are comparatively small; it is widely available in suitable strength grades, has excellent gluing properties and a clear, bright, light creamy colour. The stress (strength) grades are LA, LB and LC, with LC being the weakest of the three. Glulam members are usually composed of LB and LC grades or a combination of LB outer and LC inner laminates. The wood is cut into laminae up to 45 mm finished thickness for straight members and as thin as 13 mm for curved members. Laminates are kiln-dried to a moisture content of 12%. Individual lengths of timber are finger-jointed at the butt end. The ends of the laminae are cut or stamped to form interlocking protruding fingers that are 50 mm long. The lengths of the end jointed laminates are planed to the required thickness and a waterproof adhesive is applied to the faces to be joined. The adhesive used is, like the wood it is used to bond, resistant to chemical attack in polluted atmospheres and chemical solutions.

The adhesive-coated laminates are assembled in sets to suit the straight or curved section member they will form. Before the adhesive hardens, the laminates for curved members

Photograph 4.4   Glue-laminated timber structure.

are pulled around steel jigs to form the shape required. Both straight and curved sets are hydraulically cramped up until the adhesive is hardened. After assembly, the glulam members are cured under controlled conditions of temperature and humidity to the required moisture content. The surfaces of the straight members are then planed to remove adhesive that has been squeezed out and to reduce the section to its required dimensions and surface finish. Curved members are made oversize. The staggered ends of laminae are then cut to the required outside and inside curvature and the faces are then planed in the same manner as that for the straight members. The planed natural finish of the wood is usually left untreated to expose its natural colour and grain.

Timber decking can be used to serve as a natural wood finish to ceilings between glulam frames and rafters and as solid deck to support the roof covering and thermal insulation. The decking is laid across and screwed or nailed to roof beams and portal frames.

Photograph 4.5   Glue-laminated frame.

Photograph 4.6   Glue-laminated timber frame and roof.

Symmetrical pitch glulam timber portal frame

Symmetrical pitch glulam timber portal frames are usually fabricated in two sections for ease of transportation to the site. They are erected and bolted together at the ridge, as illustrated in Figure 4.39. The portals are spaced fairly widely apart to support timber or

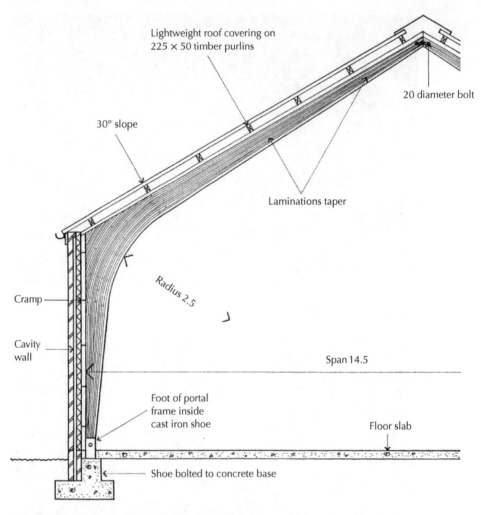

Figure 4.39   Glue-laminated timber portal frame.

steel purlins, which can be covered with sheet cladding materials, slates or tiles. Timber decking is commonly used to provide a soffit of natural timber. For buildings that require heating, the thermal insulation is placed above the timber soffit. The laminations of the timber from which the portal is made are arranged to taper so that the depth is greatest at the knee, where the frame tends to spread under load and where the depth is most needed. The portal is more slender at the apex and at the base of the post where the least section is required for strength and rigidity. The maximum radius of curve for shaped members is governed by the thickness of the laminates. A maximum radius of 5625 mm is recommended for 45 mm and 2500 mm for 20 mm thick laminae. Because of the labour involved in the assembly of curved members, they are appreciably more expensive than straight members.

Flat glulam timber portal frame

The flat portal frame illustrated in Figure 4.40 is designed for the most economic use of timber and consists of a web of small section timbers glued together with the top and bottom booms of glued laminate with web stiffeners. The portal frames are used to support metal decking on the roof and profiled sheeting on the walls. This long-span structure is lightweight and free from maintenance.

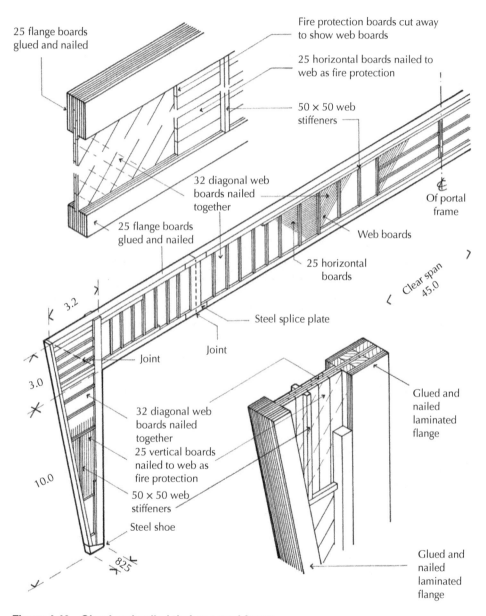

Figure 4.40   Glued and nailed timber portal frame.

Mid- to large-span glulam structures

The scale and span of glue-laminated structures has increased in recent years, as has the quality of the adhesive and structural fixings (Photograph 4.7a–d). Due to the lightweight nature of the frame and coverings, the foundations both transfer the loads to the ground and act as anchors to prevent uplift of the structure in high winds. The structural fixings are formed using angle brackets, bearing plates and splice plates, in much the same way that steel fixings are made (Photograph 4.7e–h).

Flat roof frame construction

Medium- and long-span flat roof structures are structurally less efficient and therefore less economic than truss, lattice or portal frames. The main reason for this is the need to prevent too large a deflection of the flat roof structure under load, thus leading to ponding of water on the surface of the roof. The advantage of a flat roof is that there is little unused roof space to be heated. Solid web I-section steel beams supported by steel columns may be used for industrial applications where the main beams are used to support lifting gear, but the most common form of framed flat roof construction is with lattice beam or with space frames.

Lattice beam flat roof construction

The terms beam and girder are used in a general sense to describe lattice construction. The term 'beam' is used to describe small depths associated with most roof construction and 'girder' for deeper depths associated with (e.g. bridge construction). For flat and low-pitch roofs, it is convenient to fabricate the top boom to provide a fall for the roof decking or sheeting. Lattice beams are either hot-dip galvanised, stove enamel primed or spray primed after manufacture.

Short-span beams that support relatively light loads may be constructed from cold-formed steel strip top and bottom booms with a lattice of steel rods welded between them, as illustrated in Figure 4.41. The top and bottom booms are formed as 'top hat' sections designed to take timber inserts for fixing roof decking and ceiling finishes.

The majority of lattice beams used for flat and low-pitch roofs are fabricated from hollow round and rectangular steel sections. For most low-pitch roofs to be covered with profiled sheeting, a slope of 6° is provided, as illustrated in Figure 4.42.

Space grid flat roof construction

Where there is a requirement for a large unobstructed floor area, such as exhibition areas and sports halls, a space deck roof can be used (Figures 4.43 and 4.44). A two-layer space deck comprises a grid of standard prefabricated units, each in the form of an inverted pyramid, as illustrated in Figure 4.45 and Photographs 4.8–4.10. The units are bolted together and connected with tie bars to form the roof structure. The tie bars can be adjusted to create an upward camber to the top deck to allow for deflection under load and also to provide a fall to the roof to encourage rainwater to discharge to gutters and thus avoid ponding. Photographs 4.8 and 4.9 show fixing nodes that allow different-length rods to be inserted.

A camber is formed by inserting shorter tie bars in the lower section of the structure. The roof of the structural deck may be covered with thermal insulation and steel decking or sheeting. Rooflights can be easily accommodated within the standard units and the roof can be cantilevered beyond supporting perimeter columns to provide an overhang.

Photograph 4.7 (a) Glue-laminated long-span structure, (b) Mid- to large-span glue-laminated beam anchor foundation. (c) Mid- to large-span glue-laminated beam mounted in thrust block and anchor foundation. (d) Steel rope prevents buckling and deformation of the structure. (e) Steel rope threaded through the structure. (f) Glue-laminated roof structure. (g) Beam-to-beam connection. (h) Beam to column connections.

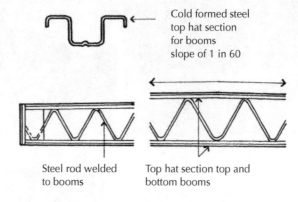

Cold formed steel
top hat section
for booms
slope of 1 in 60

Steel rod welded
to booms

Top hat section top and
bottom booms

Shallow pitch lattice beam

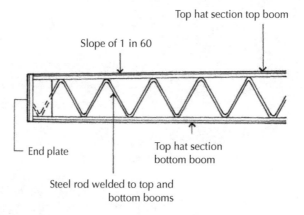

Top hat section top boom

Slope of 1 in 60

End plate

Top hat section
bottom boom

Steel rod welded to top and
bottom booms

Figure 4.41   Tapered lattice beam.

Space deck roofs may be designed as a two-way spanning structure with a square grid, or as a one-way spanning structure with a rectangular grid. Economic grid sizes are 12 × 12 m, 18 × 18 m and 12 × 18 m. The main advantage of the space deck roof is the wide spacing of the supporting columns and the economy of the structure in the use of standard units and the speed of erection. One disadvantage is that the members tend to attract dust and will require regular cleaning. Regular maintenance is also required to prevent rust.

Units are usually connected to the supporting steel columns at the junction of the trays of the units. Figures 4.46 and 4.47 illustrate the fixing of a space deck to perimeter and internal columns, respectively. At perimeter columns, a steel cap plate is welded to the cap of the column, to which a seating is bolted. This seating of steel angles has brackets welded to it into which the flanges of the trays fit and to which the trays are bolted. Similarly, a seating is bolted to a cap plate of internal columns with brackets into which the flanges of the angles of four trays fit.

The Eden project in Cornwall uses a fabricated steel dome space frame (Photographs 4.11 and 4.12). It was designed to accommodate hexagonal transparent membranes called

6° dual pitch lattice beam

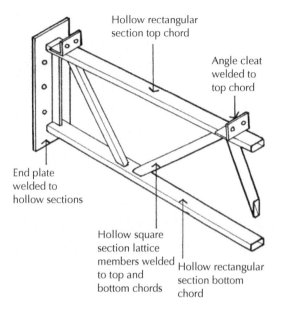

Hollow rectangular section top chord

Angle cleat welded to top chord

End plate welded to hollow sections

Hollow square section lattice members welded to top and bottom chords

Hollow rectangular section bottom chord

Figure 4.42   Lattice beam.

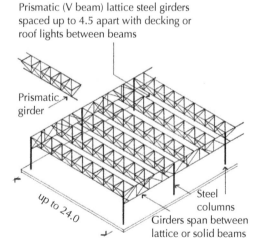

Prismatic (V beam) lattice steel girders spaced up to 4.5 apart with decking or roof lights between beams

Prismatic girder

up to 24.0

Steel columns

Girders span between lattice or solid beams

Figure 4.43   Prismatic (V beam) lattice steel roof on steel columns.

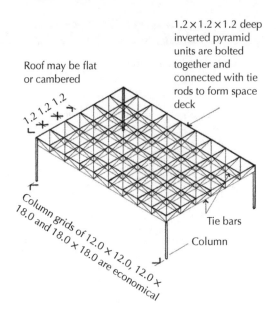

1.2 × 1.2 × 1.2 deep
inverted pyramid
units are bolted
together and
connected with tie
rods to form space
deck

Roof may be flat
or cambered

1.2 1.2 1.2

Tie bars

Column

Column grids of 12.0 × 12.0, 12.0 ×
18.0 and 18.0 × 18.0 are economical

Figure 4.44   Steel space deck root.

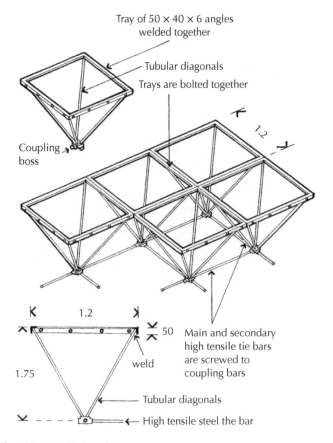

Tray of 50 × 40 × 6 angles
welded together

Tubular diagonals

Trays are bolted together

1.2

Coupling
boss

1.2

50   Main and secondary
high tensile tie bars
are screwed to
coupling bars

weld

1.75

Tubular diagonals

High tensile steel the bar

Figure 4.45   Steel space deck units.

Photograph 4.8   Space frame rods bolted to coupling nodes.

Photograph 4.9   Rods screwed into multiple fixing coupling node.

Photograph 4.10   Fully fabricated and welded spaces frame.

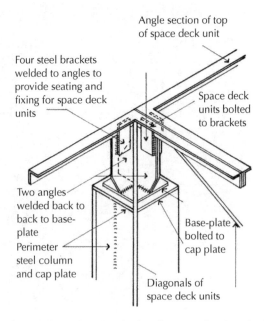

Figure 4.46   Support and fixing of space deck units to perimeter steel columns.

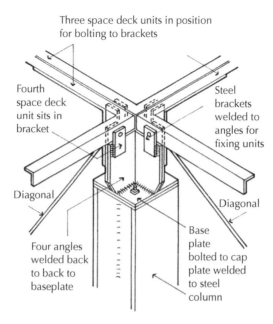

Three space deck units in position
for bolting to brackets

Fourth
space deck
unit sits in
bracket

Steel
brackets
welded to
angles for
fixing units

Diagonal

Diagonal

Four angles
welded back
to back to
baseplate

Base
plate
bolted to cap
plate welded
to steel
column

Figure 4.47   Connection of space deck units to an internal column.

Photograph 4.11   The hexagonal steel space frame.

Photograph 4.12   Hexagonal space frame with ETFE biomes.

biomes, which are made of an inflated ethylene tetrafluoroethylene (ETFE co-polymer foil), a development from the flat roof space frame technology.

Composite frame construction

A composite frame construction comprises prefabricated concrete and steel components, usually offered by one supplier as part of a design, manufacture and erection service for both single- and multi-storey-framed buildings. Precast reinforced concrete structural beams and columns are used to support lattice steel roof beams. The columns and beams are precast under carefully controlled factory conditions, with frame joints and base fixings, etc., cast in as necessary. The advantage of the composite frame construction is that the reinforced concrete columns and beams provide good fire resistance to the main structure and the lattice steel roof provides a lightweight covering. Economy of initial build cost can be made in the extensive use of prefabricated units.

Figure 4.48 is an illustration of a typical two-bay, single-storey composite frame structure. The precast reinforced concrete columns, which have fixed bases, serve as vertical cantilevers to take the major part of the loads from wind pressure. Steel brackets, cast into the column head, support the concrete and lattice steel roof beams. Concrete or lattice steel spine beams are used under the roof valley to provide intermediate support for every other roof beam. The top of the lattice steel roof beams, which are pitched at 6° to the horizontal, supports the low-pitch profiled steel roof sheeting. Fixing slots or brackets cast into the columns provide a fixing and support for sheeting rails, which in turn support the profiled steel cladding to the walls.

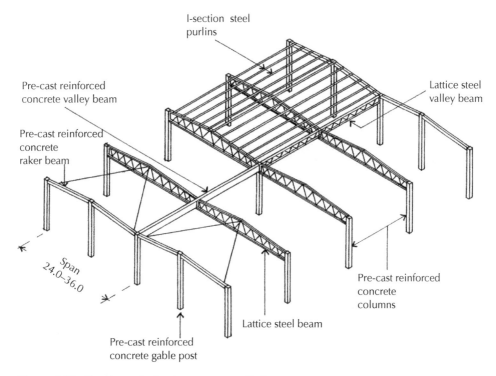

Figure 4.48 Two-bay single-storey composite frame.

4.2 Roof and wall cladding and decking

Plastic-coated profiled steel sheeting is the principal sheet material used to provide weather protection to single-storey-framed buildings . Laminated (sandwich) panels that incorporate thermal insulation are also available (see Chapter 8). Fibre cement sheet is also used.

Functional requirements

The functional requirements of roofs and walls have already been set out in *Barry's Introduction to Construction of Buildings*. In relation to wall and roof cladding, the following functional requirements need to be addressed:

❑   Strength and stability
❑   Resistance to weather
❑   Durability and freedom from maintenance
❑   Safe access during maintenance
❑   Fire safety
❑   Resistance to passage of heat
❑   Resistance to passage of sound
❑   Security
❑   Aesthetics

Strength and stability

The strength of roof and wall cladding and roof decking depends on the properties of the materials used and their ability to support the self-weight of the cladding and the imposed loads of wind and snow between the supporting purlins, rails, bearers and beams. The stability of the cladding and decking depends on the:

❑ Depth and spacing of the profiles of sheeting and decking
❑ Composition of the materials and thickness of the boards and slabs used for decking
❑ Ability of the materials to resist distortion due to wind pressure, wind uplift, snow loads and the weight of personnel engaged in fixing and maintaining the roofs

The strength and stability of the thin sheets of steel or aluminium derive principally from the depth and spacing of the profiles: shallow depth of profile for small spans to deep trapezoidal profiles and standing seams for medium to large spans between supports. Longitudinal and transverse ribs provide additional rigidity against buckling to deep profile sheeting. The comparatively thick corrugated and profiled fibre cement cladding sheets have adequate strength in depth of the profiles for anticipated loads and rigidity in the material to resist distortion and loss of stability over moderate spans between supports. Steel roof cladding sheets fixed across a structural frame act as a diaphragm, which contributes to the stability of the frames in resisting the racking effect of lateral wind forces that act on the sides and roofs of buildings. The extent of the contribution to the stability of the frames depends on the thickness of the sheets and the strength of the fasteners used to fix the sheets, as well as the ability of the sheets to resist the tearing effect of the fasteners fixed through it. Manufacturers provide guidance on the size and thickness of their sheets, minimum end lap, maximum purlin and rail centres, and maximum unsupported overhang of the sheets, as well as guidance on the type and spacing of fixings to match the exposure of the site.

Resistance to weather

Sheet steel and aluminium cladding resist the penetration of rainwater through the material's impermeability to water and the ability of the side and end laps to keep water out. The lowest allowable pitch of the roof is dictated by the end lap of the sheets. Thermal and structural movement is accommodated by the profiles, the end lap and designed tolerances at the fixings. Where long sheets are used, the secret fixing of the standing seam will allow for movement. Profiled metal sheets are usually fixed with screws, driven through the sheets into steel purlins and rails. Integral steel and Neoprene washers on the screw head effectively seal the perforation of the sheet against water penetration. Fixing is through the troughs of the profiles (where the rainwater runs) or (preferably) at the ridge of the trough, which takes a little more care and skill. Top fixing is preferred to bottom fixing because the perforation of the sheet is less exposed to water. Profiled cladding for walling is usually fixed through the troughs of the profile for ease of fixing and where the screw heads will be least visible. Standing seams to the edges of long sheets provide a deep upstand as protection against rain penetration, particularly with very low-pitch roofs.

Fibre cement sheets will resist water penetration through the density of the material, the slope of the roof and the end laps. The sheets will absorb some rainwater and should be laid at a pitch of 10° or more to avoid the possibility of frost damage. The sheets will accommodate moisture, thermal and structural movement through the end and side laps, as well as through the relatively large fixing holes for screws or hook bolts.

Flat roof membranes which resist the penetration of rainwater through the impermeability of the two-, three- or single-ply membranes and the sealed joints will, in time, harden and no longer retain sufficient elasticity or tensile strength to resist the thermal movements common to flat roof coverings laid over insulation materials.

Durability and freedom from maintenance

Coated profiled steel sheeting is easily damaged and so its durability depends to a certain extent on the care in handling and fixing on the building site. Damage to protective coatings can lead to corrosion of exposed steel, especially around fixing holes, and fixings driven home too tightly can easily distort the thin metal. Durability also depends on the climate and the colour of the coating material. Sheeting on buildings close to marine environments and in polluted industrial areas will deteriorate more rapidly than those in more sheltered, less polluted areas. Light-coloured coatings tend to be more durable than dark coatings, due to the effect of ultraviolet light on dark hues and the increased heat released from solar radiation on the more absorptive dark coatings. Organic-coated sheeting is a relatively short-lived material with a service life of around 25 years in favourable conditions and as low as 10 years in more aggressive climates.

Fibre cement sheeting does not corrode or deteriorate for many years, provided it is laid at a sufficiently steep pitch to shed water. The material is, however, relatively brittle and is liable to damage from impact and pressure from people accidentally walking over its surface. Reinforced fibre cement sheets are available that have a higher impact strength. These sheets tend to attract dirt because of the coarse texture of the surface, which is not easily washed away; thus, the sheets can become unsightly quite quickly.

Flat roof membranes, laid directly over thermal insulation material, will experience considerable temperature variations between day and night. In consequence, there is considerable expansion and contraction of the membrane, which in time may cause the membrane to tear. Solar radiation also causes oxidation and brittle hardening of bitumen-saturated or coated materials, which in time will no longer be impermeable to water. The durability of a roofing membrane in an inverted roof (upside down roof) is much improved by the layer of thermal insulation laid over the membrane, which helps protect it from the destructive effects of solar radiation and less extreme variations in temperature. The useful life of bitumen-impregnated felt membranes is from 10 years, for organic fibre felts up to 20 years and for high-performance felts up to 25 years: this can be extended by using an inverted roof construction. Mastic asphalt will oxidise and suffer brittle hardening over time which, combined with thermal movements, will give the material a useful life of around 20 years.

Safe maintenance

The Construction (Design and Management) (CDM) Regulations 2007 require that buildings should be designed so that they can be constructed, maintained and demolished safely. One in five construction-related accidents is caused by falls from, or through, roofs (HSE 1998). Care should be taken when designing structures to ensure that falling through sheeting materials and from the roof is recognised as a hazard and the risks of such occurrence are reduced. Provision should be made to prevent falls, including adequate access for plant and equipment. Safety rails should be used to prevent falls over the edges of roof structures.

Harnesses, fall arrest systems and safety nets do not prevent falls but do reduce the risks of injury in the event of a fall. Inclement weather poses a significant risk to those working in exposed positions and at heights. Work at heights should not continue during high winds or conditions that make the risks unacceptable. Debris netting (as well as safety netting) or birdcage scaffolds may be used to offer protection from falling objects and allow work to continue in the zone below the roof area. Debris shoots should also be used to ensure that waste, which presents a hazard if it falls, is quickly removed from the roof. Consideration must be given to maintenance operations once the roof structure is complete. Guarded walkways, access platforms, safety rails, etc., will be needed to ensure safe access.

Fire safety
Particular attention should be given to the internal and external fire spread characteristics of sheet materials in relation to the overall design of the building. A further cause for concern in framed buildings is concealed spaces, such as voids above suspended ceilings, roof and wall cavities. Cavity barriers and smoke stops should be fitted in accordance with current regulations and manufacturers guidance.

Resistance to passage of heat and ventilation
Resistance to the passage of heat is provided by thermal insulation materials, either separate from the sheeting material or as an integral part of the sheet in composite panels. Consideration must be given to thermal bridging in steel-framed buildings, especially at junctions, and care is required to avoid condensation. The principles of condensation, or rather the manner in which it can be avoided within the roof and wall structures, were discussed in *Barry's Introduction to Construction of Buildings*. Sheet metal may, in time, suffer corrosion from heavy condensation on the underside of the sheet. Ventilation to the space between the sheeting and the insulation, combined with a vapour check to the lining sheets, is the most effective way of minimising the risk of condensation. Fibre cement sheet is permeable to water vapour and thus provides less of a risk from condensation.

Resistance to passage of sound
The thin metal skin of profiled metal sheeting affords no appreciable resistance to sound penetration; thus, insulation must be provided, usually via the thermal insulation materials and effective seals around the opening parts of doors and windows. If sound insulation is a primary performance requirement, it may be advantageous to adopt a denser form of enclosure, such as brick or concrete to help provide the necessary sound reduction.

Security
Many single-storey-framed buildings are only occupied during working hours and are vulnerable to damage by vandalism and forced entry unless adequately protected through passive and active security measures. Apart from the obvious risk of forced entry through doors, windows and rooflights, there is a risk of entry by prising thin profiled sheeting from its fixing and so making an opening large enough to enter. Given that many buildings clad with steel sheeting are for warehousing purposes, this presents a serious challenge to the owners. Where the cost of the goods contained within is high and the likelihood of theft also high, it is wise to use a more solid form of wall construction, such as brick. Roofs are more difficult to protect, and some form of secondary protection is often used, such as a secondary steel cage under the roof (this is outside the scope of this book).

Aesthetics

Choice of an appropriate cladding for the building frame will also be determined by the appearance of the sheeting used and its ability to withstand weathering for a given timescale. Sheet profile and colour will be primary concerns, and a wide range of profiles and colours are available from manufacturers.

Profiled steel sheeting

The advantages of steel as a material for roof and wall sheeting are that its favourable strength-to-weight ratio and ductility make it both practical and economic to use comparatively thin, lightweight sheets that can be cold, roll formed to profiles with adequate strength and stiffness (Figures 4.49 and 4.50). The disadvantage of steel as a sheeting material is that it suffers rapid and progressive corrosion unless protected. The corrosive process is a complex electrochemical action that depends on the characteristics of the metal, atmosphere and temperature, and is most destructive in conditions of persistent moisture, atmospheric pollution and where different metals are in contact. Typically, steel is protected with a zinc coating by the hot-dip galvanising process.

Organic (plastic)-coated profiled steel sheets

The majority of profiled sheets used today are coated with an organic plastic coating to provide a protective coating and to provide an attractive finish. The plastic coating is applied to the galvanised zinc-coated steel sheets to serve as a barrier to atmospheric corrosion of the zinc, the erosive effect of wind and rain, and some degree of protection to damage during handling, fixing and in use. Colour is applied to the coated steel sheets by the addition of pigment to the coating material. There will be loss of colour, which tends not to be uniform over the whole sheet, especially on south-facing slopes over time. This spoils the appearance of the building, and cladding sheets may need to be replaced long before there is any danger of corrosion of the steel sheet. Light colours tend to exhibit better colour retention than darker colours. Four organic coatings are available, as described further.

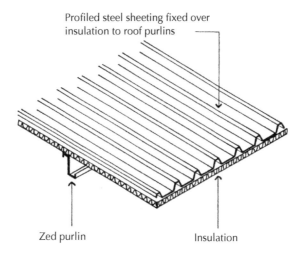

Profiled steel sheeting fixed over insulation to roof purlins

Zed purlin

Insulation

Figure 4.49 Cladding.

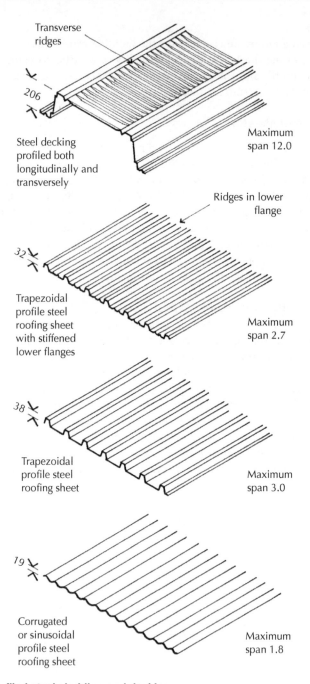

Transverse ridges

206

Steel decking profiled both longitudinally and transversely

Maximum span 12.0

Ridges in lower flange

32

Trapezoidal profile steel roofing sheet with stiffened lower flanges

Maximum span 2.7

38

Trapezoidal profile steel roofing sheet

Maximum span 3.0

19

Corrugated or sinusoidal profile steel roofing sheet

Maximum span 1.8

Figure 4.50   Profiled steel cladding and decking.

Polyvinyl chloride coatings
Polyvinyl chloride (PVC) coatings are the cheapest and most used of the organic plastic coatings (known as 'plastisol'). The comparatively thick (200 µm) coating that is applied over the zinc coating provides good resistance to normal weathering agents. The material is ultraviolet stabilised to retard the degradation by ultraviolet light and the inevitable loss of colour. The durability of the coating is good as a protection for the zinc coating below, but the life expectancy of colour retention is between 10 and 20 years. PVC is an economic, tough, durable, scratch-resistant coating but has poor colour retention.

Acrylic-polymethyl methacrylate coatings
Acrylic-polymethyl methacrylate is an organic plastic that is applied with heat under pressure, as a laminate to galvanised zinc steel strip to a thickness of 75 µm. It forms a tough finish with high strength, good impact resistance and good resistance to damage in handling, fixing and in use. It has excellent chemical resistance and its good resistance to ultraviolet radiation gives a life expectancy of colour retention of up to 20 years. The hard smooth finish of this coating is particularly free from dirt staining. It costs about twice as much as PVC coatings (unplasticated polyvinyl chloride [uPVC]).

Polyvinylidene fluoride coatings
Polyvinylidene fluoride (PVF) is a comparatively expensive organic plastic coating for profiled steel sheets, which is used as a thin (25 µm) coating for its excellent resistance to weathering, excellent chemical resistance, durability and resistance to all high-energy radiation. Because the coating is thin, careless handling and fixing may damage it. Durability is good and colour retention can be from 15 to 30 years.

Silicone polyester
Silicone polyester is the cheapest of the organic coatings used for galvanised steel sheet. It has a short life of between 5 and 7 years in a temperate, non-aggressive climate. Galvanised sheets are primed and coated with stoved silicone polyester to a thickness of 25 pm The coating provides reasonable protection against damage in handling and fixing.

Profiled steel cladding systems for roofs and walls
The term 'cladding' is a general description for materials, such as steel sheets, used to clothe or clad the external faces of framed buildings to provide weather protection. Thermal insulation is fixed under or behind the cladding sheets to provide the required thermal insulation to roofs and walls, respectively. A wide range of profiles are available, some of which are illustrated in Figure 4.51.

Single-skin cladding
The simplest system of cladding consists of a single skin of profiled steel sheeting fixed directly to purlins and sheeting rails without thermal insulation. This cheap form of construction is only used for buildings that do not need to be heated, such as warehouses and stores.

Over purlin insulation
The most straightforward and economic system of supporting insulation under cladding is to use semi-rigid or rigid insulation boards laid across roof purlins and sheeting rails, as shown in Figure 4.52. Timber spacers are used to provide an airspace for passive ventilation between the cladding sheets and the insulation. This system of cladding is suitable for buildings with low to medium levels of humidity and where the appearance of the insulation board is an acceptable finish.

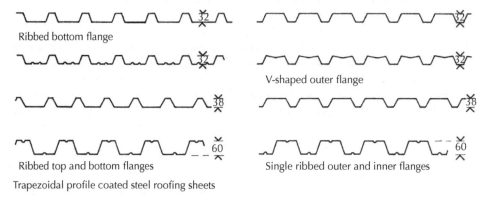

Ribbed bottom flange

V-shaped outer flange

Ribbed top and bottom flanges

Single ribbed outer and inner flanges

Trapezoidal profile coated steel roofing sheets

Figure 4.51   Trapezoidal profile-coated steel wall cladding sheets.

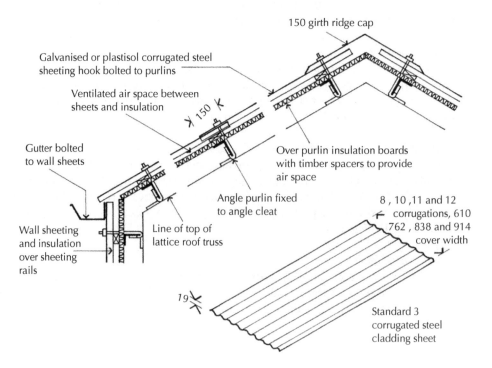

150 girth ridge cap

Galvanised or plastisol corrugated steel
sheeting hook bolted to purlins

Ventilated air space between
sheets and insulation

Gutter bolted
to wall sheets

Over purlin insulation boards
with timber spacers to provide
air space

Wall sheeting
and insulation
over sheeting
rails

Angle purlin fixed
to angle cleat

Line of top of
lattice roof truss

8 , 10 ,11 and 12
corrugations, 610
762 , 838 and 914
cover width

Standard 3
corrugated steel
cladding sheet

Figure 4.52   Corrugated steel cladding sheets.

Over purlin insulation with inner lining
Where mineral fibre mat insulation is used and where more rigid forms of insulation will not be self-supporting between widely spaced purlins, it is necessary to use profiled inner lining sheets (or trays) to provide support for the insulation. The lining sheets also help to provide a more attractive finish to the interior.

Linings are cold, roll-formed, steel strips with shallow depth profiles adequate to support the weight of the insulation. The sheets are hot-dip galvanised and coated with a protective and decorative organic plastic coating. To prevent compression of the loose mat or quilt, its thickness is maintained by Zed section spacers fixed between cladding and lining panels, as illustrated in Figures 4.53 and 4.54. The space between the top of the sheeting and the insulation is passively ventilated to minimise condensation, and a breather paper is usually spread over the top of the insulation. The breather paper protects the insulation from any rain or water condensate, yet allows moisture vapour to penetrate it. Some manufacturers also manufacture 'structural' trays, which provide a stronger internal lining and thus help to improve security to the roof.

Over-purlin composite (site assembled)
The over-purlin composite (site assembled) system comprises a core of rigid preformed lightweight insulation (or mineral wool and spacer), shaped to match the profile of the sheet and the inner lining tray. The separate components are assembled on site and fixed directly to purlins and lining sheets with self-tapping screws (Photograph 4.4). Side and

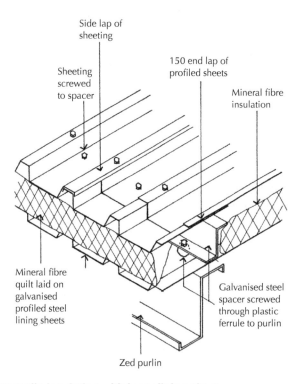

Figure 4.53   Over purlin insulation with inner lining sheets.

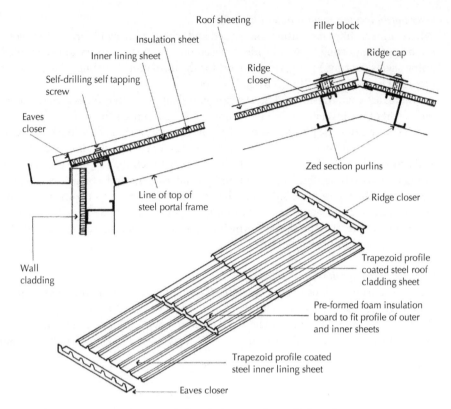

Figure 4.54   Profiled steel cladding, insulation and inner lining sheets.

end laps are sealed against the penetration of moisture vapour. Factory-formed composite panels have largely replaced this system.

Over-purlin composite (factory formed)
Factory-formed composite panels consist of a foamed insulation core enclosed and sealed by profiled sheeting and inner lining tray. The two panels and their insulating core act together structurally, to improve load-bearing characteristics. Panels have secret fixings to improve their visual appearance. Figure 4.55 is an illustration of factory-formed panels.

Standing seams
Standing seams are principally used for low and very low-pitch roofs to provide a deep upstand as weathering to the side joints of sheeting and to allow space for secret fixings. Sheets usually run from ridge to eaves to avoid the complication of detailing at the end laps with standing seams. The standing seam allows some tolerances for thermal movement of the long sheets and also provides some stiffness to the sheets, thus allowing a shallower profile to be used. Figure 4.56 illustrates a standing seam.

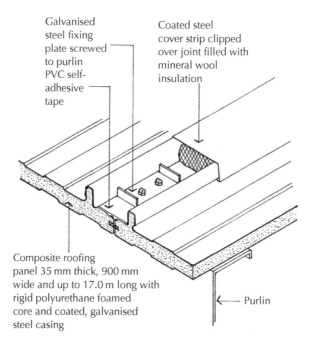

Galvanised steel fixing plate screwed to purlin PVC self-adhesive tape

Coated steel cover strip clipped over joint filled with mineral wool insulation

Composite roofing panel 35 mm thick, 900 mm wide and up to 17.0 m long with rigid polyurethane foamed core and coated, galvanised steel casing

Purlin

Figure 4.55   Composite roofing panels.

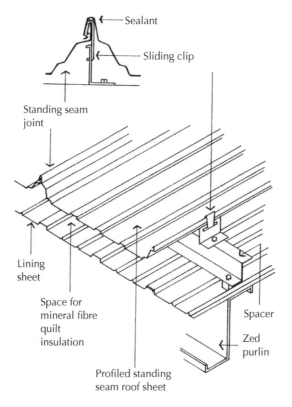

Sealant

Sliding clip

Standing seam joint

Lining sheet

Space for mineral fibre quilt insulation

Spacer

Zed purlin

Profiled standing seam roof sheet

Figure 4.56   Profiled standing seam roofing.

Fasteners

Steel cladding, lining sheets and spacers are usually fixed with coated steel or stainless steel self-tapping screws, illustrated in Figure 4.57. The screws are mechanically driven through the sheets into purlins or spacers. These primary fasteners for roof and wall sheeting may have coloured heads to match the colour of the sheeting. Secondary fasteners, which have a shorter tail, are used for fixing sheet to sheet and also flashing to sheet.

Gutters

Gutters are usually made from cold-formed, organic-coated steel and are laid at a slight fall to rainwater pipes. Gutters are supported on steel brackets screwed to eaves purlins. Valley gutters and parapet wall gutters usually have the inside of the gutter painted with bitumen as additional protection against corrosion.

Ridges

Ridges are covered with a cold-formed steel strip that is coated to provide the same finish as the roof sheeting. The ridge may be profiled to match the roof profile, or flat with a shaped filler piece to seal the space between sheet and ridge.

Wall cladding

Profiled steel sheeting is usually fixed to walls with the profile vertical, for convenience of fixing to horizontal sheeting rails fixed across the columns. Horizontal fixed sheeting can also be used for a different appearance, although some additional steel support may be required for widely spaced columns. The wall cladding is usually the same profile as that used for the roof. Figures 4.58 and 4.59 illustrate a typical section through a steel-framed building with steel sheeting above a lower wall of masonry. A drip flashing helps to keep the top of the wall dry by shedding the rainwater as it runs down the sheets. To provide a flush

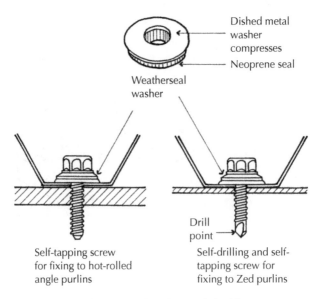

Figure 4.57   Fasteners for profiled steel sheeting and decking.

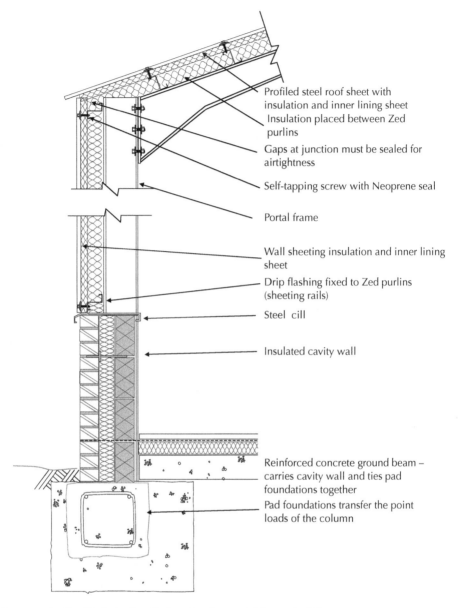

Profiled steel roof sheet with insulation and inner lining sheet

Insulation placed between Zed purlins

Gaps at junction must be sealed for airtightness

Self-tapping screw with Neoprene seal

Portal frame

Wall sheeting insulation and inner lining sheet

Drip flashing fixed to Zed purlins (sheeting rails)

Steel cill

Insulated cavity wall

Reinforced concrete ground beam – carries cavity wall and ties pad foundations together

Pad foundations transfer the point loads of the column

Figure 4.58   Profiled steel wall sheeting for portal frame building.

soffit to the roof cladding, the inner lining and insulation can be fitted under the purlins between the roof frames, as illustrated in Figure 4.60.

Profiled aluminium roof and wall cladding

On exposure to the atmosphere, aluminium corrodes to form a thin coating of oxide on its surface. This oxide coating, which is integral with the aluminium, adheres strongly and, being insoluble, protects the metal below from further corrosion so that the useful

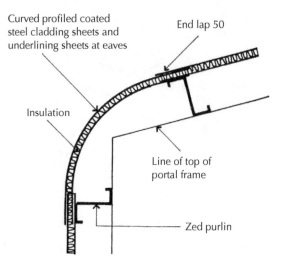

Figure 4.59   Curved profiled steel cladding.

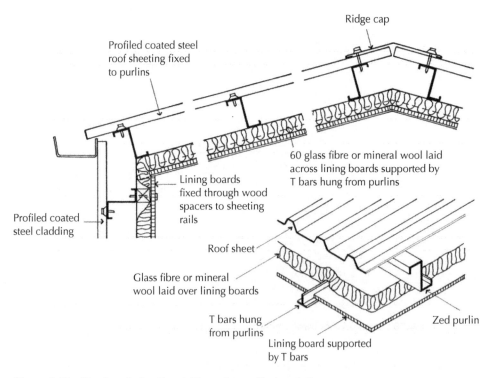

Figure 4.60   Steel roof sheeting with under-purlin insulation.

life of aluminium is 40 years or more. Aluminium is a lightweight, malleable metal with poor mechanical strength, which can be cold-formed without damage. Aluminium alloy strip is cold rolled as corrugated and trapezoidal profile sheets for roof and wall cladding. The sheets are supplied as metal mill finish, metal stucco embossed finish, pre-painted or organically coated.

Mill finish is the natural untreated surface of the metal from the rolling mill. It has a smooth, highly reflective metallic silver grey finish, which dulls and darkens with time. Variations in the flat surfaces of the mill finish sheet will be emphasised by the reflective surface. A stucco embossed finish to sheets is produced by embossing the sheets with rollers to form a shallow, irregular raised patterned finish that reduces direct reflection and sun glare and so masks variations in the level of the surface of the sheets. A painted finish is provided by coating the surface of the sheet with a passivity primer and a semi-gloss acrylic or alkyd-amino coating in a wide range of colours. A two-coat PVF acrylic finish to the sheet is applied by roller to produce a low-gloss coating in a wide range of colours.

Aluminium sheeting is more expensive than steel sheeting and is used for its greater durability, particularly where humid internal atmospheres might cause early deterioration of coated steel sheeting. The material also offers some more interesting architectural features and has been used instead of steel sheet for its attractive natural mill finish. Figure 4.61

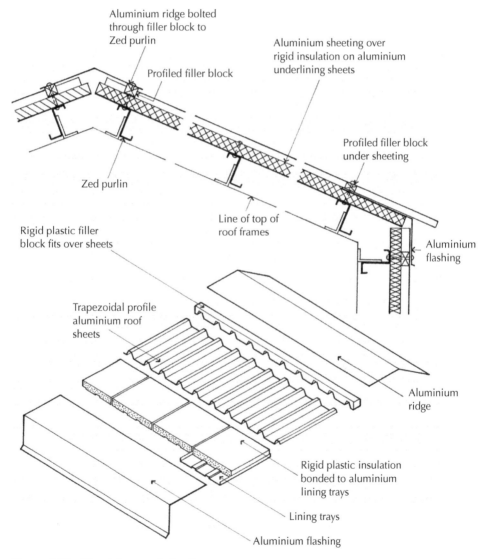

Figure 4.61   Aluminium roof sheeting.

is an illustration of profiled aluminium roof and wall sheeting, fixed over rigid insulation boards bonded to steel lining trays, to a portal steel frame.

Fibre cement profiled cladding

Fibre-reinforced cement sheets are manufactured from cellulose and polymeric fibres, cement and water, and pressed into a range of profiles. High-strength fibre-reinforced cement sheets are made with polypropylene reinforcement strips inserted along precisely engineered locations along the length of the sheet, which provides greater impact strength without affecting the durability of the product.

Sheets are usually finished in a natural grey colour, especially when used for industrial and agricultural buildings, although a range of natural colours and painted finishes are also available from some manufacturers. Fibre cement sheets are vapour permeable, which greatly reduces the risk of condensation. The sheets are a Class 0 material, provide excellent acoustic insulation, have a high level of corrosion resistance, are easy to fix and are maintenance-free. Manufacturers provide guarantees of up to 30 years. The reinforced sheets should comply with the requirements for roof safety, as set out by the Health and Safety Executive.

Fibre cement sheets are heavier than steel sheets and so require closer centres of support from purlins and sheeting rails. Corrugated fibre cement sheet may be pitched as low as 5° to the horizontal in sheltered locations, although upwards of 10° is more common. The detail shown in Figure 4.62 is typical of the type of construction used

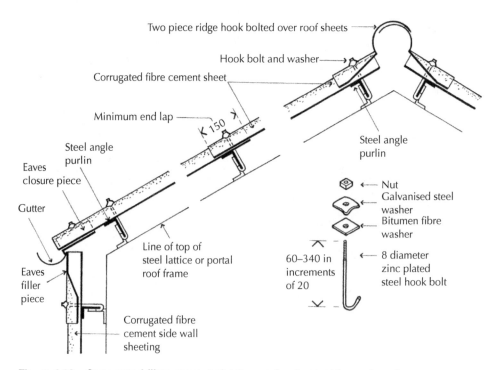

Figure 4.62   Corrugated fibre cement sheet covering to steel-framed roof.

in unheated outbuildings such as garages and tool sheds clad with fibre sheets. Typical fixings for fibre cement sheets are illustrated in Figures 4.62–4.64 is a typical section through a steel structure with profiled fibre cement sheets, insulation and underlining sheets.

Manufacturers of fibre cement sheets offer bespoke systems that combine profiled fibre cement weathering sheets with thermal insulation and an underlining sheet of fibre cement or coated steel. These are offered with a proprietary support bar system, which both supports the roof cladding sheets and helps to maintain a clear cavity into which the insulation blanket is placed. The system is built up on site in accordance with the manufacturer's guidance to provide a highly durable roofing system with tested performance characteristics, giving very good acoustic and thermal insulation as well as high resistance to condensation. Recommended pitch ranges from between 5° and 30°. A typical system is illustrated in Figure 4.65.

Roof ventilation to agricultural buildings
Fibre cement sheeting is used extensively in agricultural buildings, many of which have very specific ventilation requirements (e.g. cattle sheds or pig pens). A number of profiled prefabricated ridge fittings, including open ridges, are available that provide high levels of ventilation to the covered area given in Figure 4.66.

Ridge ventilation is usually used in combination with a spaced roof or a breathing roof. A breathing roof is constructed using Tanalised 50 × 25 mm timber battens or strips of nylon mesh to form a spacer between the courses of profiled sheets, thus providing a simple,

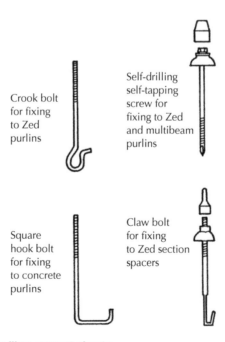

Crook bolt for fixing to Zed purlins

Self-drilling self-tapping screw for fixing to Zed and multibeam purlins

Square hook bolt for fixing to concrete purlins

Claw bolt for fixing to Zed section spacers

Figure 4.63   Fixings for fibre cement sheets.

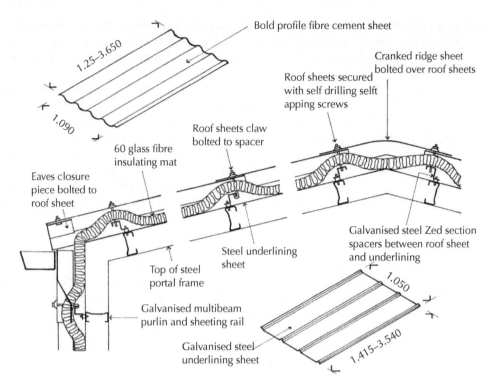

Figure 4.64   Fibre cement sheeting with insulation and steel sheet underlining.

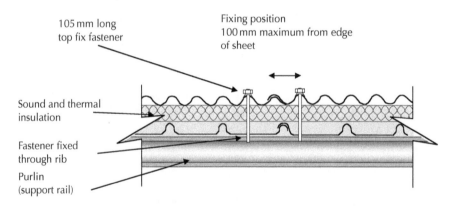

Figure 4.65   Fibre cement roofing sheets.

cheap and effective means of ventilation (Figure 4.67). A spaced roof is used for buildings that house high unit intensive rearing, which require high levels of natural ventilation. In this roof, the profiled sheets are positioned to create a gap of between 15 mm and 25 mm between the sheets; this provides excellent ventilation but also allows some rain and snow penetration (Figure 4.68).

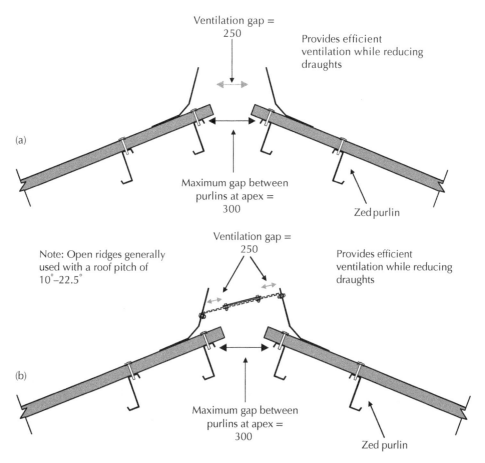

Figure 4.66 **(a) Agricultural roof ventilation: unprotected open ridges (*Source*: adapted from www.marleyeternit.co.uk.). (b) Agricultural roof ventilation: protected open ridges (*Source*: adapted from http://www.eternit.co.uk.).**

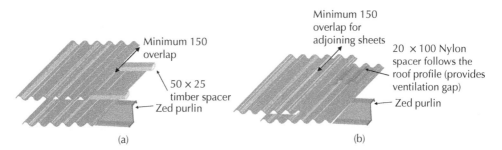

Figure 4.67 **(a) Fibre-based agricultural roof: breathing roof with timber spacer. (b) Fibre-based agricultural roof: breathing roof with nylon mesh.**

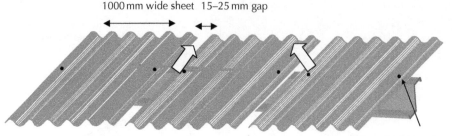

1000 mm wide sheet   15–25 mm gap

Small gap left between each sheet allows for ventilation and minimises weather ingress

Fixing should be placed in first corrugation from edge

Figure 4.68   Fibre-based agricultural roof: spaced breathing roof.

Decking

Decking is the general term used for the material or materials used and fixed across roofs to serve as a flat surface on to which one of the flat roof weathering membranes is laid. The decking is also used to support the thermal insulation, thus creating a warm roof construction. The decking is designed to support the weight of the materials of the roof and imposed loads of wind and snow, and is laid to a shallow fall to encourage rainwater run-off. Decking is sometimes applied to low-pitch lattice beam and portal frames. The most common form of decking is constructed from profiled steel sheeting. Decking can also be made of timber (for timber structures) or lightweight concrete slabs (for steel or concrete frames).

Profiled steel decking

The most commonly used form of decking is constructed from galvanised profiled steel sheeting, which is fixed with screws across beams or purlins. The underside of the decking may be primed ready for painting or be manufactured with a coated finish. Typical spans between structural frames or beams are up to 12 m for 200 mm deep trapezoidal profiles. The decking provides support for rigid insulation board, which is laid on a vapour check. The weathering membrane is then bonded to the insulation boards, as illustrated in Figure 4.69. Manufacturers produce a range or proprietary composite steel decking systems for long spans that provide high thermal insulation values.

Flat roof weathering

There is no economic or practical advantage in the use of a flat roof structure unless the roof is to be used (e.g. for leisure). A flat roof structure is less efficient structurally than a pitched roof, and there is little saving on unused roof space compared with the profiled metal sheeting, which can be laid to pitches as low as 2.5° to the horizontal. The roof surface must be constructed to create falls to rainwater outlets to avoid ponding of water on the roof surface, so it is not entirely 'flat'. In the UK climate, flat roofs have not performed particularly well; however, improvements in flat roof weathering membranes and careful detailing may help to make flat roofs a viable alternative to profiled sheet metal. See Chapter 6 of *Barry's Introduction to Building* for further details of materials, insulation and ventilation for flat roofs.

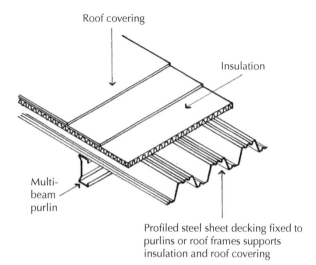

Roof covering

Insulation

Multi-
beam
purlin

Profiled steel sheet decking fixed to
purlins or roof frames supports
insulation and roof covering

Figure 4.69   Roof decking.

Drainage and falls

Given the importance of removing water from flat roofs, it is important to consider how and in which direction the water will fall to eaves, valley and/or central outlets, as illustrated in Figure 4.70. A one-direction fall is the simplest to construct, for example from a lattice beam with sloping top boom or with firring pieces of wood or tapered insulation boards laid over the structure to provide the necessary falls. A two-directional fall is more complicated and hence more time-consuming to construct, because of the need to mitre the ends of the tapered materials. A wet screed of concrete can be laid and finished with cross falls without difficulty.

Flat roof coverings are laid so that they fall directly to rainwater outlets, usually at a fall of 1 in 40. A typical straight-fall rainwater gutter is illustrated in Figure 4.71, where the roof falls to a central valley and rainwater pipes are positioned to run down against the web of structural columns.

Built-up roof coverings to roof decks

The first layer of built-up roof sheeting has to be attached to the surface of the roof deck to resist wind uplift. The manner in which this is done will depend on the nature of the roof deck. Full and partial bond methods were described in *Barry's Introduction to Construction of Buildings*. Particular attention should be given to the detailing and quality of the work to vulnerable areas such as eaves and verges, skirtings and upstands and joints. At control (expansion) joints in the structure, it is necessary to make some form of upstand in the roof on each side of the joint (Figure 4.72). The roofing is dressed up on each side of the joint as a skirting to the upstands. A plastic-coated metal capping is then secured with secret fixings to form a weather capping to the joint.

Single-ply roofing

Single-ply roofing materials provide a tough, flexible, durable lightweight weathering membrane, which is able to accommodate thermal movements without fatigue. To take the maximum advantage of the flexibility and elasticity of the membrane, the material should be loose laid over roofs so that it is free to expand and contract independently of the roof deck. To resist wind uplift, the membrane is held down either by loose ballast, a system

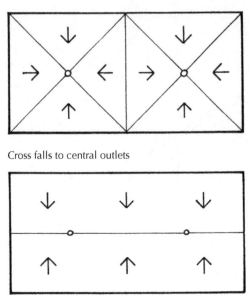

Cross falls to central outlets

Straight falls to outlets

Figure 4.70   Falls and drainage of flat roofs.

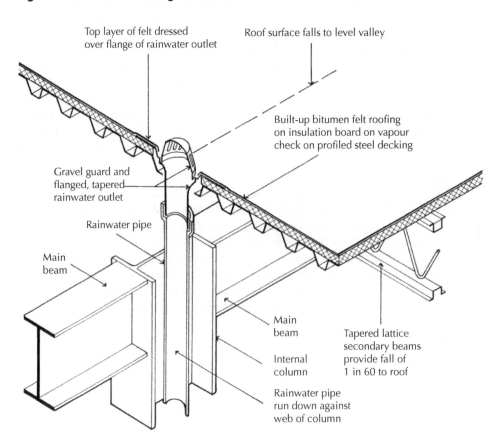

Top layer of felt dressed over flange of rainwater outlet

Roof surface falls to level valley

Built-up bitumen felt roofing on insulation board on vapour check on profiled steel decking

Gravel guard and flanged, tapered rainwater outlet

Rainwater pipe

Main beam

Main beam

Internal column

Tapered lattice secondary beams provide fall of 1 in 60 to roof

Rainwater pipe run down against web of column

Figure 4.71   Rainwater outlet in built-up bitumen felt roof.

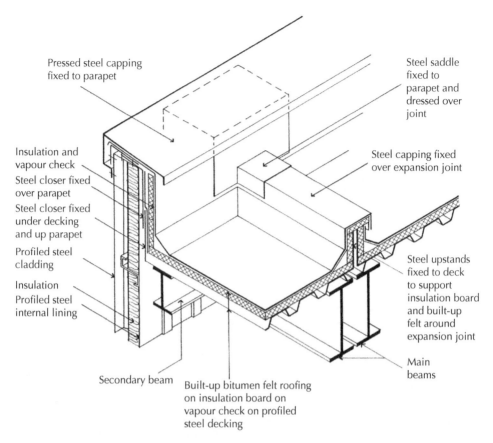

Pressed steel capping
fixed to parapet

Steel saddle
fixed to
parapet and
dressed over
joint

Insulation and
vapour check

Steel capping fixed
over expansion joint

Steel closer fixed
over parapet

Steel closer fixed
under decking
and up parapet

Profiled steel
cladding

Steel upstands
fixed to deck
to support
insulation board
and built-up
felt around
expansion joint

Insulation

Profiled steel
internal lining

Main
beams

Secondary beam

Built-up bitumen felt roofing
on insulation board on
vapour check on profiled
steel decking

**Figure 4.72   Parapet and expansion joint to profiled steel decking covered with built-up
bitumen felt roofing.**

of mechanical fasteners or adhesives. The materials used in the manufacture of single-ply
membranes are grouped as thermoplastic, plastic elastic and elastomeric.

❏   Thermoplastic materials include PVC, chlorinated polyethylene (CPE), chlorosulpho-
nated polyethylene (CSM) and vinyl ethylene terpolymer (VET). The materials are
tough with good flexibility. All of these materials can be solvent or heat-welded.
❏   Plastic elastic materials include polyisobutylene (PIB) and butyl rubber (IIR). PIB can
be solvent or heat welded; IIR is joined with adhesive.
❏   Elastomeric, ethylene propylene diene monomer (EPDM). Materials are flexible and
elastic with good resistance to oxidation, ozone and ultraviolet degradation. The materials are joined with adhesives.

These single-ply materials are impermeable to water, moderately permeable to moisture
vapour, flexible and maintain their useful characteristics over a wider range of temperatures
than the materials used for built-up roofing. To enhance tear resistance and strength, these
materials may be reinforced with polyester or glass fibre fabric. Manufacturers provide
detailed guidance on fixing, exposure and durability, together with conformity to relevant
standards and product guarantees.

4.3 Rooflights

The traditional means of providing daylight penetration to the working surfaces of large single-storey buildings is through rooflights, either fixed in the slope of roofs or as upstand lights in flat roofs. With the increase in automated manufacturing and artificial illumination, combined with concerns over poor thermal and sound insulation, unwanted glare, solar heat gain, and concerns over security, the use of rooflights has become much less common.

Functional requirements

The primary function of a rooflight is to allow the admission of daylight. As a component part of the roof, the rooflight also has to satisfy the functional requirements of the roof, being strength and stability; resistance to weather; durability and freedom from maintenance; fire safety; resistance to the passage of heat; resistance to the passage of sound; and security.

Daylight
Rooflights should be of sufficient area to provide satisfactory daylight and be spaced to give reasonable uniformity of lighting on the working surface without an excessive direct view of the sky, to minimise glare or penetration of direct sunlight and to avoid excessive solar gain. The area chosen is a compromise between the provision of adequate daylight and the need to limit heat loss through the lights. In pitched roofs, rooflights are usually formed in the slope of the roof to give an area of up to one-sixth of the floor area and spaced, as indicated in Figure 4.73, to give good uniformity and distribution of light. Rooflights in flat roofs are constructed with upstand curbs to provide a means of finishing and hence weathering the roof covering, and should be designed and positioned to provide an area of up to one-sixth of the floor area. North rooflights are used to minimise solar heat gain and solar glare; the area of the rooflight may be up to one-third of the floor area, as shown in Figure 4.73. Monitor rooflights is a term used to describe vertical or sloping slides to a rooflight, as illustrated in Figure 4.73, and these should have an area of up to one-third of the floor area.

Strength and stability
The materials used for rooflights tend to be used in the form of thin sheets to obtain the maximum transmission of light and also for economy. Glass will require support at relatively close centres to provide adequate strength and stiffness as part of the roof covering. Plastic profiled sheets tend to have less strength than the metal profiled sheets and so as a general rule require support at closer centres. Plastic sheets extruded in the form of double and triple-skin cellular flat sheets have good strength and stiffness. Attention must be paid to the safety of rooflights so as to prevent the possibility of anyone falling through the covering.

Resistance to weather
Metal glazing bars, used to provide support for glass, are made with non-ferrous flashings or plastic cappings and gaskets that fit over the glass to exclude wind and rain. A minimum pitch of 15° to the horizontal is recommended. Profiled plastic sheets are designed

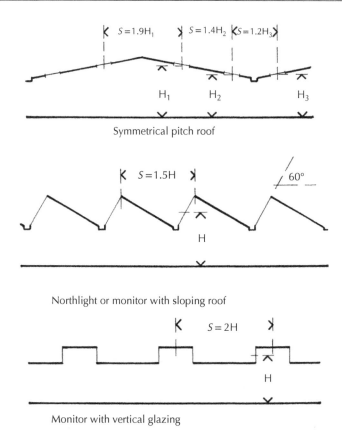

Figure 4.73 Spacing of rooflights.

to provide an adequate side lap and sufficient end lap to give the same resistance to the penetration of wind and rain as the profiled metal cladding in which they are fixed. A minimum pitch of not less than 10° to the horizontal is recommended. For lower pitches, it is necessary to seal both side and end laps to profiled metal sheeting with a silicone sealant to exclude wind and rain. Cellular flat plastic sheets are fitted with metal or plastic gaskets to weather the joints between the sheets fixed down the slope and with nonferrous metal flashings at overlaps at the top and bottom of sheets. Rooflights in flat and low-pitch roofs are fixed on a curb (upstand) to which the roof covering is dressed to exclude weather.

Durability and freedom from maintenance
Glass is the most durable of materials; however, regular washing is required to maintain adequate daylight penetration to the working surface below. Plastic materials will discolour over time and, depending on the profile of the plastic sheets, may also trap dirt. Regular cleaning is also required to maintain adequate daylight penetration and a regular replacement strategy will be required to replace the discoloured sheets. Manufacturers provide guidance as to the expected life of translucent sheets.

Fire safety

Fire safety in relation to rooflights is concerned with limiting the internal and external spread of flame. To limit the spread of fire over the surface of materials, it is necessary to limit the use of thermoplastic materials in rooflights. The Building Regulations limit the number, position and use of thermoplastic rooflights. Thermoplastic rooflights must not be used in a protected shaft. Materials for rooflights should be chosen with care and with reference to their spread of flame characteristics. To reduce the risk of a rooflight allowing fire to pass from one building to another, there are limitations on the minimum distance within which a rooflight can be placed in relation to the boundary. The distance of the rooflight from the boundary is dependent on the type of rooflight and the type of roof covering used.

Resistance to passage of heat

Limiting the number and size of rooflight can mitigate heat transfer through rooflights. Sealed double (or triple) glazed units will go some way in helping to improve the thermal resistance of the roof.

Resistance to passage of sound

The thin sheets of plastic or glass used in rooflights offer little resistance to the transfer of sound. Although some reduction in sound transfer can be achieved with double and triple-glazed units, it will be necessary to limit the size and number of rooflights for buildings that house noisy activities. In buildings where sound reduction is a critical requirement, only specifically designed acoustic rooflights should be used; normally these are triple-glazed units with a 90–150 mm cavity between the internal and external sheets of glass.

Security

Single-storey buildings clad with lightweight metal cladding to roofs and walls are vulnerable to forced entry through windows, doors, walls and roof cladding, and through glass and plastic rooflights. Security against forced entry and vandalism is best achieved via secure perimeter fencing and effective 24 hour surveillance. As a general guide, rooflights should be designed and constructed so as not to compromise the security of the roof structure.

Safety – Fragile roofs and rooflights

Rooflights and fragile roofs are a potential source of danger when constructing the roof when carrying out maintenance on the roof and to trespassers. Falls through fragile material give rise to more fatal accidents in the construction industry than any other single cause (HSE 1998). Adequate measures must be taken to prevent people from falling through fragile roofs and rooflights. Safe access to, and over, the roof surface must be provided. Platforms and staging may be provided to allow access for maintenance and inspections. Guarding should be provided to prevent persons who are on the roof from entering into the vicinity of the fragile surface. When carrying out refurbishment or maintenance staging, safety nets, birdcage scaffolds, harnesses and line system, as well as other safe means of access may need to be provided to sufficiently reduce the risk of anyone falling through the roof or rooflight. Precautions must also be taken to prevent unauthorised access to fragile roofs. Relevant legislation includes:

❑   The Health and Safety at Work etc. Act 1974.
❑   The Management of Health and Safety at Work Regulations 1999.

❏   The Construction (Health, Safety and Welfare) Regulations 1996.
❏   The CDM Regulations 2007.
❏   The Lifting Operations and Lifting Equipment Regulations 1998.

Materials used for rooflights

The traditional material for rooflights was glass laid in continuous bays across the slopes of roofs and lapped under and over slate or tile roofing. The majority of rooflights constructed today are of translucent sheets of plastic, usually formed to the same profile as the roof sheeting.

Glass

The types of glass used for rooflights are float glass, solar control glass, patterned glass and wired glass, which is used to minimise the danger from broken glass during fires. Glass has poor mechanical strength and must be supported with metal or timber glazing bars, at relatively close centres of about 600 mm for patent glazing. The principal advantage of glass is that it provides a clear view and, with regular washing, maintains a bright surface appearance.

Profiled, cellular and flat plastic sheets

Transparent or translucent plastic sheet material is used as a cheaper alternative to glass. The materials used for profiled sheeting are:

❏   *uPVC – PVC – rigid PVC*. This is one of the cheapest materials and has a light transmittance of 77%, reasonable impact resistance and good resistance to damage. The material will discolour when exposed to solar radiation.
❏   *Glass-reinforced plastic (GRP)*. GRP is usually inflammable and has good impact resistance, rigidity and dimensional stability. The material is translucent and has a moderate light transmittance of between 50% and 70%. Translucent GRP sheets comprise thermosetting polyester resins, curing agents, light stabilisers, flame retardants and reinforcing glass fibres. Three grades of GRP sheet are produced to satisfy the conditions for external fire exposure and surface spread of flame.

The materials used for flat sheet rooflights, laylights and domelights are:

❏   *Polycarbonate (PC)*. This material has good light transmittance, up to 88%, good resistance to weathering, reasonable durability and very good impact resistance. PC is the most expensive of the materials and is used principally for its high-impact resistance.
❏   *Polymethyl methacrylate (PMMA)*. This plastic is used for shaped rooflights, having good impact resistance and resistance to ultraviolet radiation, but softens and burns readily when subject to the heat generated by fires.

Rooflights

The most straightforward way of constructing rooflights in pitched roofs covered with profiled sheeting is by the use of GRP or uPVC, which is formed to match the profile of the roof sheeting. The translucent sheets are laid so that they cover the lower sheet and adjacent sheet to form an end and side lap, respectively. All side laps should be sealed with self-adhesive closed-cell PVC sealing tape to make a weather-tight joint. End laps between translucent sheets and between translucent sheet and roof sheets to roofs pitched below 20° should be sealed with extruded mastic sealant. Fixing of sheets is critical to resist wind uplift, in common with all lightweight sheeting materials used for roofing, and the fixing usually follows that used for the main roofing material.

Double-skin rooflights are constructed with two sheets of GRP, as illustrated in Figure 4.74, which have the same profile as the sheet roof covering. Profiled, high-density foam spacers, bedded top and bottom in silicone mastic, are fitted between the sheets to maintain the airspace and also to seal the cavity. Double-sided adhesive tape is fixed to all side laps of both top and bottom sheets as a seal. The double-skin rooflight is secured with fasteners driven through the sheets and foam spacers to the purlins. Stitching screws are then driven through the crown of profiles at side and end laps. Factory-formed sealed double-skin GRP rooflight units are made from a profiled top sheet and a flat underside with a spacer and sealer.

Translucent PVC (uPVC) sheets are produced in a range of profiles to match most metal and fibre cement sheeting. For roof pitches of 15° or less, the side and end laps should be sealed with sealing strips and all laps between uPVC sheets should be sealed. Fixing holes should be 3 mm larger in diameter than the fixing to allow for thermal expansion of the material. Fasteners similar to those used for fixing roofing sheets are used. Double-skin rooflights are formed in a similar manner to that shown in Figure 4.74.

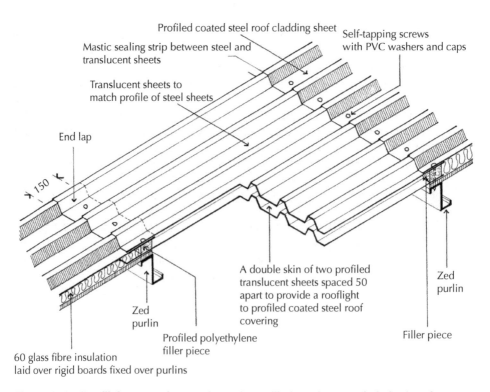

Figure 4.74   Rooflights: translucent sheets in profiled steel-covered pitched roof.

Flat cellular sheets of PC are supported by aluminium glazing bars fixed to purlins, as illustrated in Figure 4.75, to form a rooflight to a north-facing roof slope. The capping of the glazing bars compresses a Neoprene gasket to the sheets to make a watertight seal.

Patent glazing
The traditional method of fixing glass in the slopes of roofs to create a rooflight is by means of wood or metal glazing bars that provide support for the glass and form weather flashings, or cappings, to exclude water. The word 'patent' refers to the patents taken out by the original makers of glazing bars. Timber, iron and steel glazing bars have largely been replaced by aluminium and lead or plastic-coated steel bars. Likewise, single glazing has been replaced by double-glazed units and wired glass. Patent glazing is relatively labour intensive due to the provision and fixing of the glazing bars at relatively close centres; however, the result can be an attractive, durable rooflight with good light transmission.

The most commonly used glazing bars are of extruded aluminium with seatings for glass, condensation channels and a deep web top flange for strength and stiffness in supporting the weight of the glass. The glass is secured with clips, beads or cappings. Figure 4.76 illustrates aluminium glazing bars supporting single-wired glass in the slope of a pitched roof.

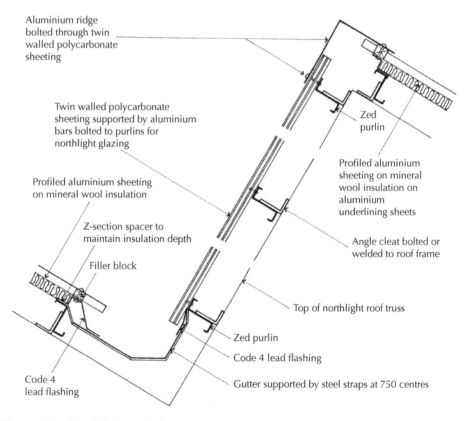

Figure 4.75 North light roof glazing.

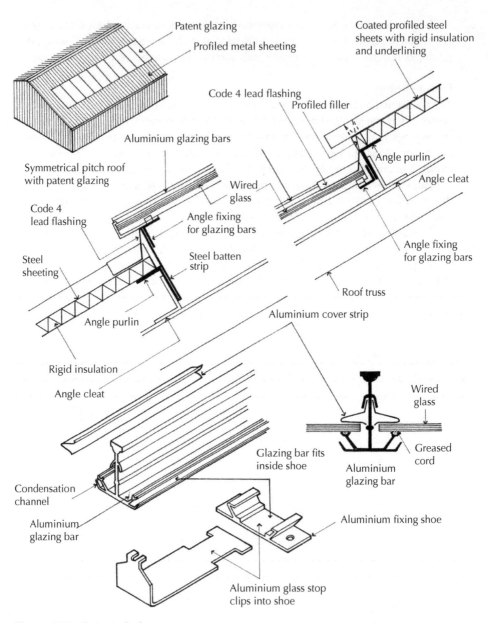

Figure 4.76   Patent glazing.

The glazing bars are secured in fixing shoes screwed or bolted to angles fixed to purlins and fitted with aluminium stops to prevent glass from slipping down the slope of the roof. Aluminium spring clips, fitted to grooves in the bars, keep the glass in place and serve as weathering between the glass and the bar. Also illustrated is a system of steel battens and angles, an angle and a purlin to provide fixing for the glass and sheeting at their overlap. Lead flashings are fixed as weathering at the overlap of the glass and sheeting.

Figure 4.77 shows six different types of glazing bar. Aluminium glazing bar for sealed double glazing (Figure 4.77a) and single glazing (Figure 4.77b) are secured with aluminium beads bolted to the bar and weathered with butyl strips. Aluminium glazing bars with bolted aluminium capping and snap-on aluminium cappings to the bars are illustrated in Figure 4.77c,d. Cappings are used to secure glass in position on steep slopes and for vertical glazing, as they afford a more secure fixing than spring clips; visually they give greater emphasis to the bars. Steel bars covered with lead and PVC sheathing as protection against corrosion are shown in Figure 4.77e,f. Steel bars are used for mechanical strength of the material and the advantage of more widely spaced supports than is possible with aluminium bars of similar depth.

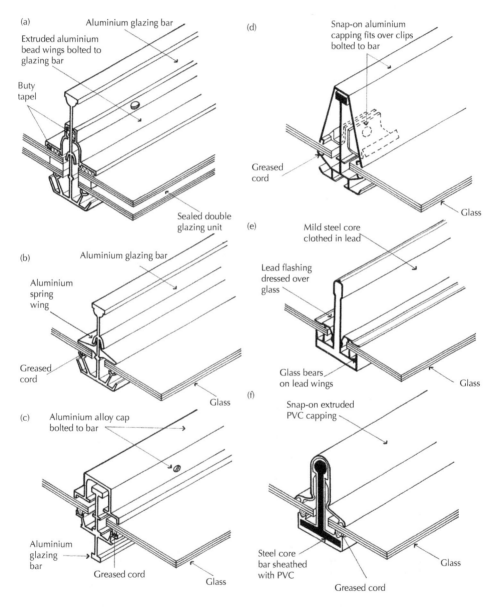

Figure 4.77   (a) Aluminium patent glazing bar with sealed double glazing. (b) Aluminium patent glazing bar for single glazing. (c) Aluminium glazing bar with aluminium cap. (d) Aluminium glazing bar with snap-on capping. (e) Lead-clothed steel core patent glazing. (f) PVC sheathed steel core glazing bar.

Rooflights in flat and low-pitch roofs

A lantern light is constructed with glazed vertical sides and a hipped or gable-ended glazed roof. The vertical sides of the lantern light are used as opening lights for ventilation. Lantern lights were often used to cover considerable areas, the light being framed with substantial timbers of iron or steel, to provide top light to large stairwells and internal rooms. Ventilation from the opening upstand sides is controlled by cord or winding gear from below to suit the requirements of the occupants of the space below. The lantern light requires relatively frequent maintenance if it is to remain sound and watertight, and many have been replaced by domelights. Figure 4.78 is an illustration of an aluminium lantern light

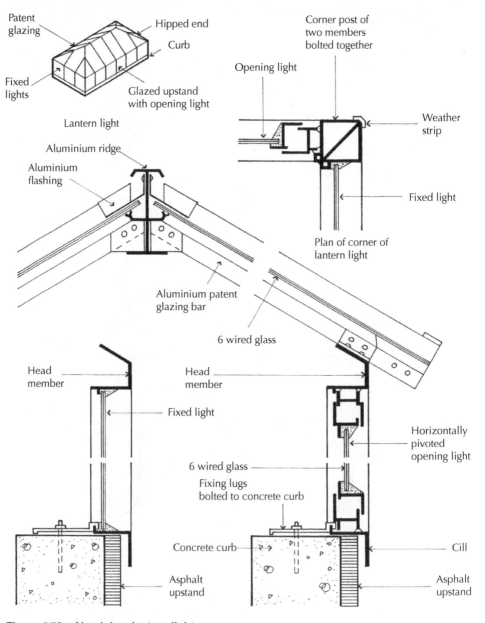

Figure 4.78   Aluminium lantern light.

constructed with standard aluminium window frame and sash sections, aluminium corner posts, aluminium patent glazing to the pitched roof and an aluminium ridge section. The lantern light is bolted to an upstand curb (in common with all rooflights fixed in a flat roof) to resist wind uplift, and the roof covering is dressed to a height of at least 150 mm above the surface of the flat roof.

Decklights are constructed as a hipped or gable-ended glazed roof with no upstand sides; thus, they provide daylight to the space below but no ventilation, as shown in Figures 4.79 and 4.80. This decklight is constructed with lead-sheathed steel glazing bars pitched and fixed to a ridge and bolted to a steel tee fixed to the upstand curb.

A variety of shapes are produced to serve as rooflights for flat roofs, as illustrated in Figure 4.81. The advantage of the square and rectangular shapes over the circular and ovoid ones is that they require straightforward trimming of the roof structure around the openings and upstands. Plastic rooflights are made as either single-skin lights or as sealed double-skin lights, which improves their resistance to the transfer of heat. Plastic rooflights are bolted or screwed to upstand curbs to resist wind uplift, and the roof covering dressed against the upstand as illustrated in Figure 4.82. To provide diffused daylight through concrete roofs, a lens light may be used, comprising square or round glass blocks (lenses) that

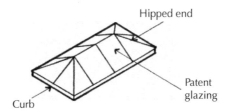

Figure 4.79. Decklight (3D view).

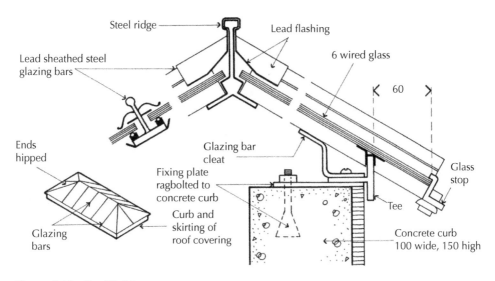

Figure 4.80 Decklight.

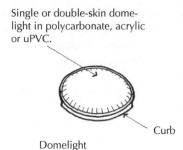

Single or double-skin dome-light in polycarbonate, acrylic or uPVC.

Curb

Domelight

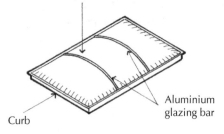

Rectangular base single or double-skin domelight in polycarbonate, acrylic or uPVC.

Aluminium glazing bar

Curb

Rectangular base domelight

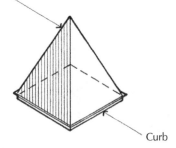

Single or double skin pyramid rooflight in polycarbonate acrylic or uPVC.

Curb

Figure 4.81   Pyramid rooflight.

are cast into reinforced concrete ribs, as illustrated in Figure 4.83. The lens light can be pre-fabricated and bedded in place on site, or it can be cast in situ. Although light transmission is poor, these rooflights are used primarily to provide resistance to fire, to improve security and to reduce sound transmission through the roof.

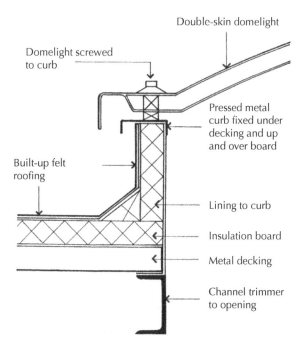

Figure 4.82   Upstand to domelight.

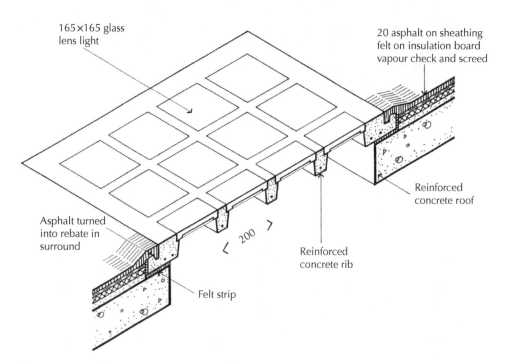

Figure 4.83   Reinforced concrete and glass rooflights.

Lightweight, tensile membrane structures

Film and fabric roof coverings are used in many different ways to create large canopies over open landscaped areas, sport facilities, and buildings. They can also be incorporated in a multi-function fabric providing a watertight, thermally efficient and light-emitting enclosure (Figure 4.84).

The Olympic stadium in Munich is a well-known example of a tensile structure, designed by Frei Otto. Because of the exposed structure, the components can clearly be seen (Photograph 4.13). The stadium was a pioneering design in creativity and scale, designed

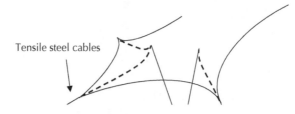

Tensile steel cables

Figure 4.84   Main components of lightweight tensile fabric structure.

Foundation and connection to ground anchor

Frame and network of cables supporting the acrylic tiles
Cables at 750 mm centres

Interior under canopy, clear open spaces and translucent panels

Masts, tension cables and acrylic panels
2.9 × 2.9 m, resting on net at 750 centres

Photograph 4.13   Lightweight tensile structure, Munich Stadium.

and built for the 1972 Olympics. The principles now form the basis of many lightweight tensile membrane structures.

The stadium is a tensile steel structure primarily designed to cover a large area and allow in natural light. The lightweight translucent skin is supported by masts, anchors and cables, to create a precise steel net covered by rigid acrylic panels. Lightweight acrylic panels are often used in construction, sometimes called acrylic glass, glass/plastic laminate or PMMA. PMMA is a transparent shatter-resistant thermoplastic used as a lightweight alternative to glass, often used in profiled cladding and rooflights.

The structure relies on central masts (columns) and ties, with a network of cables that are pinned together by connectors, which distribute the tensile forces to the ground anchors. The cylindrical welded tube masts are up to 80 m in length. The connectors are made of cast steel, which act as central nodes to resolve and distribute the tensile forces. Each cable is connected to the node by an end bracket that is linked by a large pin to the connector. The nodes and cables are literally pinned together and it is the pin that allows for rotation and movement in the structure. Tensile forces are resolved through the network of pinned cables, which are then distributed to the foundations and ground anchors.

The acrylic panels are 4 mm thick and measure $2.9 \times 2.9 \, m^2$. They are bolted to the intersection nodes laid on the cables. Neoprene gaskets are used to join, seal and accommodate 6° of movement. The net uses 750 mm aluminium clamps pressed on to all of the strands at 750 mm centres. Connections use one bolt per joint, providing a node that can freely rotate. The cable is made from 19 heavily galvanised, 2 and 3 mm diameter, steel wires. The main cables are made from five strands formed by 37–109 wires each. The cables are held under high tension to control the level deformation that could take place under snow and wind loads and the ropes are coupled to accommodate higher loadings. A combination of tension foundations is used to anchor the main cables including:

- ❏ inclined slot foundations, which act like tent pegs.
- ❏ gravity foundations, using self-weight plus the weight of the soil surcharge to anchor the foundations.
- ❏ earth anchor foundations for the masts, which allowed dynamic movement when erecting the mast and accommodates the compressive forces on the mast.

Fabric structures can be constructed as a single skin or as an inflated structure. Photograph 4.14 shows a concrete and steel structure covered by inflated transparent laminate, ETFE, film. This material can be used in single-skin lightweight fabric construction or used to trap air creating an inflated structure. The transparent laminate does not degrade under ultraviolet light and has an expected service life of 50 years.

4.4 Diaphragm, fin wall and tilt-up construction

The majority of tall, single-storey buildings that enclose large open areas, such as sports halls, warehouses, supermarkets and factories, with walls more than 5 m high are constructed with a frame of lattice steel or a portal frame covered with lightweight steel cladding and infill brick walls at a lower level. An alternative approach is to use diaphragm walls and fin walls constructed of brickwork or blockwork. Brickwork is preferred to block-work because the smaller unit of the brick facilitates bonding and avoids cutting of blocks. Some of the advantages of diaphragm and fin wall construction include durability, security, thermal insulation, sound insulation and resistance to fire. Visual appearance of the wall can be enhanced with the use of special bricks and creative design of fin walls.

A fabric tented structure

Fabric covering

Allianz Arena covered by ETFE inflated structure

Concrete and steel structure covered by translucent ethylene tetrafluoroethylene

Photograph 4.14   Fabric and foil-covered structures.

Brick diaphragm walls

A diaphragm wall is built with two leaves of brickwork bonded to brick cross ribs (diaphragms) inside a wide cavity between the leaves, thus forming a series of stiff box or I-sections structurally, as illustrated in Figure 4.85. The compressive strength of the bricks and mortar is considerable in relation to the comparatively small dead load of the wall, roof and imposed loads. Stability is provided by the width of the cavity and the spacing of the cross ribs, together with the roof, which is tied to the top of the wall to act as a horizontal plate to resist lateral forces.

Construction

The width of the cavity and the spacing of the cross ribs is determined by the size of the box section required for stability and the need for economy in the use of materials by using whole bricks. Cross ribs are usually placed four or five whole brick lengths (with mortar joints) apart and the cavity one-and-a-half or two-and-a-half whole bricks (with mortar joints) apart, so that the cross ribs can be bonded in alternate courses to the outer and inner leaves, as illustrated in Figure 4.86. Loads on the foundations are relatively slight, thus a simple strip foundation can be used in good ground conditions.

The roof is tied to the top of the diaphragm wall to act as a prop in resisting the overturning action of lateral wind pressure, by transferring the horizontal forces on the long walls to

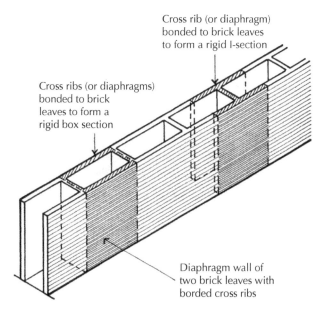

Figure 4.85   Brick diaphragm wall.

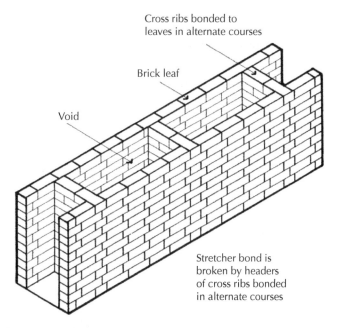

Figure 4.86   Bonding of diaphragm wall.

the end walls of the building that act as shear walls. The roof structure is tied to a reinforced concrete capping beam by bolts, as illustrated in Figure 4.87. Care is required at this junction to ensure that thermal bridging does not occur across the capping beam. Roof beams are braced by horizontal lattice steel wind girders, which are connected to roof beams, as illustrated in Figure 4.88.

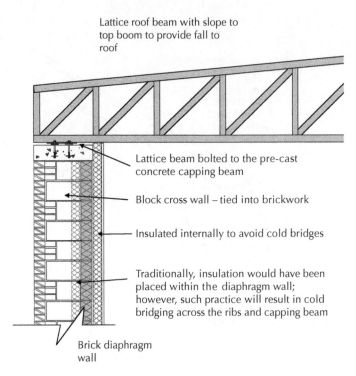

Lattice roof beam with slope to top boom to provide fall to roof

Lattice beam bolted to the pre-cast concrete capping beam

Block cross wall – tied into brickwork

Insulated internally to avoid cold bridges

Traditionally, insulation would have been placed within the diaphragm wall; however, such practice will result in cold bridging across the ribs and capping beam

Brick diaphragm wall

Figure 4.87   Connection of roof beams to diaphragm wall.

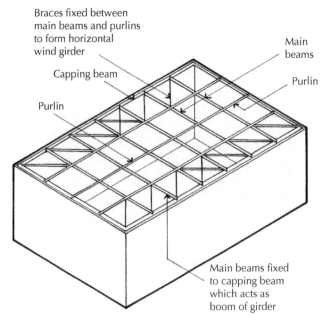

Braces fixed between main beams and purlins to form horizontal wind girder

Capping beam

Purlin

Main beams

Purlin

Main beams fixed to capping beam which acts as boom of girder

Figure 4.88   Wind girder to beam roof.

Door and window openings should be designed to fit between the cross ribs so that the ribs can form the jambs of the opening. Large door and window openings will cause large local loadings; thus, double ribs (or thicker ribs) are built to take the additional load, as illustrated in Figure 4.89. Vertical movement joints are formed by the construction of double ribs at the necessary centres to accommodate thermal movement (Figure 4.90).

Diaphragm walls built in positions of severe exposure will resist moisture penetration, although the cavity should be ventilated to assist with the drying out of the brickwork. Given the problem of thermal bridging inherent in the brick diaphragms, the most convenient method of insulation is to fix insulation to the inside face of the wall. A long, high

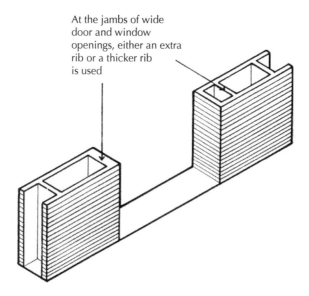

At the jambs of wide door and window openings, either an extra rib or a thicker rib is used

Figure 4.89   Openings in diaphragm wall.

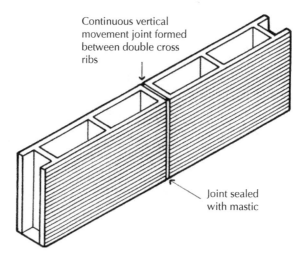

Continuous vertical movement joint formed between double cross ribs

Joint sealed with mastic

Figure 4.90   Movement joint in diaphragm wall.

diaphragm wall with flat panels of brickwork may have a rather uninspiring appearance. Variations in the depth of the cavity wall, the use of projecting brick fins and polychromatic brickwork may go some way to alleviate the monotony, although there will be cost implications.

Brick fin walls

A fin wall is built as a cavity wall buttressed with piers (fins), which are bonded to the external leaf of the cavity wall to buttress and hence stiffen the wall against overturning. A fin wall acts structurally as a series of T-sections, as illustrated in Figure 4.91. The compressive strength of the bricks and mortar is considerable in relation to the comparatively small dead load of the wall, roof and imposed loads. Stability against lateral forces from wind pressure is provided by the T-sections of the fins and the prop effect of the roof, which is usually tied to the top of the wall to act as a horizontal plate to transfer forces to the end walls. The minimum dimensions and spacing of the fins are determined by the cross-sectional area of the T-section of the wall required to resist the tensile stress from lateral pressure and by considerations for the appearance of the building. Spacing and dimensions of the fins can be varied to suit a chosen external appearance. Some typical profiles for brick fins are illustrated in Figure 4.92, with brick specials use for maximum effect.

Construction

The wall is constructed as a cavity wall, with inner and outer leaves of brick tied with wall ties and thermal insulation positioned within the cavity. The fins are bonded to the outer leaf in alternate courses. Thickness of the fin will typically be one brick thick with a projection of four or more brick lengths, with the size of the fin varying to suit structural and aesthetic requirements. The fins should be spaced to suit whole brick sizes, thus minimising

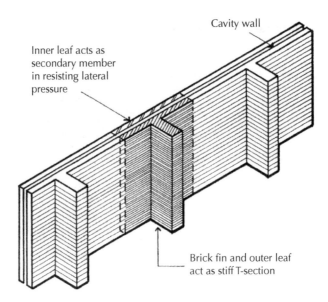

Inner leaf acts as secondary member in resisting lateral pressure

Cavity wall

Brick fin and outer leaf act as stiff T-section

Figure 4.91   Fin wall.

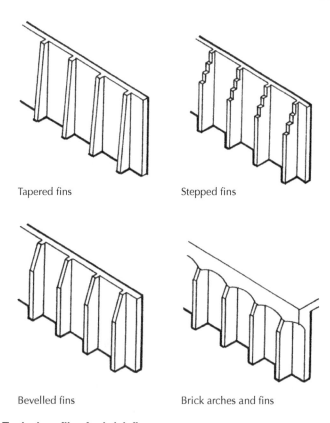

Tapered fins   Stepped fins

Bevelled fins   Brick arches and fins

Figure 4.92 Typical profiles for brick fins.

the cutting of bricks, and at regular centres necessary for stability and for appearance. The loads on the foundation of a fin wall are relatively slight, and a continuous concrete strip foundation should provide adequate support and stability on good bearing ground. The foundation will extend under the fin, as illustrated in Figure 4.93.

Roof beams are usually positioned to coincide with the centres of the fins and tied to a continuous reinforced capping beam that is cast or bedded on the top of the wall, or to concrete padstones cast or bedded on top of the fins, as illustrated in Figure 4.94. To resist wind uplift on lightweight roofs, the beams are anchored to the brick fins through bolts built into the fins, cast or threaded through the padstones and bolted to the beams. Horizontal bracing to the roof beams is provided by lattice wind girders fixed to the beams to act as a plate in propping the top of the wall.

Door and window openings should be the same width as the distance between the fins for simplicity and economy of construction. To allow sufficient cross section of brickwork at the jambs of wide openings, a thicker fin or a double fin is built, as illustrated in Figure 4.95. Movement joints are usually formed between double brick fins, as illustrated in Figure 4.96. In addition to the usual resistance to weather provided by brickwork, the projecting fins may provide some additional shelter to the wall from driving rain.

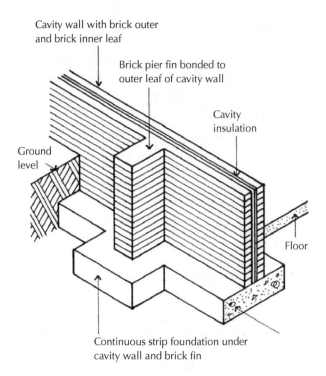

Figure 4.93   Brick pier fin bonded to outer leaf of cavity wall.

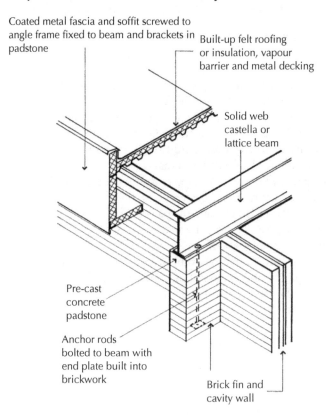

Figure 4.94   Fin wall, beams and roofing.

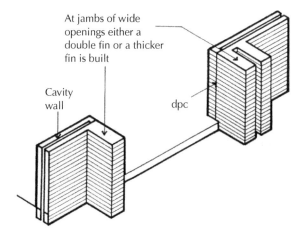

Figure 4.95   Openings in fin walls.

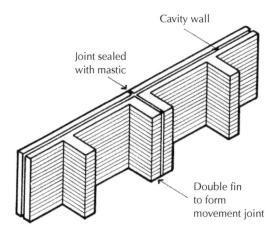

Figure 4.96   Movement joint in fin wall.

Concrete tilt-up construction

Tilt-up construction is a technique of precasting large, slender reinforced concrete wall panels on site (on a temporary casting bed or on the concrete floor slab) which, when cured, are tilted by crane into position. This technique has been used principally for the construction of single-storey commercial and industrial buildings on open sites where there is room for casting and the necessary lifting equipment. Tilt-up construction has been used extensively in the US, where it originated, and many other countries such as Australia and New Zealand, for the speed of casting and speed of erection of the panels. The technique is most economical when there is a high degree of repetitiveness in the structure and the walls are used in a loadbearing capacity.

Typical applications include low-rise warehouses, offices and factories. There are few examples of this type of construction in the UK.

The concrete panels provide good resistance to the penetration of rain and also provide good durability and freedom from maintenance. Panels also provide good fire resistance, resistance to the passage of sound and relatively good security against forced entry. The reinforced concrete panels do not provide adequate thermal insulation for heated buildings. Thermal insulation is usually applied to the internal face of the panels with a moisture vapour check between wall and insulation. Insulation boards are used to provide both insulation and an internal finish to the building, fixed to timber battens, which are shot fired to the panel.

Tilt-up concrete panels vary in size, shape and thickness, but typically will be around 7×5 m, 160 mm thick, and weigh between 20 and 30 tonnes. Panel size is limited by the strength of the reinforced concrete panel necessary to accommodate the stresses induced in the panel as it is tilted from the horizontal to the vertical and also by the lifting capacity of the cranes. Wall panels may vary in design from plain, flat slabs to frames with wide openings for glazing, provided that there is adequate reinforced concrete to carry the anticipated loads. A variety of shapes and features are made possible by repetitive use of the formwork in the casting bed, and a variety of external finishes can be produced, ranging from smooth to textured finishes.

Sequence of assembly

The sequence of operations is shown in Figure 4.97. The site slab of concrete is cast over the completed foundations, drainage and service pipework and accurately levelled to provide a level surface on which the wall panels can be cast. A bond breaker/cure coat is then applied to the concrete slab and the panels cast around reinforcement inside steel (or timber) edge shuttering, which is placed as near as possible to the final position of the wall panel. Lifting lugs and other fittings are usually cast into the upper face of the panels, which will be covered by insulation and internal finishes. Wall panels may be cast individually or as a continuous strip. If the panels are cast as a continuous strip, they are cut to size once the concrete has gone off but during the early stages of the concrete's maturity (one or two days).

Panels may also be cast as a stack, one on top of the other, separated by a bond breaker. Once cured, the hardened panels are then gently lifted or tilted into position and propped or braced ready to receive the roof deck. The panels are tilted up and positioned on the levelled foundations against a rebate in the concrete, or up to timber runners or onto a sheathing angle and then set level on steel levelling shims. A mechanical connection between the foot of the slabs to the foundation and/or floor slab is usually employed. Cast in metal, dowels projecting from the foot of the panels are set into slots or holes in the foundations and grouted into position. Alternatively, a plate welded to studs or bar anchors, cast into the foot of the panel, provides a means of welded connection to rods cast into the site slab, as illustrated in Figure 4.98. The roof deck serves as a diaphragm to give support to the top of the wall panels and to transmit lateral wind forces back to the foundation. Lattice beam roof decks are welded to seat angles, welded to a plate and cast in studs, as shown in Figure 4.98. A continuous chord angle is welded to the top of the lattice beams and to bolts cast or fixed in the panel. The chord angle serves as a transverse tie across the panels and is secured to them with bolts set into slots in the angle to allow for shrinkage movements of the panels.

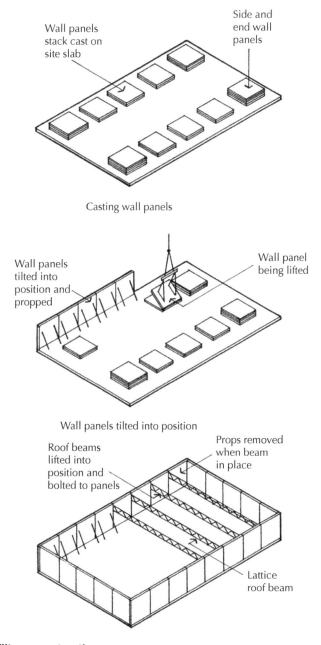

Wall panels
stack cast on
site slab

Side and
end wall
panels

Casting wall panels

Wall panels
tilted into
position and
propped

Wall panel
being lifted

Wall panels tilted into position

Roof beams
lifted into
position and
bolted to panels

Props removed
when beam
in place

Lattice
roof beam

Figure 4.97   Tilt-up construction.

4.5 Shell structures

A shell structure is a thin, curved membrane or slab, usually of reinforced concrete, that functions both as a structure and covering, the structure deriving its strength and rigidity from the curved shell form (see Photographs 4.15 and 4.16). The term 'shell' is used to

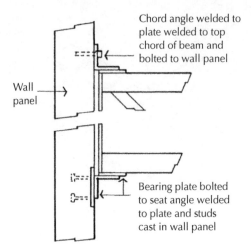

Chord angle welded to
plate welded to top
chord of beam and
bolted to wall panel

Wall
panel

Bearing plate bolted
to seat angle welded
to plate and studs
cast in wall panel

Connection of roof beams to
wall panel

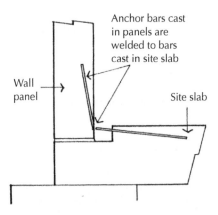

Anchor bars cast
in panels are
welded to bars
cast in site slab

Wall
panel

Site slab

Figure 4.98   Connection of wall panels to site slab.

describe these structures by reference to the considerable strength and rigidity of thin, natural and curved forms such as the shell of an egg. The material most suited to the construction of a shell structure is concrete, which is a highly plastic material when wet and that can take up any shape inside formwork (also known as centring). Small section reinforcing bars can readily be bent to follow the curvature of shells. Wet concrete is spread over the centring and around the reinforcement, and compacted to the required thickness, with the stiffness of the concrete mix and the reinforcement preventing the concrete from running down the slope of the curvature of the shell while the concrete is wet. Once the concrete has hardened, the reinforced concrete membrane or slab acts as a strong, rigid shell, which serves as both structure and covering to the building. The strength and rigidity of curved shell structures make it possible to construct single curved barrel vaults 60 mm thick and double-curved hyperbolic paraboloids 40 mm thick in reinforced concrete for clear spans up to 30 m.

Photograph 4.15   Shell roof construction.

Photograph 4.16   Curved and domed shell roof.

The attraction of shell structures lies in the elegant simplicity of the curved shell form that utilises the natural strength and stiffness of shell forms with great economy in the use of material. The main disadvantages related to their cost and poor thermal insulation properties. A shell structure is more expensive than, for example, a portal-framed structure covering the same floor area, because of the considerable labour required to construct the centring on which the shell is cast. Shell structures cast in concrete are also difficult to insulate economically, because of their geometry and so are mainly suited to unheated spaces.

Shell structures tend to be described as single or double-curvature shells. Single curvature shell structures are curved on one linear axis and form part of a cylinder in the form of a barrel vault or conoid shell; double curvature shells are either part of a sphere as a dome or a hyperboloid of revolution (see Figure 4.99). The terms are used to differentiate the comparative rigidity of the two forms and the complexity of the formwork (centring) necessary to construct the shell form. Double curvature of a shell adds considerably to its stiffness, resistance to deformation under load and reduction in the need for restraint against deformation.

Centring (or formwork) is the term used to describe the necessary temporary support on which a curved reinforced concrete shell structure is cast. The centring for a single

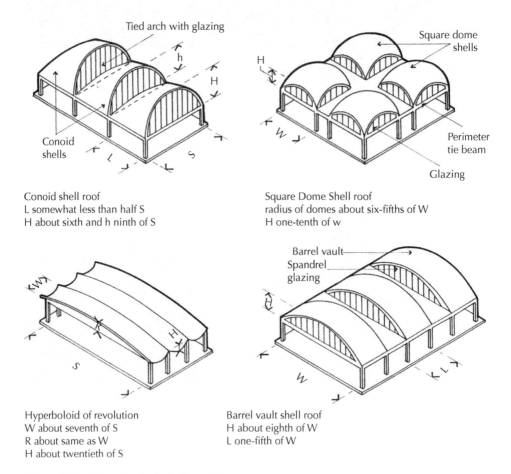

Conoid shell roof
L somewhat less than half S
H about sixth and h ninth of S

Square Dome Shell roof
radius of domes about six-fifths of W
H one-tenth of w

Hyperboloid of revolution
W about seventh of S
R about same as W
H about twentieth of S

Barrel vault shell roof
H about eighth of W
L one-fifth of W

Figure 4.99 Some typical shell roof forms.

curvature barrel vault is less complex than that for a dome, which is curved from a centre point. Advances in computer software have made the design of shell structures and the setting out of formwork much easier; however, there is still a considerable demand on labour to make and erect the centring, and the more complex the shape, the greater the amount of cutting and potential waste of material. The simplest, and hence most economic, of all shell structures is the barrel vault, constructed in concrete or timber.

Reinforced concrete barrel vaults

Reinforced concrete barrel vaults consist of a thin membrane of reinforced concrete positively curved in one direction so that the vault acts as both structure and roof surface. The most common form of barrel vault is the long-span vault, illustrated in Figure 4.100, where the strength and stiffness of the shell lie at right angles to the curvature. Typical spans range from 12 to 30 m, with the width being about half the span and the rise about one-fifth of the width. To cover large areas, multi-span, multi-bay barrel vault roofs can be used (see Figure 4.101). The concrete shell may be from 57 to 75 mm thick for spans of 12 and 30 m, respectively. The thickness of the concrete provides sufficient cover of concrete to protect the reinforcement against damage by fire and corrosion.

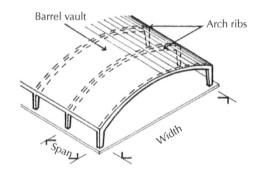

Short-span barrel vault

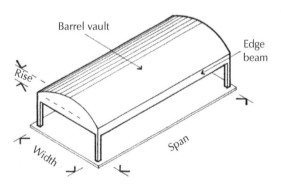

Long-span barrel vault

Figure 4.100   Reinforced concrete barrel vaults.

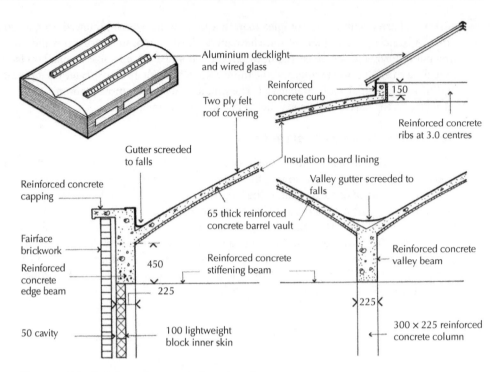

Figure 4.101   Reinforced concrete barrel vault.

Stiffening beams and arches

Under local loads, the thin shell of the barrel vault will tend to distort and lose shape and, if this distortion were of sufficient magnitude, the resultant increase in local stress would cause the shell to progressively collapse. To strengthen the shell against this possibility, stiffening beams or arches are cast integrally with the shell. Figure 4.102 illustrates the four types of stiffening members generally used, with common practice being to provide a stiffening member between the columns supporting the shell. The downstand reinforced concrete beam, which is usually 150 or 225 mm thick, is the most efficient of the four because of its depth. To avoid the interruption of the line of the soffit of the vaults caused by a downstand beam, an upstand beam is sometimes used. The disadvantage of an upstand beam is that it breaks up the line of the roof and also needs protection against the weather. Arch ribs are sometimes used because they follow the curve of the shell and therefore do not interrupt the line of the vault; however, these are less efficient structurally because they have less depth than beams.

Edge and valley beams

Reinforced concrete edge beams are cast between columns as an integral part of the shell, to resist the tendency of the thin shell to spread and its curvature to flatten out due to self-weight and imposed loads. The edge beams may be cast as dropped beams, upstand beams, or partly upstand or partly dropped beams, as illustrated in Figure 4.103. Between multi-bay vaults, the loads on the vaults are largely transmitted to adjacent shells and then to the edge beams, thus allowing the use of comparatively slender featheredge beams.

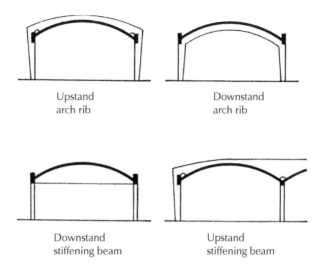

Upstand
arch rib

Downstand
arch rib

Downstand
stiffening beam

Upstand
stiffening beam

Figure 4.102   Stiffening beams and arches for reinforced concrete barrel vaults.

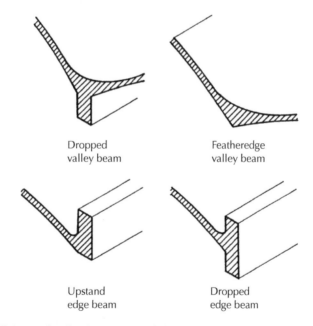

Dropped
valley beam

Featheredge
valley beam

Upstand
edge beam

Dropped
edge beam

Figure 4.103   Edge and valley beams for reinforced concrete barrel vaults.

Rooflights

Natural light through the shell structure can be provided by decklights formed in the crown of the vault, as illustrated in Figure 4.101, or by domelights. Rooflights are fixed to an upstand curb cast integrally with the shell, as illustrated in Figure 4.101. Care is required to avoid overheating and glare. One way of providing natural light and avoiding glare and overheating is to use a system of north-light barrel vaults, as illustrated in Figures 4.104 and 4.105. The roof consists of a thin reinforced concrete shell on the south-facing side of

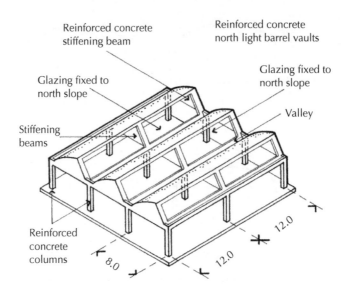

Reinforced concrete stiffening beam

Reinforced concrete north light barrel vaults

Glazing fixed to north slope

Glazing fixed to north slope

Valley

Stiffening beams

Reinforced concrete columns

8.0   12.0   12.0   12.0

Figure 4.104   Three-bay reinforced concrete north light barrel vault.

the roof, with a reinforced concrete-framed north-facing slope, and pitched at between 60° and 80°. This construction is less efficient structurally than a barrel vault because the rigidity of the shell is interrupted by the north lights.

Thermal insulation
The thin concrete shell offers poor resistance to the transfer of heat, and some form of insulating soffit lining is necessary to meet the requirements of the building regulations. This is difficult to achieve without causing thermal bridges and also avoiding interstitial condensation between the insulation and the concrete structure, which adds considerably to the cost of the shell, and combined this makes concrete shells largely unsuitable for buildings, which are to be heated.

Expansion joints
To limit expansion and contraction caused by changes in temperature, continuous expansion joints are formed at intervals of approximately 30 m along the span and across the width of multi-bay, multi-span barrel roofs. The expansion joints are formed by erecting separate shell structures, each with its own supports and with a flexible joint material between neighbouring elements (see Figure 4.106). Vertical expansion joints are made so as to form a continuous joint to the ground with double columns on either side of the joint. Longitudinal expansion joints are formed in a valley with upstands weathered with non-ferrous cappings over the joint.

Roof covering
A variety of materials may be used to cover concrete shells, the choice depending on the use of the building and to a certain extent the position of the thermal insulation. Lightweight

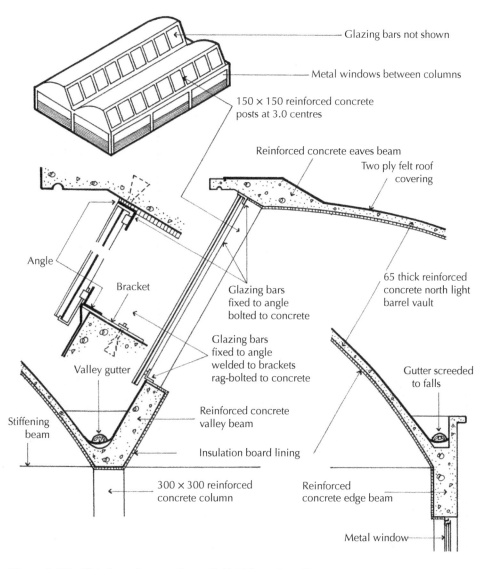

Glazing bars not shown

Metal windows between columns

150 × 150 reinforced concrete posts at 3.0 centres

Reinforced concrete eaves beam

Two ply felt roof covering

Angle

Bracket

Glazing bars fixed to angle bolted to concrete

65 thick reinforced concrete north light barrel vault

Glazing bars fixed to angle welded to brackets rag-bolted to concrete

Gutter screeded to falls

Valley gutter

Stiffening beam

Reinforced concrete valley beam

Insulation board lining

300 × 300 reinforced concrete column

Reinforced concrete edge beam

Metal window

Figure 4.105   Reinforced concrete north light barrel vault.

materials such as thin non-ferrous sheet metal, bitumen felt and plastic membranes may be used.

Walls

The walls of shell structures between the columns are non-loadbearing, their purpose being to provide shelter, security and privacy, as well as thermal and sound insulation. Thus, a variety of partition wall constructions may be used, from brick and blockwork to timber and steel studwork with facing panels.

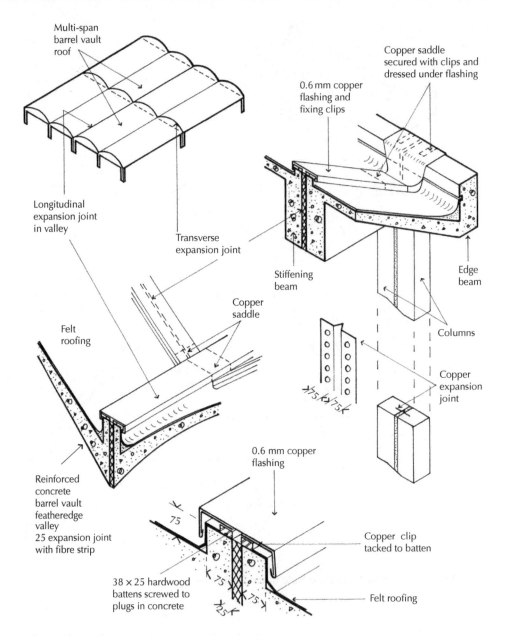

Figure 4.106   Expansion joints and flashings in reinforced concrete barrel vaults. Expansion joints at intervals of not more than 30 m.

Conoid and hyperboloid shell roofs

Reinforced concrete conoid shell

In the reinforced concrete conoid shell form, the curvature and rise of the shell increases from a shallow curve to a steeply curved end in which the north light glazing is fixed, as illustrated in Figure 4.99. The glazed end of each shell consists of a reinforced concrete or steel lattice, which serves as a stiffening beam to resist deformation of the shell. Edge beams resist spreading of the shell as previously described.

Hyperbolic paraboloid shells

The hyperbolic paraboloid shells provide dramatic shapes and structural possibilities of doubly curved shells (see Photograph 4.17). The name hyperbolic paraboloid comes from the geometry of the shape: the horizontal sections through the surface are hyperbolas and the vertical sections parabolas, as illustrated in Figures 4.107 and 4.108. The structural

Photograph 4.17   Hyperbolic paraboloid shell roof.

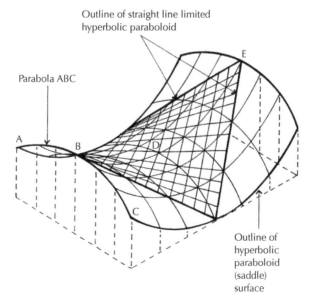

Figure 4.107   Hyperbolic paraboloid (saddle) surface.

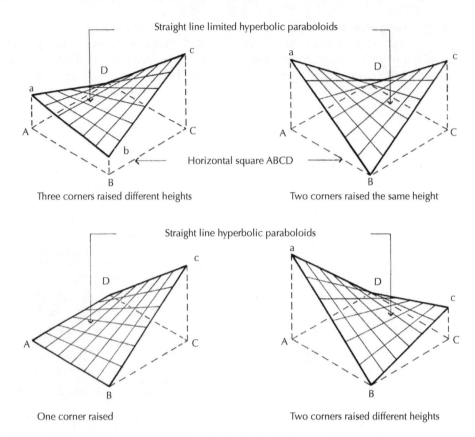

Figure 4.108 illustrations:

Straight line limited hyperbolic paraboloids

Horizontal square ABCD

Three corners raised different heights

Two corners raised the same height

Straight line hyperbolic paraboloids

One corner raised

Two corners raised different heights

Figure 4.108 Setting out straight line limited hyperbolic paraboloid surfaces on a square base.

significance of this shape is that at every point on the surface, straight lines, which lie in the surface, intersect so that in effect, the surface is made up of a network of intersecting straight lines. Thus, the centring (formwork) can consist of thin straight sections of timber, which are simple to fix and support.

Reinforced concrete hyperbolic paraboloid shell

Figure 4.109 illustrates an umbrella roof formed from four hyperbolic paraboloid surfaces supported on one column. The small section reinforcing mesh in the surface of the shell resists tensile and compressive stress, and the heavier reinforcement around the edges and between the four hyperbolic paraboloid surfaces resists shear forces developed by the tensile and compressive stress in the shell. A series of these roofs can be combined, with glazing between them, to provide shelter to the area below.

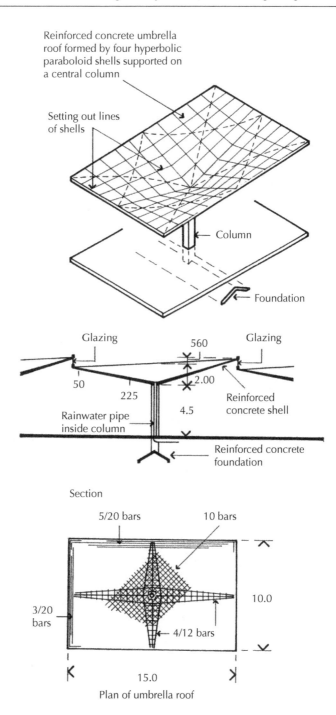

Reinforced concrete umbrella roof formed by four hyperbolic paraboloid shells supported on a central column

Setting out lines of shells

Column

Foundation

Glazing

560

Glazing

50

2.00

225

4.5

Reinforced concrete shell

Rainwater pipe inside column

Reinforced concrete foundation

Section

5/20 bars

10 bars

3/20 bars

10.0

4/12 bars

15.0

Plan of umbrella roof

Figure 4.109   Reinforced concrete hyperbolic paraboloid.

Timber shell structures

Timber barrel vaults

Single- and multi-bay barrel vaults can be constructed from small-section timber with spans and widths similar to reinforced concrete barrel vaults (Figure 4.110). The vault is formed from layers of boards glued and mechanically fixed together and stiffened with ribs at close centres. The timber ribs serve both to stiffen the shell and to maintain the boards' curvature over the vault. Glue-laminated edge and valley beams are formed to resist spreading of the vault. Timber barrel vaults have some advantage over concrete, in that the material performs better in terms of providing some thermal insulation. Indeed, it is easier to include thermal insulation within the construction while maintaining the visual integrity of the shell.

Timber hyperbolic paraboloid shell

Timber can also be used to form hyperbolic paraboloid shell structures (Figure 4.111). Laminated boards and edge beams are used. Low points of the shell are usually anchored to concrete abutments/ground beams to prevent the shell from spreading under load.

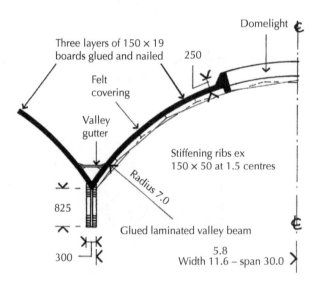

Figure 4.110   Timber barrel vault.

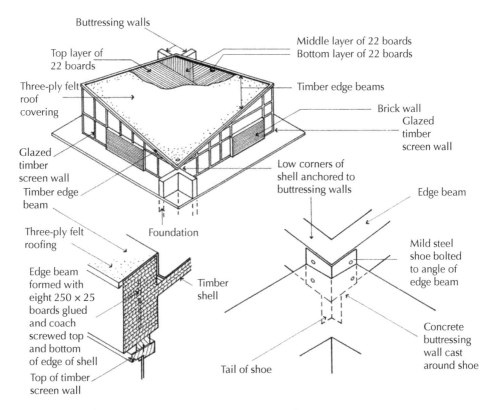

Figure 4.111   Hyperbolic paraboloid timber shell roof.

Further reading

This chapter covers quite a range of materials and structural systems. For roofing and cladding see, for example, the National Federation of Roofing Contractors: www.nfrc.co.uk.

Reflective exercises

You are designing an Olympic size swimming pool on the edge of a large town (choose one near you). The pool and surrounding area need to be covered with a clear-span single-storey structure. Your client is unsure what materials to use.

❑ How would you advise your client? What would be the most appropriate structural system and why?
❑ You client would also like to use local materials and labour. How does this influence your reasoning?

Structural Timber Frames: Chapter 5
AT A GLANCE

What? Structural timber frames are assembled from timber components, engineered timber products or a combination of both. Hardwood, such as oak, is used where the frame will form a visual feature. Softwood is used where the frame is to be covered, and this is common in timber-framed housing. Softwood is also used to produce engineered timber components, such as I-beams and cross-laminated timber (CLT) panels that help to provide additional strength and rigidity to structural timber frames. Timber is a renewable resource, is carbon neutral, and there is little to no waste in the manufacturing process. Subject to appropriate detailing and fixing, the timber components may be reclaimed and reused once the building has reached the end of its service life. Timber has inherent fire resistance and has an attractive appearance. The structural timber frame must be kept dry and ventilated to avoid the risk of decay.

Why? Framed buildings offer a dry and rapid form of construction. The framed structure allows freedom for positioning of internal partition walls and reasonable clear spans. Structural timber frames are a popular choice for relatively low-rise buildings of two to four storeys in height, although it is possible to build higher. Framed construction allows considerable design freedom for the building envelope. External walls and glazing may be positioned between the structural timber frame to express the frame, and this is common in oak-framed construction. Alternatively, the external envelope may be positioned to completely cover the frame, which is common in timber framed construction.

When? The structural timber frame will be erected after the foundations are complete. The rapid construction sequence provides a structural framework onto which the roof may be positioned and the external envelope fixed. This quickly provides a weatherproof envelope within which internal works can proceed unhindered by wind and rain.

How? The components of a timber frame will be produced under factory conditions to a specific design. Components are available off the shelf or they can be manufactured to suit bespoke design requirements. Components may be delivered to site as individual units, for example for self-build housing projects, or more commonly, they are pre-assembled into large wall and floor elements or modular units. The size of the preassembled elements is limited by transportation restrictions and physical access to the construction site. The structural timber frame is erected onto pre-prepared foundations using cranes and site personnel to help position the timber members. Members are fixed in position by using timber screws and or bolting through pre-drilled holes. The ease of fixing helps to aid the speed of erection. Similarly, the ease of fixing helps to facilitate the ease of alterations, replacement and recovery of components at a future date.

5 Structural Timber Frames

Structural timber frames are a popular choice for relatively low-rise building of one to two storeys and increasingly popular for mid-rise buildings (three to five storeys). More recently, structural timber frames have been used for buildings of up to 9 and 10 storeys high. The advantages of the structural timber frame are the speed of erection of the pre-cut timber members and ease of fixing. Timber has inherent fire resistance and it is a sustainable material (carbon neutral). Structural timber frames may comprise seasoned timber, such as oak columns and beams, engineered timber components or a mix of timber and engineered timber.

5.1 Functional requirements

The functional requirements of a structural frame are:

- ❏ Strength and stability.
- ❏ Durability and freedom from maintenance.
- ❏ Fire safety.

Strength and stability

The strength of timber varies with species and is generally greater with dense hardwoods than less dense softwoods. Strength is also affected by defects in timber such as knots, shakes, wane and slope of the grain of the wood. The strength and stability of a structural timber frame depends on the combination of elements, such as columns and beams and the rigidity of the connections. Simple calculations will help to establish the size of columns and beams. Connections tend to follow a tradition of using a combination of nails, screws and bolts. Factory-produced assemblies will be joined by mechanical connections.

Stress grading of timber

There are two methods of stress grading: visual grading and machine grading:

(1) *Visual grading*. Trained graders or computer software determine the grade of a timber by visual examination in the timber mill. This is to assess the effect on strength of observed defects such as knots, shakes, wane and slope of grain. There are two visual

Barry's Advanced Construction of Buildings, Fifth Edition. Stephen Emmitt.
© 2023 John Wiley & Sons Ltd. Published 2023 by John Wiley & Sons Ltd.

grades, general structural (GS) and special structural (SS). In SS, the allowable stress is higher than that in GS timber.

(2) *Machine grading.* Timbers are subjected to a test for stiffness by measuring deflection under load in a machine that applies a specified load across overlapping metre lengths to determine the stress grade. This mechanical test, premised on the fact that strength is proportional to stiffness, is a more certain assessment of the true strength of a timber than a visual test. The machine grades, which are comparable to the visual grades, are machine general structural (MGS) and machine special structural (MSS). There are, in addition, two further machine grades: M50 and M75. The stress of M50 lies between MGS and MSS, and M75 is the highest stress grade in the series.

Stress-graded timbers are marked GS and SS at least once within the length of each piece for visually graded timber, together with a mark to indicate the grader or company. Machine-graded timber is likewise marked MGS, M50, MSS and M75 together with the BS Kitemark and the number of the British Standard, 4978. Approved Document A provides practical guidance to meet the requirements of the Building Regulations for small buildings, and includes tables of the sizes of timber required for floors and roofs, related to load and span.

Durability and freedom from maintenance

The durability of timber is determined by careful specification and selection of seasoned timber together with careful detailing to avoid the possibility of rot. As a general rule, hardwood timbers are more durable than softwood timbers. Softwood will usually require some form of surface protection to increase its durability, such as paint, stain or impregnation with a preservative. Some hardwoods, such as oak, can be left to weather naturally without surface protection. Careful detailing can also help to avoid the possibility of wet or dry rot, as discussed later.

Fire safety

There is a misconception that timber-framed buildings are not safe in a fire; this is incorrect. Structural fire safety is provided by passive protection, relating to the sizing, spacing and protection of timber. The reaction of timber in a fire has been extensively tested over the years; thus, the behaviour is well known and buildings can be designed accordingly. Timber will char when exposed to flame, forming a protective layer and helping to preserve structural integrity. Large section timber beams and columns may be left without additional protection; however, the smaller section timber members of low-rise timber framed houses will need to be protected. As a general rule, 30 minutes of fire protection can be achieved from components made of solid timber and glulam, by protecting it with a 12.5 mm (or 15 mm) thick layer of gypsum plasterboard. Additional fire resistance, for example, 60 minutes, can be achieved by using 25 mm thick gypsum plasterboard. For exposed timber structures, the fire resistance will depend on the charring characteristics of the timber components and the protection of fixings and joints. Exposed timber frames can be designed to incorporate sacrificial timber so that the timber exposed to flame will char and protect the inner, loadbearing material from damage. The biggest threat to timber frame construction occurs during construction and refurbishment work, when unprotected elements are more susceptible to fire.

5.2 Timber

The word 'timber' describes wood that has been cut for use in building. Timber has many advantages as a building material. Timber is a comparatively lightweight material that is easy to cut, shape and join using hand- or power-operated tools. As a structural material, it has good weight-to-cost, weight-to-strength and weight-to-modulus of elasticity ratios and coefficients of thermal expansion, density and thermal resistance. With sensible selection, fabrication and fixing, and adequate weather protection, timber is a reasonably durable material in relation to the life of most buildings. When sourced from sustainable forests, timber has excellent environmental credentials and is considered carbon neutral during its lifetime. Softwood and hardwood are the main terms used to classify timber, with engineered timber covering timber products that have been modified to improve strength and or durability.

It is rare to use timber in its natural state; however, it is possible to do so. The term 'roundwood construction' refers to buildings that employ timber logs. The seasoned timber is used without any form of processing, other than cutting the tree trunks to size (and removing the bark) to create logs of the required length. Building with logs will create a unique low-impact building when combined with natural walling materials such as unfired bricks and hemp lime. This form of construction tends to be used infrequently, as the timber logs vary in shape and it is not easy to calculate the strength of the structure other than in general terms. However, this is a sustainable and relatively thermally efficient way of building single storey domestic scale properties; ideally suited to naturalistic settings.

Properties of timber

Up to two-thirds of the weight of growing wood is due to water in the cells of the wood. When the tree is felled and the wood is cut into timber, this water begins to evaporate, and the wood gradually shrinks as water leaves the cell walls. As the shrinkage in timber is not uniform, the timber may lose shape and it is said to warp. It is essential that before timber is used in buildings, either it should be stacked for a sufficient time in the open air to allow most of the water in it to dry out or it should be artificially dried in a kiln. If unseasoned or wet timber is used in building, it will dry out and shrink, causing twisting of doors and windows and cracking at joints with plastered walls. This means that only seasoned timber should be used and that the timber should be protected from moisture on the building site, when stored and as work proceeds. The process of allowing, or causing, newly cut wood to dry out is called seasoning, and timber that is ready for use in building is said to have been properly seasoned.

Natural dry seasoning

When logs have been cut into timbers they are stacked either in the open or in an open-sided shed. Timbers are stacked with battens between them to allow air to circulate around them and are left stacked until most of the moisture in the wood has evaporated. Softwoods are stacked for a year or two before they are sufficiently dried or seasoned. Hardwoods need to be stacked for much longer, for up to 10 years before they are seasoned. The lowest moisture content of timber that can be achieved by this method of seasoning is about 18%.

Artificial or kiln seasoning

Because of the great length of time required for natural dry seasoning and because sufficiently low moisture contents of wood cannot be achieved, artificial seasoning is used. After the wood has been converted to timber it is stacked in a kiln with battens between the timber. Air is blown through the kiln, the temperature and humidity of the air being regulated to affect seasoning more rapidly than with natural seasoning, but not so rapidly as to cause damage to the timber. If the timber is seasoned too quickly, it shrinks and is liable to crack and lose shape badly. To avoid this, it is common to allow timber to season naturally for a time and then to complete the process artificially.

Moisture content of timber

It is necessary to specify the moisture content of timber. Moisture content is stated as a percentage of the dry weight of the timber. The dry weight of any piece of timber is its weight after it has been so dried that further drying causes it to lose no more weight. This dry weight is reasonably constant for a given cubic measure of each type of wood and is used as the constant against which the moisture content can be assessed. Table 5.1 sets out moisture contents for timber. The moisture content of timber should be such that the timber will not appreciably gain or lose moisture in the position in which it is fixed.

Conversion of wood into timber

The method of converting wood to timber affects the timber in two ways: (i) by the change of shape of the timber during seasoning, and (ii) in the texture and differences in colour on the surface of the wood. Because the springwood is less dense than the summerwood, the shrinkage caused when the wood is seasoned (dried) occurs mainly along the line of the annual rings. Circumferential shrinkage is greater than the radial shrinkage. Because of this, the shrinkage of one piece of timber cut from a log may be quite different from that cut from another part of the log. This can be illustrated by showing what happens to the planks of a log converted by the 'through and through' cut method shown in Figure 5.1. When the planks have been thoroughly seasoned, their deformation due to shrinkage can be compared by putting them together in the order in which they were cut from the log, as in Figure 5.1.

If the face of a timber is cut on a radius of the circle of the log, the cells of the medullary rays may be exposed where the cut is made. With many woods, this produces very pleasing texture and colour on the surface of the wood and it is said that the 'figure' of the wood has been exposed (Figure 5.1). Radial cutting of boards as shown is very expensive and is

Table 5.1   Moisture content of timber

	1	2
Position of timber in building	%	%
External uses fully exposed	20 or more	–
Covered and generally unheated	18	24
Covered and generally heated	15	20
Internal and continuously heated building	12	20

Source: Column 1: Average moisture content likely to be attained under service conditions.
Column 2: Moisture content which should not be exceeded in individual pieces at time of
erection. Data from BS 5268: Part 2:1996 (issue 2, May 1997).

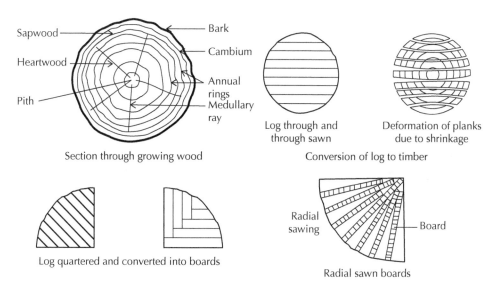

Section through growing wood

Log through and through sawn

Conversion of log to timber

Deformation of planks due to shrinkage

Log quartered and converted into boards

Radial sawn boards

Figure 5.1   Conversion of wood into timber.

employed only for high-class cabinet-making and panelling timbers where the wood will be used decoratively.

Surface finishes for timber

Timber can be left unfinished or a protective decorative finish can be applied to the surface. Unfinished timber is sometimes used on buildings for aesthetic effect and to reduce the environmental impact and maintenance costs of applying finishes to the timber at regular intervals. Care is required in the selection and detailing of suitable timbers for a given location and degree of exposure. There are three types of applied surface finish for wood: paint, varnish and stains (see also Chapter 10). Paint and varnish are protective and decorative finishes that afford some protection against water and provide a decorative finish that can easily be cleaned. Paints are opaque and hide the surfaces of the wood, whereas varnishes are sufficiently transparent for the texture and grain of the wood to show. There is a wide range of stains available, from those that leave a definite film on the surface to those that penetrate the surface. Stains range from gloss through semi-gloss to matt finish. The purpose of this finish is to give a selected uniform colour to wood without masking the grain and texture of the wood. Most stains contain a preservative to inhibit fungal surface growth. These stains are most effective on rough-sawn timbers.

Decay in timber

Decay in timber is caused by fungal decay and insect infestation, both of which can be prevented through careful detailing and specification as well as through careful work both off and on the construction site. Insect attack will depend upon location, moisture content and temperature. Any one of a number of wood-destroying fungi may attack timber that is persistently wet and has a moisture content of over 20%. Fungal decay is classified as being 'dry' or 'wet' rot.

Dry rot and its prevention

Dry rot is the most serious form of fungal decay and is caused by *Serpula lacrymans*, which can spread and cause extensive destruction of timber. Airborne spores of this fungus settle on timber, and if its moisture content is greater than 20%, they germinate. The spore forms long thread-like cells that pierce the wood cells and use the wood as a food. The thread-like cells multiply, spreading out long white thread-like arms called mycelium, which feed on other wood cells. This fungus can spread many metres from the point where the spore first began to thrive, and is capable of forming thick greyish strands that can find their way through lime mortar and softer bricks. Timber that is affected by this fungus turns dark brown and shrinks and dries into a cracked powdery dry mass, hence the name dry rot. It is generally accepted that there is little risk of fungal decay in softwood if the timber is maintained at a moisture content of 20% or less.

Dry rot can be prevented through careful detailing to allow adequate ventilation and regular maintenance to prevent the moisture content of the timber from rising due to, for example, leaks in the building fabric or service pipes. Unseasoned timber must not be used and effort must be made during the construction work to keep seasoned timber dry. Buildings should be detailed to include horizontal and vertical damp-proof courses (dpcs) and all underfloor areas to suspended timber floors and roof spaces must be adequately ventilated. Manufacturers and suppliers of timber-framed buildings will detail their bespoke systems to minimise the risk of dry and wet rot occurring.

The cause of the persistent dampness that has raised the moisture content of timber above 20%, such as leaking gutters or water pipes, must be corrected. This may involve providing a new dpc, preventing moisture penetration through walls, repairing leaking roofs and plumbing, eliminating condensation, clearing the bridging of dpcs and improving ventilation, including the provision of additional airbricks, and so on. Timber, plaster, dust and debris that have been affected by the fungus or are in close proximity to it should be taken out of the building and be burnt immediately. The purpose of burning the affected timber is to ensure that none of it is used in the repairs and to kill any spore that might cause further rot. All walls and surrounding masonry should be cleaned and treated by surface application of masonry biocide; preservative plugs or pastes should be inserted into holes drilled into localised problem areas and/or irrigation of fungal solution inserted via holes drilled into the wall (PCA 2008). Although the use of heat has been suggested as a means of sterilising masonry adjacent to the affected area, there is little reported evidence of the use and effectiveness of this. Fungicide fluids for treatment against dry rot in masonry are water-based and normally require total saturation; due to the health and safety risks caused by the preservative, this should be avoided wherever possible. Preference should be given to the use of preservative plugs, paste or organic solvent-based fungicidal products. Old lime plaster on which, or through which, the rot has spread should be hacked off and renewed. New timber used to replace affected timber should be treated with a wood preservative before it is fixed or built in.

Because dry rot is capable of spreading through other materials and attacking dry timber, eradication involves the replacement of all affected timber with pre-treated timber and the treatment of adjoining timbers, brickwork, plasterwork and adjacent areas well away from the point of decay.

The following operations must be carried out to ensure that the fungus is completely eradicated:

❏   Locate and eliminate the dampness responsible for the attack.
❏   Cut out, remove from the site and burn all defective timbers in the vicinity of the attack. All sound timber within 300 to 600 mm of the defective timber should be cut out, removed and burnt.
❏   Note that surface spraying of timbers will not prevent the spread of dry rot from infected timber.
❏   Carry out a thorough check of all other timbers within the building to which the fungus might have spread. This may involve removing flooring, ceilings, skirtings, architraves, wall panels and so on.
❏   Strip off all wall plaster that may contain fungal strands to 300 mm beyond the observed limit of growth (HSE 2001; PCA 2008).
❏   Clean down all exposed masonry by wire brushing. Remove all stripped plaster, debris and dust from the site.
❏   Surrounding masonry should be cleaned and treated by surface application of masonry biocide, preservative plugs or pastes inserted into holes drilled into localised problem areas and/or irrigation of fungal solution inserted via holes drilled into the wall (PCA 2008).
❏   Replace all timber that has been cut out with new, well-seasoned timber that has been pre-treated with fungicidal preservative.

Wet rot and its prevention
Wet rot is caused principally by *Coniphora puteana*, the cellar fungus, which occurs more frequently, but is less serious, than dry rot. Decay of timber due to wet rot is confined to timbers that are in damp situations such as cellars, ground floors without dpcs and timbers and joinery that is frequently in contact with moisture. The rot causes darkening and longitudinal cracking of timber and there is often little or no visible growth of fungus on the surface of timber.

Timber should not be built into or in contact with any part of the structure that is likely to remain damp. dpcs and damp-proof membranes (dpms) above and at ground level and sensibly detailed flashings and gutters to roofs and chimneys will prevent the conditions suited to the growth of wet rot fungus.

Wet rot is not as prolific as dry rot and the outbreaks are usually more localised; thus, the treatment and eradication are simpler. It is usually only necessary to replace the affected timber with new pre-treated timber. This combined with the elimination of the wet rot attack and removal of the original cause is normally sufficient.

A wet rot attack is remedied by cutting out the affected wood from the point of decay with a safety clearance, and splicing-in new, preservative-treated timber.

Attacks of wet rot are often caused by the breakdown of the protective paint finish, which allows rainwater to penetrate the timber and create suitable conditions for attack. The cause of wet rot on internal elements is usually water penetration through the roof, walls and leaking plumbing. It is essential that the source of the water or moisture is identified and eliminated.

Insect attack on wood
In the UK, the three types of insects that most commonly cause damage to timber are the furniture beetle, the death-watch beetle and the house longhorn beetle. Timber which the larvae of these beetles have affected should be sprayed or painted with a preservative

that contains an insecticide during early summer and autumn. These preservatives prevent the larvae changing to beetles at the surface of the wood, which helps to prevent further infestation.

Wood preservatives

Wood may be preserved as a precaution against fungal decay or insect attack. Current practice is based on the premise that prevention is better than cure. The two types of preservative in general use are waterborne or organic solvent formulations, where water or a volatile solvent serves as a vehicle for the active fungicide or insecticide components. If timber decay is discovered, it is advisable to consult one of the many companies that specialise in its diagnosis and treatment. They will carry out a detailed survey and diagnosis of the timber decay and provide a treatment and eradication package backed by guarantee.

Treatment of insect attack

Where there are only a few insect exit holes and the insect attack is not advanced, it is recommended that the timber be treated with water-based insecticide. The insecticide should be applied so that it minimises the exposure to people, pets and the environment. If the insect attack goes to the heart of the timber, the amount of organic solvent-based treatment must match the degree of penetration. If the attack is widespread, a paste or injected solvent-based formulations should be used and applied to all timbers. If the attack is localised, treatment need only be applied 300 mm beyond the area of the affected timber. The cure to both insect attack and mould growth is to remove the source of the damp condition and environment. Without the water content, the insects cannot survive and the mould cannot grow. Timbers that have been structurally weakened by insect attack must be treated with insecticide and strengthened. If the timber is beyond use and repair, it should be cut out, burnt and completely replaced with new, pre-treated timber.

The primary control method to prevent insect infestation and fungal attack is to keep timber dry. Excessive use of even modern water-based treatment is not recommended.

5.3 Modified and engineered timber products

Modified and engineered timber products are manufactured under factory conditions to produce products that are superior to natural wood. These comprise modified wood products (MWPs), a range of wood structural panels, glue-laminated beams (see Chapter 4), CLT panels and engineered beams. MWPs primarily enhance the durability of timber. Engineered products involve the bonding together of timber using permanent adhesives (glues) to create products that offer structural advantages over natural timber. The primary functional requirements are durability, strength and stability, but some of these products also claim to have strong environmental credentials.

Modified wood products (MWPs)

It is possible to modify the properties of timber through impregnation, chemical treatment and/or heat treatment. This allows cheaper, less durable timbers to be used and modified into highly durable products that are less sensitive to moisture, UV, dimensional change, fungal and insect attack. The timbers most used include pine, poplar, eucalyptus and other fast growing varieties. Modified timber products may be used for flooring in wet areas and externally for cladding buildings. There are many processes and products on the market, but the main approaches are:

❏ *Impregnation*: is achieved by using a liquid catalyst and vacuum pressure impregnation, with a drying temperature above 100 °C.
❏ *Thermal modification*: involves heating the timber to temperatures over 200 °C in an oxygen-free environment. The lack of oxygen prevents the wood from burning. The heat treatment causes a chemical reaction within the timber, which 'bakes' the sugars and tannins in the wood. This makes the wood darker and harder and also makes it inedible to microbes and insects. This tends to result in a better performing product compared to pressure-treated timber.
❏ *Chemical modification*: involves a variety of processes, the most well-known being a process called acetylation. The wood is impregnated with acetic anhydride and reacts at high temperature. The process turns free hydroxyls, which absorb and release moisture and which are naturally present in the wood, into acetyl groups. This reduces the timber's ability to absorb water, which makes it more dimensionally stable and very durable. Silicon-based compounds are also used to similar effect.

Wood structural panels

Wood structural panels are precision-engineered panels, manufactured under factory conditions. Wood structural panels comprise chipboard, medium-density fibreboard (MDF), plywood and orientated strand board (OSB). Plywood and OSB have similar characteristics in terms of strength and stiffness, but their composition is different. As a general rule, all of these products need to be protected from moisture during construction, and also after construction, to prevent expansion and damage from moisture/water ingress. Wood structural panels must be correctly installed and kept dry and ventilated. As a general rule, they should not be used in areas of high humidity.

Chipboard
Wood particles, sawdust, timber chips and shavings are glued together with an adhesive and hot pressed to create a rigid board. Chipboard, also known as particleboard, is a relatively cheap product, but tends to have limited use in construction, because it retains water and is prone to moisture damage. Three densities are available, normal-, medium- and high-density chipboard. Chipboard is usually protected with a decorative laminate, melamine or wood veneer to produce countertops.

Medium-density fibreboard (MDF)
MDF is formed of fine wood dust and resin, which is then hot pressed into a board or formed into specific shapes, such as skirting board profiles. MDF can be manufactured to be fire and water resistant with the addition of specific resins. The material is easy to saw, drill and work with; however, this needs to be done in a well-ventilated area and workers should wear eye protection and a face mask because the product may be harmful to health. The product is knot free and therefore well suited to internal joinery.

Plywood

Thin sheets of softwood or hardwood, called veneers, are positioned at right angles to one another and glued together using a hot press to form a cross-laminated panel. The grain of each veneer is perpendicular to the adjacent layer. Plywood always comprises an odd number of layers so that it is balanced and hence less prone to bending, warping or shrinking. Plywood is used for roof and wall sheathing, as well as for flooring. The material will splinter when cut or sawn.

Orientated strand board (OSB)

OSB is manufactured in a similar way to plywood, although it is considered to be a more versatile alternative to plywood. Timber logs are ground into thin wood strands and dried. These strands are then mixed with adhesive and wax, formed into a mat and then hot pressed into panels. Wood fibres are positioned with opposing orientation to create a rigid and structurally strong board. OSB tends to be used for roofing and wall sheathing and also for flooring. OSB is easy to saw, drill, nail, screw and sand. Panels are available in a range of standard sizes (e.g. 1220×2440 mm and different thicknesses, 9, 11 and 18 mm). Panels can also be produced to bespoke sizes. OSB tends to react slower to moisture compared to plywood, but it is also slower to shed the moisture and hence prone to rot. OSB will expand quicker around the edges than the middle if it gets wet. There are two classifications, OSB/2 and OSB/3. The latter is designed and manufactured to cope with damp conditions.

Structurally insulated panels (SIPs)

Structurally insulated panels (SIPs) have been developed around the concept of composite (or sandwich) panels. The SIPs are composed of two faces, usually OSB/3, that 'sandwich' a rigid core of expanded polystyrene (EPS) or polyurethane (PU). This forms a lightweight panel that is easy to erect on site. The panels offer very good thermal insulation and airtightness in addition to inherent structural properties. SIPs offer a simple, dry form of construction that includes the insulation and structure in one panel, rather than building the structure then adding the insulation. Panels tend to be around 140–175 mm thick and are manufactured in standard widths from 200 to 1200 mm, and lengths of up to 7.5 m. Bespoke widths can be made to order. The panels are used in floor, wall and roof construction, usually in conjunction with timber I beams and joists. A variety of proprietary products are available, some of which are designed specifically as cladding panels. SIPs may be used to achieve *Passivhaus* standards and feature in some of the offsite housing systems.

Cross-laminated timber (CLT) panels

CLT panels are engineered timber products using kiln-dried small sections of timber. Panels are manufactured in a similar way to glue-laminated timber beams in a quality-controlled factory environment. Small sections of timber are bonded together in layers with permanent adhesives to form large panels. Although the glues tend to have a high environmental impact, the glue content is very low, accounting for around 0.6% of the panel. Layers of timber (lamellas) are bonded perpendicular to one another (longitudinal and transverse layers), hence the term cross-laminated (or 'crosslam' or 'X-LAM'). The cross-lamination gives the panel strength in two directions (dimensions) and helps to ensure the product retains its dimensional integrity when subjected to changing levels of humidity. Panels comprise a minimum of three layers of timber, rising to a maximum of seven layers, depending on the desired structural properties and overall thickness. The timber most commonly used is spruce, but other softwoods such as larch and some of the fir family are also used. The majority of the timber is kiln-dried C24, with a small amount of C16-grade timber. The moisture content should be 12% (plus or minus 2%). Panel depth depends upon the number of layers used, ranging from 60 mm to around 340 mm.

Panels are manufactured in a range of sizes from 1.2 to 3.5 m wide, typically 2.4 m wide for floor panels and 2.95 m for wall panels. Length may be up to 22 m, but more usually around 13 m to facilitate ease of transportation. The CLT panels are transported to site, craned into position and secured using screws and/or brackets. Panels should be kept dry on site, and if they do get wet (e.g. from a rain shower), they should be allowed to dry out prior to fixing and/or covering with another material. The size of the panels tends to be restricted by transportation to the site and access restrictions. This provides the potential for fast, dry on site assembly.

Three grades of surface finish (visual quality) are available, depending on building function, whether or not the panels will be covered or left fair face and the desired aesthetic. The quality classifications are:

(1) *Visible quality AB.* Panels comprise a mixture of A and B quality lamellas that are planed and sanded on one side to give a visual quality suitable for residential buildings, offices and schools. The inherent finish can be left as supplied or a decorative finish (wash or varnish) may be applied.
(2) *Visible quality BC.* One side of the panel is planed and sanded. Typically used in industrial and commercial buildings where the visible quality may not be a major concern.
(3) *Non-visible quality C.* This is the lowest visual quality and therefore the panels are commonly used in situations where they will be covered by, for example, a roof covering, plasterboard or render.

CLT panels are vapour permeable, which means that they can be used as part of a 'breathing wall' assemblage. They have a low environmental impact (carbon is stored within the product during its service life) and when the raw material is sourced from ecological, economical and socially responsible forests, they make a positive contribution to a green supply chain. The thermal conductivity is around 0.13 W/mK, which is similar to a lightweight concrete block. CLT has very good fire resistance, retaining its load-bearing capacity in a fire. The surface will char when exposed to flame and hence protect the inner core. The product can also positively contribute to acoustic insulation and airtightness, subject to appropriate detailing and fixing.

CLT panels are used in the construction of floors, walls and roofs. The smooth finish of the panels provides an inherent surface finish that does not require further treatment (unless required, such as a coloured wash). The panels are easy to fix into for first and second fix, using self-tapping wood screws. CLT panels are used in the construction of high-rise timber framed buildings, because of their strength and lightweight.

All CLT panels must be kept dry during storage on site and during installation. They must be positioned above the level of the dpc or dpm. CLT panels will require some form of protection, such as external cladding or render if used for the external leaf of a wall.

However, the panels are more commonly used for the inner leaf of walls and for internal partition walls. When used to form a floor, it is possible to achieve clear spans of around 8 m. Given the structural properties of CLT panels, it is possible to construct comparatively slim floors compared to more traditional approaches.

Engineered timber beams

Engineered timber beams are usually provided as an I section. The horizontal sections of the I are called flanges and the vertical element is the web. They are manufactured using wood composites and therefore can utilise young, easy to cut, sustainably sourced trees. Compared to traditional timber beams and joists, engineered timber beams are much lighter and use about one-sixth of the material for the same structural strength. They are also dimensionally more precise and stable, as they do not dry and shrink as much as solid timber once the building is occupied. Engineered timber beams are used for floor joists (I-joists) to the longer spanning I-beams (e.g. for roof structures). They are also becoming common components of a range of prefabricated timber assemblies, such as wall, floor and roof panels. Manufacturers provide typical examples of spans and some provide interactive clear span tables for typical applications, such as floors.

Engineered bamboo products

Bamboo is a member of the grass family. As a fast-growing and sustainable material, it offers potential benefits over the comparatively slower-growing timber for use in construction. Bamboo exhibits similar characteristics to timber and it has long been used in vernacular construction, especially in countries where bamboo is abundant. The bamboo canes are referred to as culms. In its natural state, the irregularities in the bamboo culm make it challenging to connect individual sections for building. Engineered products offer greater potential for construction. The bamboo culm is processed in a factory to form a laminate composite. The culm is split, planed, bleached, caramelised and then glued and pressed to form a cross-laminated sheet product, known as cross-laminated bamboo (CLB). Alternatively, it is possible to produce products from woven strands of bamboo, such as engineered flooring. This involves the crushing of the bamboo culms, combining it with resin and pressing it into dense panels, which are then cut to form flooring and beams. Engineered bamboo appears to have comparable structural properties to timber and there are also ongoing developments into hybrid products that comprise bamboo and timber. These aim to strengthen low-grade structural timber with bamboo to create an engineered product that can be used in construction. Research and development is ongoing, along with efforts to harmonise and approve certification of engineered bamboo products for structural applications. The majority of bamboo is sourced in China (Moso bamboo) and increasingly from Colombia (Guadua bamboo).

5.4 Timber-framed walls

The construction of a timber framed wall is a rapid, clean, dry operation. Timbers can be cut and assembled with simple hand- or power-operated tools, and once the wall is raised into position and fixed, it is ready to receive wall finishes. A timber-framed wall has adequate stability and strength to support the floors and roofs of small buildings, such as houses. Covered with wall finishes, such as plasterboard, it has sufficient resistance to damage by fire, good thermal insulating properties and reasonable durability, providing it is well constructed and protected from decay. Timber-framed houses can be constructed very quickly. Using offsite prefabrication allows the erection of a house within a day, with roofing and external cladding completed soon afterwards. Alternatively, timber-framed houses can be built from timber by carpenters or by self-builders using a variety of proprietary systems (Photograph 5.1). There are a number of websites that provide information on the design and specification of timber frame housing.

- ❏   *TRADA*: Timber Research and Development Association provides details, guidance, reports and updates; http://www.trada.co.uk
- ❏   *Virtual site*: provides details, 360 images and information on low-carbon construction; http://www.leedsmet.ac.uk/teaching/vsite/
- ❏   *Woodspec*: guidance on designing and specifying timber frame building; http://www.woodspec.ie

Modern methods of timber frame construction were introduced into the UK in the 1960s. Timber frame construction offers flexible planning, energy-efficient construction,

Photograph 5.1   Self-build timber frame house under construction.

economic use of materials and a wide range of finishes. Timber frame construction, especially when light cladding is used, is lighter than a comparable masonry structure, and in some instances the foundations can be designed to be smaller and hence less wasteful of materials. The dry construction is fast and there is no need to wait for wet trades to dry out before decorating. The high levels of thermal insulation make timber frame an attractive option, given stringent thermal requirements. The structure of timber frame buildings must be designed by a structural engineer to demonstrate structural stability of the structure and compliance with the Building Regulations. It is common to erect buildings to a height of two or three storeys, although it is possible to build higher (six to eight storeys) and still satisfy the Building Regulations.

Stability of timber-framed buildings

The stability of a timber-framed wall depends on a sound foundation on which a stable structure can be constructed. Figure 5.2 is an illustration of the base of a timber-framed wall set on a brick upstand raised from a strip foundation. The $150 \times 50\,mm$ timber sole plate is bedded on a horizontal dpc; shot fired nail with sufficient anchorage is used to locate the sole plate, in exposed positions 13 mm bolts at 2 m centres built into the foundation to anchor the plate against wind uplift. As an alternative, the bolts may be shaped so that the bottom flange is built into the wall, run up on the inside face of the wall with a top flange turned over the top of the plate. Angle brackets may also be used to secure the sole plate.

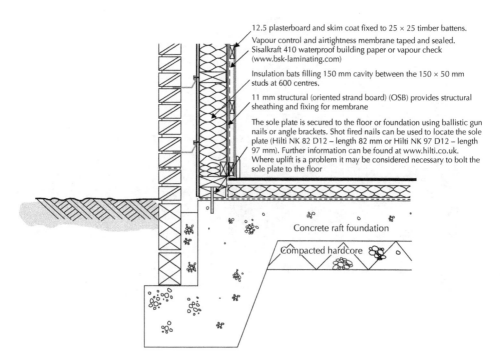

12.5 plasterboard and skim coat fixed to 25 × 25 timber battens.

Vapour control and airtightness membrane taped and sealed. Sisalkraft 410 waterproof building paper or vapour check (www.bsk-laminating.com)

Insulation bats filling 150 mm cavity between the 150 × 50 mm studs at 600 centres.

11 mm structural (oriented strand board) (OSB) provides structural sheathing and fixing for membrane

The sole plate is secured to the floor or foundation using ballistic gun nails or angle brackets. Shot fired nails can be used to locate the sole plate (Hilti NK 82 D12 – length 82 mm or Hilti NK 97 D12 – length 97 mm). Further information can be found at www.hilti.co.uk. Where uplift is a problem it may be considered necessary to bolt the sole plate to the floor

Concrete raft foundation

Compacted hardcore

Figure 5.2   Base of timber frame wall: fixing detail.

The upstand kerb of a concrete raft foundation serves as a base for the timber wall, with the anchor bolts set into the concrete kerbs and turned over the top of the sole plate. The vertical 150×50 mm studs are nailed to the sole plate at 400–600 mm centres with double studs at angles to facilitate fixing finishes (illustrated in Figure 5.3). Photograph 5.2 shows the sole plate, a packing piece of timber and the timber stud panel fixed to the floor.

A timber stud wall consists of small section timbers fixed vertically between horizontal timber head and sole plates, as illustrated in Figure 5.4 and Photograph 5.3. The vertical stud members are usually spaced at centres of 400–600 mm to support the anticipated loads and to provide fixing for external and internal linings. The horizontal noggins fixed between studs are used to stiffen the studs against movement that might otherwise cause finishes to crack. A face of plywood sheeting is often applied to both sides of the insulated stud panel to provide considerable lateral stability.

A timber stud wall has poor structural stability along its length because of the non-rigid nailed connection of the studs to the head and sole plate, which will not strongly resist racking deformation. It must be braced (stiffened) against racking. As an internal wall or partition, a timber stud frame may be braced by diagonal timbers or by being wedged between solid brick or block walls. As an external wall, a timber stud frame may be braced between division walls and braced at angles where one wall butts to another, as illustrated in Figure 5.4. Diagonally fixed boarding or plywood sheathing fixed externally as a background for finishes, braces an external stud frame wall.

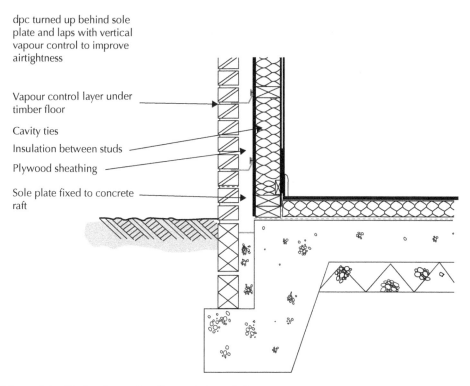

dpc turned up behind sole plate and laps with vertical vapour control to improve airtightness

Vapour control layer under timber floor

Cavity ties

Insulation between studs

Plywood sheathing

Sole plate fixed to concrete raft

Figure 5.3   Timber-framed wall on raft foundation.

Photograph 5.2   Timber frame mounted on sole plate.

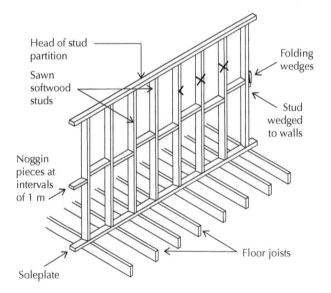

Figure 5.4   Timber stud frame.

Photograph 5.3   Timber stud frame.

Because of its small mass, a timber frame wall has poor lateral stability against forces such as wind, which tend to overturn the wall. For stability along the length of the wall, connected external and internal walls or partitions will serve as buttresses. For stability up the height of the wall, timber upper floors and the roof connected to the wall will serve as buttresses. Buttressing to timber walls that run parallel to the span of floor joists and roof frames is provided by steel straps that are fixed across floor joists and roof rafters and fixed to timber walls in the same way that straps are used to buttress solid walls as previously described.

The usual method of supporting and fixing the upper floor joists to the timber wall frame is by using separate room height wall frames. The heads of the ground floor frames provide support for the floor joists on top of which the upper floor wall frame is fixed, as illustrated in Figure 5.5. The roof rafters are notched and fixed to the head of the upper floor wall frame. As an alternative, a system of storey height wall frames may be used with the top of the head of the lower frame in line with the top of the floor joists that are supported by a timber plate nailed to the studs, as illustrated in Figure 5.6. The upper frame is formed on the lower frame. The advantage of this system is that there is continuity of the wall frame and the disadvantage is that there is a less secure connection.

Figure 5.7 is an illustration of a two-storey house with timber walls, floor and roof with a brick outer leaf, with the insulation filling the cavity rather than contained within the

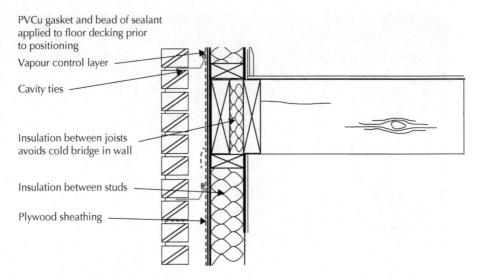

PVCu gasket and bead of sealant applied to floor decking prior to positioning

Vapour control layer

Cavity ties

Insulation between joists avoids cold bridge in wall

Insulation between studs

Plywood sheathing

Figure 5.5   Support for floor joists.

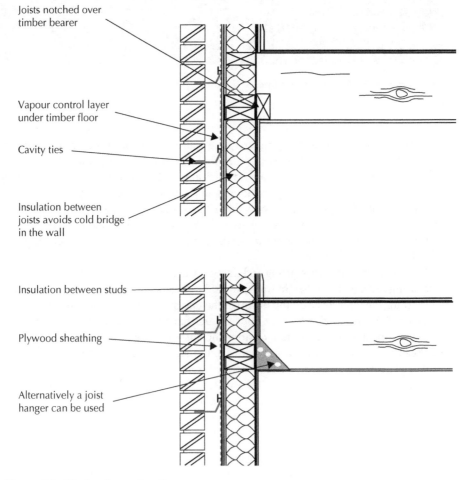

Joists notched over timber bearer

Vapour control layer under timber floor

Cavity ties

Insulation between joists avoids cold bridge in the wall

Insulation between studs

Plywood sheathing

Alternatively a joist hanger can be used

Figure 5.6   Timber-framed wall.

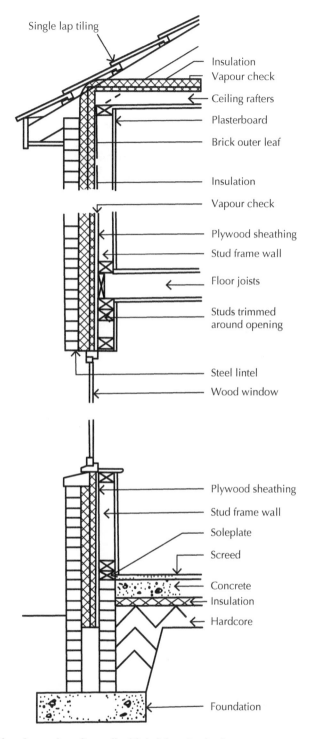

Figure 5.7   Timber-framed cavity wall with brick outer leaf.

timber stud panels. It is rare to see timber frames constructed in this manner, but there is no reason why it cannot be done. Most timber frame panels are prefabricated; therefore, site assembly is much quicker if the units come with insulation installed.

Timber or tile-clad timber frame

The traditional weather envelope for timber walls is timber weatherboarding nailed horizontally across the stud frame. The weatherboards are shaped to overlap to shed water. Some typical sections of boarding are illustrated in Figure 5.8. The wedge section, feather edge boarding, is either fixed to a simple overlap or rebated to lie flat against the studs as illustrated. The shaped chamfered and rebated and tongued and grooved shiplap boarding is used for appearance, particularly when the boarding is to be painted. To minimise the possibility of boards twisting, it is usual practice to use boards of narrow widths, such as 100 and 150 mm.

As protection against rain and wind penetrating the weatherboarding, it is usual to fix sheets of breather paper behind the weatherboarding. Breather paper serves to act as a

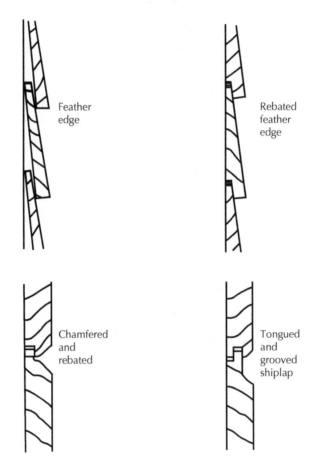

Feather edge

Rebated feather edge

Chamfered and rebated

Tongued and grooved shiplap

Figure 5.8 Timber weatherboarding.

barrier to water and at the same time allows the release of moisture vapour under pressure to move through the sheet. Instead of nailing weatherboarding directly to the studs of the wall frame, it is usual to fix either diagonally fixed boarding or sheets of plywood across the external faces of the stud frame. The boarding and ply sheets serve as a brace to the frame and as a sheath to seal the frame against weather. Figure 5.9 is an illustration of weatherboarding fixed to plywood sheathing with insulation fixed between studs.

Around openings to windows and doors, the weatherboarding and ply sheath may be butted to the back of window and door frames fixed to project beyond the stud frames for the purpose. At the head of the opening, the head of the frame may be reduced in depth so that the boarding runs down over the face of the frame. The weatherboarding butts up to the underside of a projecting cill. For extra protection, sheet lead may be fixed behind the weatherboarding and nailed and welted to window and door frames. At external angles, the weatherboarding may be mitred or finished square edged. An effective weathering is to fix a strip of lead behind the weatherboarding to form a sort of secret gutter.

Tile, slate hanging or rendering may be used to provide more durable protection, especially in exposed positions. In the UK, it has been common practice in speculative house construction to provide the weather protection with a brick outer leaf, which provides protection to the timber in most situations. A sensible argument for this form of construction could be speed of erection and completion of building work by combining the rapid framing of a timber wall, floor and roof structure that could be completed and covered in a matter of a few days, with a brick outer leaf and speedy installation of electrical, water and heating services and dry linings.

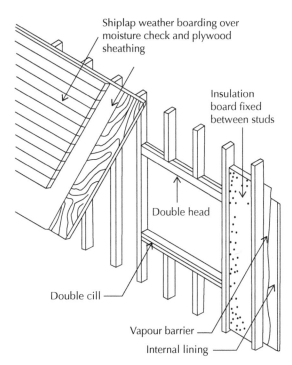

Figure 5.9 Weather envelope.

Vapour control and airtightness membranes should be lapped and taped to ensure airtightness at all points of the structure (Figure 5.11). Internal plasterboard linings to the timber-framed walls, the soffit of the first floor and the ceiling will provide a sufficient period of fire resistance to meet the requirements for a two-floor house. The requirement for barriers in external cavity walls to small houses applies only to the junction of a cavity and a wall separating buildings.

Prefabricated timber frames

There are three types of prefabricated timber frame systems that can be used: stick build, platform frame and balloon frame. Choice is often determined by the desired amount of work to be carried out on the building site, which has to be balanced against the cost of the timber frame. For example, some wall panels are delivered to site complete with sheathing and insulation, whereas other manufacturers provide the frame only, with the sheathing and insulation to be fixed on site.

Stick build
Using the traditional technique of stick build, the timber walls and floors are simply assembled from the individual members and components. Although not pre-assembled, the timber members are often delivered to site pre-cut and identified for ease of assembly. This type of construction is used to construct bespoke timber-framed houses, for example, oak post and beam structures. Apart from small, self-build and/or complicated structures, stick build is rarely used.

Platform frame
The advantage of using frames that are fabricated either on or offsite complete with outer and inner finishes is speed of erection. Where a number of houses are to be built, it is possible to complete a building in a matter of days. The systems most used are either platform or storey frames. The term 'platform' frame equally applies to light steel frame house construction made from cold-formed steel sections.

The platform frame system of construction employs prefabrication frames that are floor-to-ceiling level high, with the sole of the lower stud frame bearing on the foundation and the head of the frame supporting first-floor joists, as illustrated in Figure 5.10. The first floor can then be used as a working platform from which the upper frames are set on top of the lower. The wall frames or panels may be the full width of the front and rear walls of narrow terrace houses or made in two or more panels. The first-floor joists and roof provide sufficient bracing up the height and the separating wall will brace across the width of the wall. The wall frames may be prefabricated as stud frames sheathed with plywood or made complete with finishes on both sides.

Platform frame is currently the most common form of timber frame construction. The panel sizes will fit easily onto standard road transport and can be easily lifted into position.

Balloon frames
Storey frames ('balloon frames') are made the height of a storey, floor to floor, so that the top of the head of a frame is level with the top of the floor joists. A bearer fixed to the stud frame supports the joists. This arrangement provides continuity of the stud framing up the

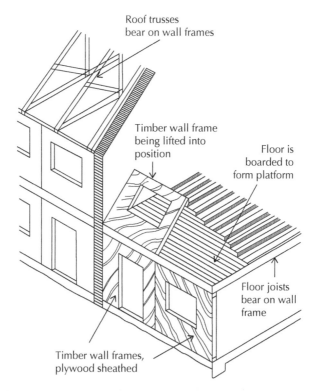

Roof trusses
bear on wall frames

Timber wall frame
being lifted into
position

Floor is
boarded to
form platform

Floor joists
bear on wall
frame

Timber wall frames,
plywood sheathed

Figure 5.10   Platform frame.

height of the wall at the expense of some loss of secure anchor of floor joists to wall. A balloon wall frame is fabricated as one continuous panel, the height of the two floors of small houses, as illustrated in Figure 5.11. The advantage of the balloon frame is speed of fabrication and erection, and the least number of joints between frames that have to be covered and weathered externally. The term 'balloon frames' also applies to steel frame housing (light steel, cold-formed sections).

Functional requirements specific to timber-framed buildings

Fire safety

Specifying a minimum period of fire resistance for the elements of structure restricts the premature failure of the structural stability of a building. A timber framed wall covered with plasterboard internally satisfies the requirement for houses of up to two storeys. To prevent the spread of fire between buildings, limits to the size of 'unprotected areas' of walls and finishes to roofs close to boundaries are set out in the Building Regulations. By reference to the boundaries of the site, the control will limit the spread of fire. Unprotected areas are those parts of external walls that may contribute to spread of fire and include glazed windows, doors and those parts of a wall that may have less than a notional fire resistance. Limits are

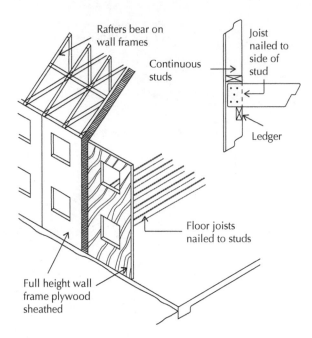

Rafters bear on wall frames

Continuous studs

Joist nailed to side of stud

Ledger

Floor joists nailed to studs

Full height wall frame plywood sheathed

Figure 5.11   Balloon frame.

set on the use of roof coverings that will not provide adequate protection against the spread of fire across their surface to adjacent buildings.

The passage of fire through connecting voids in timber frame construction is as significant as it is in masonry (loadbearing) construction. The connecting external and party wall cavity barriers must be sealed to prevent the passage of smoke and fire. The addition of edge seals reduces the passage of heat and improves airtightness. Typically, a 'T' section barrier is used to improve the thermal insulating properties of the cavity, reduce the passage of sound and prevent the passage of fire within a specified period.

Resistance to the passage of heat

Timber is a comparatively good insulator. However, the sections of a timber frame do not by themselves generally afford sufficient insulation and a layer of some insulating material has to be incorporated in the construction. The layer of insulation is fixed either between the vertical studs of the frame or on the outside or inside of the framing. The disadvantage of fixing the insulation between the studs is that there may be a great deal of wasteful cutting of insulation boards to fit them between studs; and to the extent that the U-value of the timber stud is less than that of the insulation material, there will be a small degree of thermal bridge across the studs. The advantage of fixing the insulation across the outer face of the timber frame is simplicity in fixing and the least amount of cutting. Also, the void space between the studs will augment insulation and provide space in which to conceal service pipes and cables. The disadvantage of external insulation is that the weathering finish such as weatherboarding has to be fixed to vertical battens screwed or nailed through the insulation to the studs. Unless the insulation is one of the rigid boards, it may be difficult to make a fixing for battens sufficiently firm to nail the battens to. Internal insulation is usually in the form of one of the insulation boards that combine insulation with a plasterboard finish.

Inorganic materials are most used for insulation between the studs, because there is no advantage in using the more expensive organic materials. Rolls of loosely felted fibres or compressed semi-rigid batts or slabs of glass fibre or rockwool are used. The material in the form of rolls is hung between the studs, where it is suspended by top fixing and a loose friction fit between studs, which generally maintain the insulating material in position for the comparatively small floor heights of domestic buildings. The friction fit of semi-rigid slabs or batts between studs is generally sufficient to maintain them, close butted, in position. For insulating lining to the outside face of studs, one of the organic insulants such as XPS or PIR provides the advantage of least thickness of insulating material for a given resistance to the transfer of heat. The more expensive organic insulants, in the form of boards, are fixed across the face of studs for ease of fixing and to save wasteful cutting. A vapour check should be fixed on to, or next to, the warm inside faces of insulants against penetration of moisture vapour. Organic insulants, such as XPS, which are substantially impervious to moisture vapour, can serve as a vapour check, particularly when rebated edge boards are used and the boards are close butted together.

Vapour check

A high level of insulation required for walls may well encourage moisture vapour held by warm inside air, particularly in bathrooms and kitchens, to find its way due to moisture vapour pressure into a timber-framed wall and condense to water on the cold side of the insulation. Condensate may damage the timber frame; so, as a barrier to warm moist air, there should be some form of vapour check fixed on the warm side of the insulation. Closed cell insulating materials such as XPS, in the form of rigid boards, are impermeable to moisture vapour and will act as a vapour check. The boards should either be closely butted together or supplied with rebated or tongued and grooved edges so that they fit tightly and serve as an efficient vapour check.

Where insulation materials that are pervious to moisture vapour, such as mineral fibre, are used for insulation between studs, a vapour check of polythene sheet must be fixed across the warm side of the insulation. The polythene sheet should be lapped at joints and continued up to unite with any vapour check in the roof and should, as far as practical, not be punctured by service pipes.

The vapour check is now also being incorporated as an airtightness control membrane. To ensure airtightness and vapour control, timber-framed buildings are now externally wrapped and sealed internally.

To prevent overheating of electrical cables that run through insulation, the cables should be de-rated by a factor of 0.75 by using larger cables than specified, which will generate less heat. So that cables are not run through insulation, it is wise to fix the inside dry lining to timber frames that are filled with insulation onto timber battens nailed across the frame, so that there is a void space in which cables can be safely run.

Resistance to the passage of sound

The small mass of a timber-framed wall affords little resistance to airborne sound but does not readily conduct impact sound. The thermal insulation necessary for the conservation of heat will give some reduction in airborne sound and the use of a brick or block outer leaf will appreciably reduce the intrusion of airborne sound from outside the property. Where sound pollution is a problem, acoustic plasterboard may be used to reduce sound transmission. The acoustic plasterboard is designed to reduce sound transmission and is fixed to studwork in a similar manner to plasterboard. To ensure that the acoustic board is effective, the joints should be staggered and sealed, holes should not be cut into the board and services should not pass through the board.

5.5 High-rise structural timber frames

High-rise buildings have relied on steel and concrete since their introduction and widespread use in our urban centres. These tried and tested materials are now being challenged by timber, or more specifically engineered timber and hybrid construction (comprising timber, concrete and steel). Traditionally associated with low-rise domestic developments, the rapid technological advances in timber engineering, combined with a greater awareness of the environmental impact of building materials, have led to the development of high-rise commercial buildings and apartments. These are marketed as being sustainable and a low-carbon alternative to steel and concrete framed buildings.

Until quite recently it was common to restrict framed timber buildings to relatively modest height, typically low-rise (one to two storeys) and mid-rise (three to five storeys). With advances in manufacturing techniques, a move to performance-based regulations, and increased awareness of the environmental impact of buildings, interest has turned to the construction of tall buildings using structural timber frames. New timber technologies, such as engineered timber, have helped make it possible to build taller in timber. Buildings are now being erected that are higher than six storeys, with some extending to ten or more and pioneering developments going higher, for example, up to 18 storeys in height and comprising mixed use. These are classified as high-rise timber buildings and there are many proposals for very high timber-framed buildings. This dry, offsite construction can exploit the benefits of prefabrication and modularisation, resulting in short construction programmes. Standardisation is another economic benefit that may be applied to the production of columns, beams, floors and walls. Post and beam construction, comprising glulam and laminated timber beams are used, which are sufficiently strong to replace the more carbon-intensive steel and concrete. CLT loadbearing wall panels buildings are also being used, although these tend to be more modest in height. The CLT panels are used for the walls onto which the floor is added to form the platform for the walls for the next storey. While such technologies offer fast and low-carbon construction, there are technical issues that require further research and development, and perceptions in relation to safety that need to be addressed.

Fire safety

Fire safety relates to retaining structural integrity in the event of a fire, and hence preventing collapse. Fire safety also relates to surface spread of flame. Although there are perceived concerns about fire safety and structural safety, the small number of pioneering tall timber-framed buildings have proved that it is possible to build tall in timber and maintain stability and safety. High-rise timber frames rely on heavy (massive) timber members to create the frame. The timber used is usually large-section engineered timber to form the columns and beams of the building superstructure. Solid-sawn timber sections may also be used for the structural frame. These large sections perform differently compared to light timber sections in a fire and provide inherent fire resistance. The charring to the surface in a fire protects the element and helps to maintain the structural integrity of the structural frame in a fire. The design of the columns and beams will need to allow for a charring rate in the event of a fire that will provide two-hour fire protection. As with all buildings, it will be necessary to design buildings with compartmentation to help prevent the spread of fire and provide sprinkler systems to all floors. Similar to steel and concrete framed construction, the integrity of the building in a fire is determined by the interaction of many components and their fixings, together with the provision of fire and smoke stops and fire-resistant finishes. This has to be demonstrated at the design and engineering phase, to ensure compliance with prevailing legislation.

Strength and stability

One of the advantages of being relatively lightweight also provides some structural challenges. Compared to steel and concrete framed buildings, the timber-framed buildings are more likely to move under external forces, such as wind loading. This can be addressed by using large columns and trusses, which may be designed to form part of the building's interior aesthetic. Glulam columns in the lower floors tend to be quite thick, around $1\,m^2$. The clear span of beams tends to be around $8\,m$. Steel cross bracing is usually contained within the floor construction to add additional strength to the structure.

Timber can retain a high moisture content, and as this dries it may lead to differential movement and shrinkage or shortening of columns. This can be mitigated by combining the timber frame with a secondary concrete structure, for example to the building core. The central service core, comprising the main lifts and staircase will be designed to add additional lateral restraint. The central service core may be constructed in concrete to provide lateral stability to the structural timber frame construction. Alternatively, a CLT lift/service core may be used. Other approaches rely on framed wall construction that has evolved from the low-rise timber framed housing techniques. This is an area of construction in which the boundaries of height and construction techniques are very much at an early stage, as engineers and manufacturers continue to explore possibilities.

Further reading

The Timber Research and Development Association (TRADA) has merged with the Timber Trade Federation (TTF) to form a new organisation Timber Development UK. Both organisations provide technical guidance to their membership and additional information on low-carbon construction using timber can be accessed via: www.timberdevelopment.uk. See also the Structural Timber Association: www.structuraltimber.co.uk.

Reflective exercises

Why would you choose a structural timber frame over a structural concrete or structural steel frame? What factors do you need to consider and how would you justify your decision to your client?

Your client has already 'experimented' with the use of straw bale and hemp lime render as part of a desire to have a more sustainable building portfolio. Given the positive experience, a decision has been taken to utilise as much bio-based building material as possible on the latest building project. This is a mixed-use development in the centre of a medium-sized town (pick one near you). Your client wishes to capture as much carbon as possible through the use of timber frame construction and bio-based products such as bamboo. The development is a mixed-use development comprising retail units on the ground floor and apartments on the upper floors. The building is designed to be eight storeys high.

❏   What environmental benefits does a structural timber frame offer compared to alternative materials and approaches?
❏   What are the environmental implications of importing timber frame components and engineered timber and bamboo products from around the world?

Structural Steel Frames: Chapter 6
AT A GLANCE

What? Structural steel frames are assembled from mild steel components and sections. A variety of steel sections are used to form a column and beam structure. Steel production is highly demanding in terms of energy consumption. However, once manufactured, it is possible to recycle steel components with minimal additional energy consumption, thus helping to balance out the material's environmental impact over its life. Subject to appropriate detailing and fixing, the steel components may be reclaimed and reused once the building has reached the end of its service life. Steel frames require protection from fire, which is usually achieved by encasing the steel framework with dry lining materials or painting (spraying) the frame with an intumescent coating.

Why? Framed buildings offer a dry and rapid form of construction and are a popular form of construction for tall buildings, as well as low-rise commercial and industrial units. The framed structure allows long clear spans and hence freedom for positioning of internal partition walls. The framed construction allows considerable design freedom for the building envelope. For low-rise buildings, a variety of claddings, from sheet materials to brickwork, may be used. Taller buildings tend to be designed to utilise curtain walling and cladding to form the external envelope.

When? The structural steel frame will be erected after the foundations are complete. The rapid construction sequence provides a structural framework onto which the roof may be positioned and onto which the floors and external envelope are fixed. This quickly provides a weatherproof envelope within which internal works can proceed unhindered by the wind and rain. Tall buildings are usually scheduled so that the external envelope is fixed as the steel frame is erected, thus providing protection to workers and materials. Such techniques are common in fast-track projects.

How? The components of a structural steel frame will be produced under factory conditions to the engineer's design. Components are available off the shelf for relatively simple, low-rise buildings or they can be manufactured to suit bespoke design requirements. The strength and stability of the structure will be determined by the interaction of columns and beams and the rigidity of the connections. Components may be delivered to site as individual components and/or as pre-assembled wall and floor elements. The length of the columns and beams is limited by transportation restrictions and physical access to the construction site. The structural steel frame is erected on to pre-prepared foundations, by using cranes and site personnel to help position the steel members. Members are fixed in position by using bolts through pre-drilled holes and welds. The ease of fixing helps to aid the speed of erection and facilitate the ease of alterations, replacement and recovery of components at a future date.

6 Structural Steel Frames

Structural steel frames are a popular choice for tall buildings. The advantages of the structural steel frame are the speed of erection of the ready-prepared steel members and the accuracy of setting out and connections, which is a tradition in engineering works. Accurate placing of steel members, with small tolerances, facilitates the fixing of cladding materials and curtain walls. With the use of sprayed-on or dry-lining materials to encase steel members to provide protection against damage by fire, a structural steel frame may be more economical than a reinforced concrete structural frame, because of speed of erection and economy in material and construction labour costs. Steel components can also be reclaimed, reused and recycled when the building is deconstructed at the end of its life.

6.1 Functional requirements

The functional requirements of a structural frame are:

❑ Strength and stability.
❑ Durability and freedom from maintenance.
❑ Fire safety.

Strength and stability

The requirements from the Building Regulations are that buildings must be constructed so that the loadbearing elements, foundations, walls, floors and roofs have adequate strength and stability to support the dead loads of the construction and anticipated imposed loads on roofs, floors and walls, without such undue deflection or deformation as might adversely affect the strength and stability of parts or the whole of the building. The strength of the loadbearing elements of the structure is assumed either from knowledge of the behaviour of similar traditional elements, such as walls and floors under load or by calculations of the behaviour of parts or the whole of a structure under load, based on data from experimental tests, with various factors of safety to make allowance for unforeseen construction or design errors. The strength of individual elements of a structure may be reasonably accurately assessed by taking account of tests on materials and making allowance for variations of strength in both natural and man-made materials.

Barry's Advanced Construction of Buildings, Fifth Edition. Stephen Emmitt.
© 2023 John Wiley & Sons Ltd. Published 2023 by John Wiley & Sons Ltd.

The strength of combinations of elements such as columns and beams depends on the rigidity of the connection and the consequent interaction of the elements. Simple calculations, based on test results, of the likely behaviour of the joined elements or a more complex calculation of the behaviour of the parts of the whole of the structure, can be made. Various factors of safety are included in calculations to allow for unforeseen circumstances. Calculations of structural strength and stability provide a mathematical justification for an assumption of a minimum strength and stability of structures in use.

Imposed loads are those loads that it is assumed the building or structure is designed to support, taking account of the expected occupation or use of the building or structure. Assumptions are made of the likely maximum loads that the floors of a category of building may be expected to support. The load of the occupants and their furniture on the floors of residential buildings will generally be less than that of goods stored on a warehouse floor.

The loads imposed on roofs by snow are determined by taking account of expected snow loads in the geographical location of the building. Loads imposed on walls and roofs by wind (wind loads) are determined by reference to the situation of the building on a map of the UK on which basic wind speeds have been plotted. These basic wind speeds are the maximum gust speeds averaged over three-second periods, which are likely to be exceeded on average only once in 50 years. In the calculation of the wind pressure on buildings, a correction factor is used to take account of the shelter from wind afforded by obstructions and ground roughness.

The stability of a building depends initially on a reasonably firm, stable foundation. The stability of a structure depends on the strength of the materials of the loadbearing elements in supporting, without undue deflection or deformation, both concentric and eccentric loads on vertical elements and the ability of the structure to resist lateral pressure of wind on walls and roofs.

The very considerable deadweight of walls of traditional masonry or brick construction is generally sufficient, by itself, to support concentric and eccentric loads and the lateral pressure of wind. Generally, the deadweight of skeleton-framed multi-storey buildings is not, by itself, capable of resisting lateral wind pressure without undue deflection and deformation. Some form of bracing is required to enhance the stability of skeleton-framed buildings. Unlike the joints in a reinforced concrete structural frame, the normal joints between vertical and horizontal members of a structural steel frame do not provide much stiffness in resisting lateral wind pressure.

Disproportionate collapse
A requirement from the Building Regulations is that a building shall be constructed so that, in the event of an accident, the building will not suffer collapse to an extent disproportionate to the cause. This requirement applies only to a building having five or more storeys (each basement level being counted as one storey), excluding a storey within the roof space, where the slope of the roof does not exceed 70° to the horizontal.

Durability and freedom from maintenance

The members of a structural steel frame are usually inside the wall fabric of buildings so that in usual circumstances the steel is in a comparatively dry atmosphere, which is unlikely to cause progressive, destructive corrosion of steel. Structural steel will, therefore, provide reasonable durability for the expected life of the majority of buildings and require no maintenance. Where the structural steel frame is partially or wholly built into the enclosing masonry or brick walls, the external wall thickness is generally adequate to prevent such penetration of moisture as is likely to cause corrosion of steel. Where there is some likelihood of penetration of moisture to the structural steel, it is usual practice to provide protection by the application of paint or bitumen coatings or the application of a damp-proof layer. Where it is anticipated that moisture may cause corrosion of the steel, either externally or from a moisture-laden interior, weathering steels, which are much less subject to corrosion, are used.

Fire safety

The application of the Regulations, as set out in the practical guidance given in Approved Document B, is directed to the safe escape of people from buildings in case of fire, rather than the protection of the building and its contents. Insurance companies that provide cover against the risks of damage to the building and contents by fire may require additional fire protection such as sprinklers.

Internal fire spread (structures)

The requirement from the Regulations relevant to structure is to limit internal fire spread (structure). As a measure of ability to withstand the effects of fire, the elements of a structure are given notional fire resistance times, in minutes, based on tests. Elements are tested for their ability to withstand the effects of fire in relation to:

❏ Resistance to collapse (loadbearing capacity), which applies to loadbearing elements.
❏ Resistance to fire penetration (integrity), which applies to fire-separating elements.
❏ Resistance to the transfer of excessive heat (insulation), which applies to fire-separating elements.

The notional fire-resisting times, which depend on the size, height, number of basements and use of buildings, are chosen as being sufficient for the escape of occupants in the event of fire. The requirements for the fire resistance of elements of a structure do not apply to:

❏ A structure that supports only a roof unless:
   (a) The roof acts as a floor (e.g. car parking), or as a means of escape.
   (b) The structure is essential for the stability of an external wall, which needs to have fire resistance.
❏ The lowest floor of the building.

6.2 Methods of design

There are a number of established approaches to the method of design of structural steel frames, as described later. It is also possible for engineers and designers to use some simple 'rules of thumb' to quickly establish the depth of, for example, a steel beam for a specific span to give an indication as to the depth of the structure.

Rules of thumb

Simple 'rules of thumb' can be used to quickly establish an approximate size for steel columns and beams given typical loading conditions. This may be useful in the very early design stages when a number of options are being explored to quickly establish what may or may not be feasible. This method has largely been superseded by BIM and computer software programs, although rules of thumb can also be used to check the validity of results from software programs. For steel I beams, the quick way to size a steel beam is to divide the span by 20. For a 12 m clear span, the beam would be approximately 600 mm deep. The column size will depend on the floor area it is supporting and the imposed loads on the floor area. Calculations will be required to coordinate the dimensions of beams and columns to ensure an efficient design that facilitates ease of construction.

Permissible stress design method

With the introduction of steel as a structural material in the late nineteenth and early years of the twentieth centuries, the permissible stress method of design was accepted as a basis for the calculation of the sizes of structural members. Having established and agreed yield stress for mild steel, the permissible tensile stress was taken as the yield stress divided by a factor of safety, to allow for unforeseen overloading, defective workmanship and variations in steel. The yield stress in steel is that stress at which the steel no longer behaves elastically and suffers irrecoverable elongation, as shown in Figure 6.1, which is a typical stress/strain curve for mild steel.

The loads to be carried by a structural steel frame are dead, imposed and wind loads. Dead loads comprise the weight of the structure including walls, floors, roof and all permanent fixtures. Imposed loads include all movable items that are stored on or usually supported by floors, such as goods, people, furniture and movable equipment. Wind loads are those applied by wind pressure or suction on the building. Dead loads can be accurately calculated. Imposed loads are assumed from the common use of the building to give reasonable maximum loads that are likely to occur. Wind loads are derived from the maximum wind speeds.

Having determined the combination of loads that are likely to cause the worst working conditions the structure is to support, the forces acting on the structural members are calculated by the elastic method of analysis to predict the maximum elastic working stresses in the members of the structural frame. Beam sections are then selected so that the maximum predicted stress does not exceed the permissible stress. In this calculation, a factor of safety is applied to the stress in the material of the structural frame. The permissible compressive stress depends on whether a column fails due to buckling or yielding and is determined from the slenderness ratio of the column, Young's modulus and the yield stress divided by a factor of safety. The permissible stress method of design provides a safe and reasonably economic method of design for simply connected frames and is the most commonly used method of design for structural steel frames.

A simply connected frame is a frame in which the beams are assumed to be simply supported by columns, to the extent that while the columns support beam ends, the beam is not fixed to the column and in consequence when the beam bends (deflects) under load, bending is not restrained by the column. Where a beam bears on a shelf angle fixed to a column and the top of the beam is fixed to the column by means of a small top cleat designed to

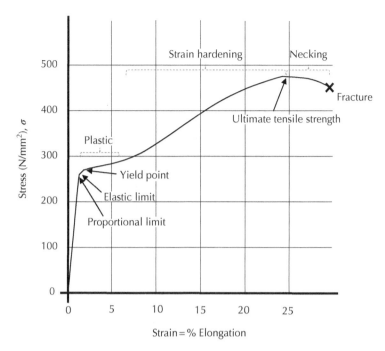

Figure 6.1   Stress/strain curve for mild steel.
Yield point: the point at which the material under stress no longer behaves elastically; point at which permanent deformation (plastic deformation or flow) begins.
Elastic limit: the maximum stress or force per unit area that can be applied to a material without causing permanent deformation; the highest point on the graph before the straight line changes to a curve.
Tensile strength: the maximum stress that a material can sustain before tearing or failing while the material is being stretched.
Fracture: the point at which the material breaks.

maintain the beam in a vertical position, it is reasonable to assume that the beam is simply supported and will largely behave as if it had a pin-jointed connection to the column.

Collapse or load factor method of design

Where beams are rigidly fixed to columns and where the horizontal or near-horizontal members of a frame, such as the portal frame, are rigidly fixed to posts or columns, then beams do not suffer the same bending under load that they would if simply supported by columns or posts. The effect of the rigid connection of beam ends to columns is to restrain simple bending, as illustrated in Figure 6.2. The fixed-end beam bends in two directions, upwards near fixed ends and downwards at the centre. The upward bending is termed 'negative bending' and the downward bending termed 'positive bending'. It will be seen that bending at the ends of the beam is prevented by the rigid connections that take some of the stress due to loading and transfer it to the supporting columns. Just as the rigid connection of beam to column causes negative or upward bending of the beam at the ends, so a comparable, but smaller, deformation of the column will occur.

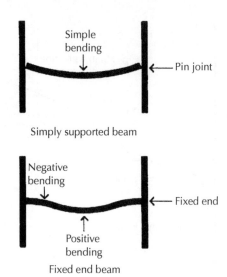

Simply supported beam

Fixed end beam

Figure 6.2   Comparison of pin-jointed and fixed end beams.

Using the elastic method of analysis to determine working stress in a fixed-end beam, to select a beam section adequate for the permissible stress, the design method produces a section greater than is needed to provide a reasonable factor of safety against collapse, because in practice the permissible stress is not reached and in consequence, the beam could safely support a greater load.

The collapse or load factor method of design seeks to provide a load factor, that is, a safety factor, against collapse applied to particular types of structural frame for economy in the use of materials, by using the load factor, which is applied to the loads instead of stress in materials. The load factor method was developed principally for use in the design of reinforced concrete and welded connection steel frames with rigid connections as an alternative to the permissible stress method, as a means to economy in the selection of structural sections. In the use of the load factor method of design, plastic analysis is used. In this method of analysis of the forces acting in members, it is presumed that extreme fibre stress will reach or exceed yield stress and the fibres behave plastically. This is a valid assumption as, in practice, the fibres of the whole section play a part in sustaining stress, and under working loads, extreme fibre stress would not reach yield point.

Limit state method of design

The purpose of structural analysis is to predict the conditions applicable to a structure that would cause it to become either unserviceable in use or unable to support loads to the extent that members might fail.

In the permissible stress method, a limit is set on the predicted working stress in the members of the frame, by the use of a factor of safety applied to the predicted yield stress of the materials used. In the load factor method of design, a limit is set on the working loads to ensure that they do not exceed a limit determined by the application of a factor of safety to the loads that would cause collapse of the structure. The limit state method of design seeks to determine the limiting states of both materials and loads that would cause a particular structure to become unserviceable in use or unsafe due to excessive load. The limiting conditions that are considered are serviceability during the useful life of the building and the ultimate limit state of strength.

Serviceability limit states set limits on the behaviour of the structure to limit excessive deflection, excessive vibration and irreparable damage due to material fatigue or corrosion that would otherwise make the building unserviceable in use. Ultimate limit states of strength set limits to strength in resisting yielding, rupture, buckling and transformation into a mechanism, and stability against overturning and fracture due to fatigue or low-temperature brittleness.

During use, the limit state method of design sets characteristic loads and characteristic strengths, which are those loads and strengths that have an acceptable chance of not being exceeded during the life of the building. To take account of the variability of loads and strength of materials in actual use, a number of partial safety factors may be applied to the characteristic loads and strengths to determine safe working loads and strengths.

The limit state method of design has not been accepted wholeheartedly by structural engineers because they say, it is academic, highly mathematical, increases design time and does not lead to economic structures. There is often little reward in employing other than the permissible stress method of design for the majority of buildings so that the use of the limit state method is confined in the main to larger and more complex structures where the additional design time is justified by more adventurous and economic design.

6.3 Steel sections

Mild steel is the material generally used for constructional steelwork. It is produced in several basic strength grades of which those designated as 43, 50 and 55 are most commonly used. The strength grades 43, 50 and 55 indicate minimum ultimate tensile strengths of 430, 500 and 550 N/mm², respectively. Each strength grade has several subgrades indicated by a letter between A and E; the grades that are normally available are 43A, 43B, 43C, 43D, 43E, 50A, 50B, 50C, 50D and 55C. In each strength grade, the subgrades have similar ultimate tensile strengths, and as the subgrades change from A to E, the specification becomes more stringent, the chemical composition changes and the notch ductility improves. The improvement in notch ductility (reduction in brittleness), particularly at low temperatures, assists in the design of welded connections and reduces the risk of brittle and fatigue failure, which is of particular concern in structures subject to low temperatures.

Properties of mild steel

Strength
Steel is strong in both tension and compression with permitted working stresses of 165, 230 and 280 N/mm² for grades 43, 50 and 55, respectively. The strength-to-weight ratio of mild steel is good, so that mild steel is able to sustain heavy loads with comparatively small self-weight.

Elasticity
Under stress induced by loads, a structural material will stretch or contract by elastic deformation and return to its former state once the load is removed. The ratio of stress to strain, which is known as Young's modulus (the modulus of elasticity), gives an indication of the resistance of the material to elastic deformation. If the modulus of elasticity is high, the deformation under stress will be low. Steel has a high modulus of elasticity, $200\,kN/mm^2$, and is therefore a comparatively stiff material, which will suffer less elastic deformation than aluminium, which has a modulus of elasticity of $69\,kN/mm^2$. Under stress induced by loads, beams bend or deflect, and in practice this deflection under load is limited to avoid cracking of materials fixed to beams. The sectional area of a mild steel beam can be less than that of other structural materials for a given load, span and limit of deflection.

Ductility
Mild steel is a ductile material that is not brittle and can suffer strain beyond the elastic limit through what is known as plastic flow, which transfers stress to surrounding material so that at no point will stress failure in the material be reached. Because of the ductility of steel, the plastic method of analysis can be used for structures with rigid connections, which makes allowance for transfer of stress by plastic flow and so results in a section less than would be determined by the elastic method of analysis, which does not make allowance for the ductility of steel.

Resistance to corrosion
Corrosion of steel occurs as a chemical reaction between iron, water and oxygen to form hydrated iron oxide, commonly known as rust. Because rust is open-grained and porous, a continuing reaction will cause progressive corrosion of steel. The chemical reaction that starts the process of corrosion of iron is affected by an electrical process through electrons liberated in the reaction, whereby small currents flow from the area of corrosion to unaffected areas and so spread the process of corrosion. In addition, pollutants in air accelerate corrosion as sulphur dioxides from industrial atmospheres and salt in marine atmospheres increase the electrical conductivity of water and so encourage corrosion. The continuing process of corrosion may eventually, over the course of several years, affect the strength of steel. Mild steel should therefore be given protection against corrosion in atmospheres likely to cause corrosion.

Fire resistance
Although steel is non-combustible and does not contribute to fire, it may lose strength when its temperature reaches a critical point in a fire in a building. Therefore, some form of protection against fire is required.

Weathering steels (COR-TEN)

The addition of small quantities of certain elements modifies the structure of the rust layer that forms. The alloys encourage the formation of a dense, fine-grained rust film and also react chemically with sulphur in atmospheres to form insoluble basic sulphate salts that block the pores on the film and so prevent further rusting. The thin, tightly adherent film that forms on this low alloy steel is of such low permeability that the rate of corrosion is reduced almost to zero. The film forms a patina of a deep brown colour on the surface of steel, giving a rust-like appearance. The low-permeability rust film forms under normal wet/dry cyclical conditions. In conditions approaching constant wetness and in conditions exposed to severe marine or salt spray conditions, the rust film may remain porous and not prevent further corrosion. Weathering steels are produced under brand names, the most well-known being 'COR-TEN', which is a particular favourite of artists, architects and engineers for its appearance and durability. Typical uses include large urban art installations, exposed frames and surfaces of buildings and bridges. Weathering steels require special welding techniques to ensure the weld points weather at the same rate as the steel alloy. Care is also required to ensure that the initial runoff from the steel does not discolour and stain adjacent materials as it weathers.

Standard rolled steel sections

The steel sections mostly used in structural steelwork are standard hot-rolled steel universal beams and columns together with a range of tees, channels and angles illustrated in Figure 6.3. Universal beams and columns are produced in a range of standard sizes and weights designated by serial sizes. Within each serial size, the inside dimensions between flanges and flange edge and web remain constant, and the overall dimensions and weights vary, as illustrated in Figure 6.4. This grouping of sections in serial sizes is convenient for production within a range of rolling sizes and for the selection of a suitable size and weight by the designer. The deep web-to-flange dimensions of beams and the near similar flange-to-web dimensions of columns are chosen to suit the functions of the structural elements. Because of the close similarity of the width of the flange to the web of column sections, they are sometimes known as 'broad flange sections'.

A range of comparatively small section 'joists' are also available, which have shallowly tapered flanges and are produced for use as beams for small to medium spans. The series of structural T's is produced from cuts that are half the web depth of standard universal beams and columns. The range of standard hot-rolled structural steel angles and channels has tapered flanges. The standard rolled steel sections are usually supplied in strength grade 43A material with strength grades 50 and 55 available for all sections at an additional cost per tonne. All of the standard sections are available in Cor-Ten B weathering steel.

Castella beam

An open web beam can be fabricated by cutting the web of a standard section mild steel beam along a castellated line, as illustrated in Figure 6.5. The two cut sections of the beam are then welded together to form an open web, castellated beam, which is one and a half times the depth of the beam from which it was formed. Because of the increase in depth, the castella beam will suffer less deflection (bending) under light loads. The castella beam is no stronger than the beam from which it was cut but will suffer less deflection under load. The increase in cost due to fabrication and the reduction in weight of the beam as compared to a solid web beam of the same depth and section justify the use of these beams for long-span, lightly loaded beams, particularly for roofs. The voids in the web of these beams are convenient for housing runs of electrical and heating services.

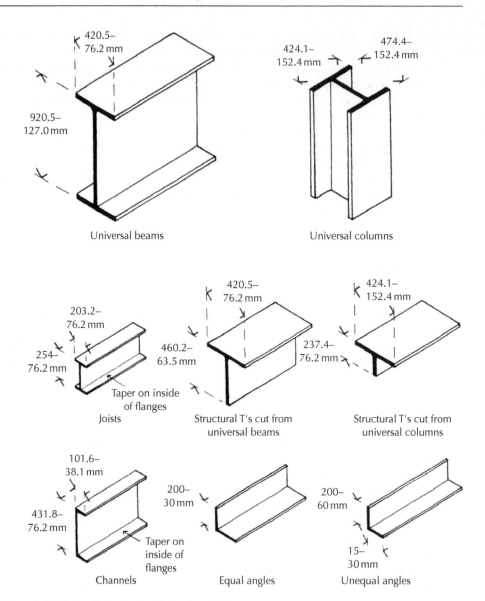

420.5–
76.2 mm

920.5–
127.0 mm

Universal beams

424.1–
152.4 mm

474.4–
152.4 mm

Universal columns

203.2–
76.2 mm

254–
76.2 mm

Taper on inside
of flanges

Joists

420.5–
76.2 mm

460.2–
63.5 mm

Structural T's cut from
universal beams

424.1–
152.4 mm

237.4–
76.2 mm

Structural T's cut from
universal columns

101.6–
38.1 mm

431.8–
76.2 mm

Taper on
inside of
flanges

Channels

200–
30 mm

Equal angles

200–
60 mm

15–
30 mm

Unequal angles

Figure 6.3 Hot-rolled structural steel sections.

Steel tubes

A range of seamless and welded seam steel tubes is manufactured for use as columns, struts and ties. The use of these tubes as columns is limited by the difficulty of making beam connections to a round section column. These round sections are the most efficient and compact structural sections available and are extensively used in the fabrication of lattice girders, columns, frames, roof decks and trusses for economy, appearance and comparative freedom from dust traps. Connections are generally made by scribing the ends of the tube

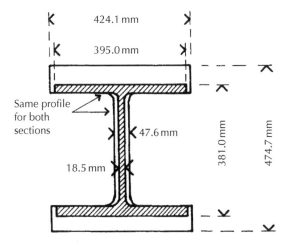

Figure 6.4 Universal columns.

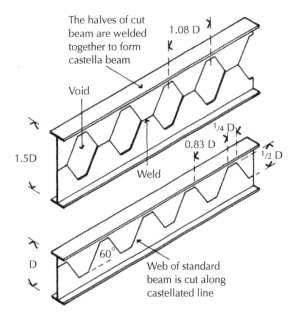

Figure 6.5 Castella beam.

to fit around the round sections to which they are welded. For long-span members such as roof trusses, bolted plate connections are made at mid-span for convenience in transporting and erecting long-span members in sections.

Hollow rectangular and square sections

Hollow rectangular and square steel sections are made from hollow round sections of steel tube, which are heated until they are sufficiently malleable to be deformed. The heated tubes are passed through a series of rollers, which progressively change the shape of the tube to

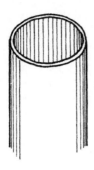 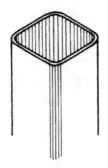 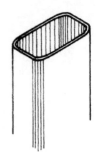

Circular hollow sections from 21.3 to 457 mm outside diameter

Square hollow sections from 20 × 22 to 400 × 400 mm

Rectangular hollow sections from 50 × 30 to 450 × 250 mm

Figure 6.6   Hollow steel sections.

square or rectangular sections with rounded edges, as illustrated in Figure 6.6. To provide different wall thicknesses, the heated tube can be gradually stretched. The advantage of these sections is that they are ideal for use as columns as the material is uniformly disposed around the long axis, and the square or rectangular section facilitates beam connections.

Hollow square and rectangular sections are much used as the members of lattice roof trusses and lightly loaded framed structures with the square sections for columns and the rectangular sections as beams. The economy in material and the neat appearance of the sections, which with welded connections have a more elegant appearance than angle connections, recommend their use. These sections are also much used in the fabrication of railings, balustrades, gates and fences with welded connections for the neat, robust appearance of the material. Prefabricated sections can be hot dip galvanised to inhibit rusting prior to painting.

Cold roll-formed steel sections

Cold roll-formed structural steel sections are made from hot-rolled steel strip, which is passed through a series of rollers. Each pair of rollers progressively takes part in gradually shaping the strip to the required shape. As the strip is cold formed, it has to be passed through a series of rollers to avoid the thin material being torn or sheared in the forming process, which produces sections with slightly rounded angles to this end. There is no theoretical limit to the length of steel strip that can be formed. The thickness of steel strip commonly used is from 0.3 to 0.8 mm and the width of strip up to about 1 m. A very wide range of sections is possible with cold-rolled forming, some of which are illustrated in Figure 6.7.

The advantage of cold-rolled forming is that any shape can be produced to the exact dimensions to suit a particular use or design. Figure 6.8 is an illustration of cold-formed sections, spot welded back-to-back to form structural steel beam sections, and sections welded together to form box form column sections. Connections of cold-formed sections are made by welding self-tapping screws or bolts to plate cleats welded to one section. Because of the

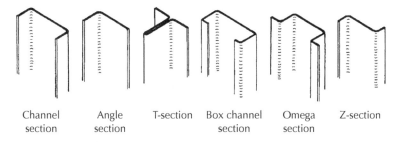

Channel section   Angle section   T-section   Box channel section   Omega section   Z-section

Figure 6.7   Cold roll-formed sections.

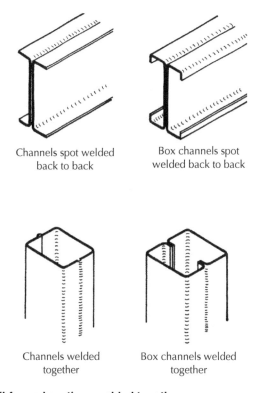

Channels spot welded back to back

Box channels spot welded back to back

Channels welded together

Box channels welded together

Figure 6.8   Cold roll-formed sections welded together.

comparatively thin material from which the sections are formed, it is necessary to use some coating that will inhibit corrosion and some form of casing as protection against early damage by fire where regulations so require. Cold-formed steel sections are extensively used in the manufacture of roof trusses and lattice beams and frames. The fabricated sections are protected with a galvanised coating where practical. Cold-formed, pressed steel sections are much used for floor and roof decking for the floors of framed buildings and also for metal doors, frames and metal trim such as skirtings.

Steel frame wall construction for domestic houses

Although steel is traditionally associated with large commercial and industrial buildings, light gauge steel (cold-formed steel sections) is used in domestic house construction. The sections of steel are used in a similar way to timber sections to form a structural frame. Because the steel frame is a good conductor of heat, it is important that the insulation is not placed between the metal frame, as this would create thermal bridges. Instead, the insulation should be applied to the inside of the cavity. Rolled steel sections can be assembled using either stick, panel or balloon construction.

When using stick construction, the individual members are delivered to site pre-cut and pre-punched for holes to be cut and self-tapping screws to be used. Advantages of stickbuild construction are that small modifications and adjustments can be made on site to accommodate site tolerances and very minor design changes. Structural members can be packed and transported in small, tightly packed, loads. However, stick-build construction is labour intensive and is not widely used in the UK. Panel and balloon frame construction tend to be more attractive alternatives.

Panel construction has the advantage that the wall subframes, panels, floors and roof trusses are prefabricated under carefully controlled factory conditions and then delivered to site as ready-assembled elements. The subframes and panels are connected on site using bolts or self-tapping screws. Quality control is achieved via factory production, and the accuracy of the components and panels make them easy and quick to assemble on site.

In balloon construction the panels are much larger, comprising floor-to-roof elements, although the components are manufactured in much the same as panel frame construction.

6.4 Structural steel frames

The earliest structural steel frame was erected in Chicago in 1883 for the Home Insurance building. A skeleton of steel columns and beams carried the whole of the load of floors, and solid masonry or brick walls were used for weather protection and appearance. Since then, the steel frame has been one of the principal methods of constructing multi-storey buildings (Photograph 6.1).

Skeleton frame

The conventional steel frame is constructed with hot-rolled section beams and columns in the form of a skeleton designed to support the whole of the imposed and dead loads of floors, external walling or cladding and wind pressure (Photograph 6.2). The arrangement of the columns is determined by the floor plans, horizontal and vertical circulation spaces and the requirements for natural light to penetrate the interior of the building. Figure 6.9 is an illustration of a typical rectangular grid skeleton steel frame. In general, the most economic arrangement of columns is on a regular rectangular grid with columns spaced at 3.0–4.0 m apart, parallel to the span of floors which bear on floor beams spanning up to 7.5 m with floors designed to span one way between main beams. This arrangement provides the smallest economic thickness of floor slab and least depth of floor beams, and therefore least height of building for a given clear height at each floor level.

Photograph 6.1   Steel frame office building under construction.

Photograph 6.2   Skeleton frame with concrete lift shaft core.

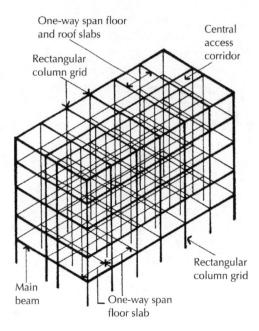

One-way span floor
and roof slabs

Central
access
corridor

Rectangular
column grid

Rectangular
column grid

Main
beam

One-way span
floor slab

Figure 6.9   Rectangular grid steel frame.

Figure 6.10 is an illustration of a typical small skeleton steel frame designed to support one-way span floors on main beams and beams to support solid walls at each floor level on the external faces of the building. This rectangular grid can be extended in both directions to provide the required floor area.

Where comparatively closely spaced columns may obstruct internal floor space, a larger rectangular or square grid of columns is used. The columns support main beams, which in turn support secondary beams spaced at up to 4.5 m apart to carry one-way span floor slabs, as illustrated in Figure 6.11 and Photograph 6.3. This arrangement allows for the least span and thickness of floor slab and the least weight of construction.

A disadvantage of this layout is that the increased span of main beams requires an increase in their depths so that they will project below the underside of the secondary beams. Heating, ventilating and electrical services, which are suspended and run below the main beams, are usually hidden above a suspended ceiling. The consequence is that the requirement for comparatively unobstructed floor space causes an increase in the overall height of a building because of the increase in depth of floor from floor finish to suspended ceiling at each floor level. Where there is a requirement for a large floor area unobstructed by columns, either a deep long-span solid web beam, a deep lattice girder or Vierendeel girders are used.

The advantage of using deep lattice girders or Vierendeel girders is that they may be designed so that their depth occupies the height of a floor and does not, therefore, increase the overall height of construction. The Vierendeel girder illustrated in Figure 6.12 is fabricated from mild steel plates, angles, channels and beam sections, which are cut and welded together to form an open web beam. The advantage of the open web form is that it can accommodate both windows externally and door openings internally, unlike the diagonals

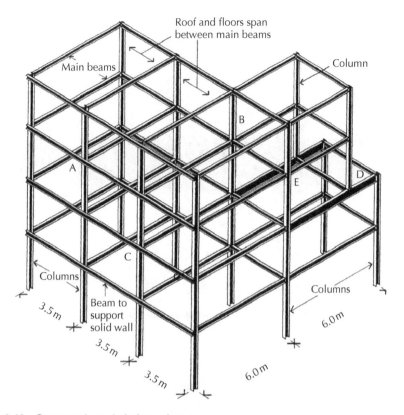

Figure 6.10   Structural steel skeleton frame.

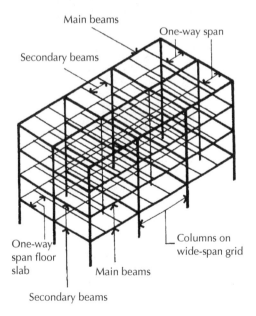

Figure 6.11   Wide-span column grid.

Photograph 6.3   Skeleton frame with wide-span column grid.

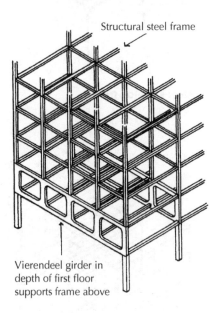

Structural steel frame

Vierendeel girder in
depth of first floor
supports frame above

Figure 6.12   Vierendeel girder.

of a lattice girder of the same depth. The solid parts of the web of this girder are located
under the columns they are designed to support. The specialist fabrication of this girder
together with the cost of transporting and hoisting into position involves considerable cost.

The conventional structural steel frame comprises continuous columns, which support
short lengths of beam that are supported on shelf angles bolted to the columns. Where there

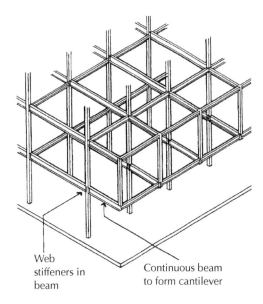

Web
stiffeners in
beam

Continuous beam
to form cantilever

Figure 6.13   Continuous beam to form cantilever.

is a requirement, for example, for the structural frame to overhang a pavement, the frame
has to be cantilevered out, as illustrated in Figure 6.13. To support the columns on the external face of the cantilever, it is necessary to use shorter lengths of the main external column
carried on continuous cantilever beams. The continuous cantilever beam is carried back
and connected to an internal column. To support the cantilever, a short length of column
is connected under the cantilever beam to which web stiffeners are welded to reinforce the
beam against web buckling.

Parallel beam structural steel frame

The parallel beam structural steel frame uses double main or spine beams fixed on each side
of internal columns to support secondary rib beams that support the floor. The principal
advantage of this form of structure is improved flexibility for the services, which can be
located in both directions within the grid between the spine beams in one direction and the
rib beams in the other. The advantage of using two parallel main spine beams is simplicity of connections to columns and the use of continuous long lengths of beam independent of column grid, which reduces fabrication and erection complexities and the overall
weight of steel.

The most economical arrangement of the frame is a rectangular grid with the more
lightly loaded rib beams spanning the greater distance between the more heavily loaded
spine or main beams. Where long-span ribs are used, for reasons of convenience in internal
layout or for convenience in running services or both, a square grid may be most suitable.
The square grid illustrated in Figure 6.14 uses double spine or main beams to internal columns with pairs of rib beams fixed to each side of columns with profiled steel decking and
composite construction structural concrete topping fixed across the top of the rib beams.
The spine beams are site bolted to end plates welded to short lengths of channel section steel

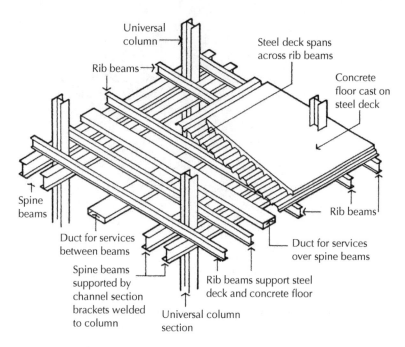

Figure 6.14   Parallel beam structural steel frame.

that are shop welded to the columns. At the perimeter of the building, a single spine beam is bolted to the end plate of channel sections welded to the column.

The parallel beam structural frame may be used, with standard I-section beams and columns or with hollow rectangular section columns and light section rolled steel sections or cold-formed strip steel beams and ribs, for smaller buildings supporting moderate floor loads in which there is a need for provision for the full range of electric and electronic cables and air conditioning. Although the number of steel sections used for each grid of the framework in this system is greater than that needed for the conventional steel frame, there is generally some appreciable saving in the total weight and, therefore, the cost of the frame, and appreciable saving in the erection time due to the simplicity of connections. The overall depth of the structural floor is greater than that of a similar conventional structural steel frame. Services may be housed within the structural depth, rather than being slung below the structural floor of a conventional frame above a suspended ceiling, which can help to reduce the overall height of the building for a given clear height between the finished floor and ceiling level.

Pin-jointed structural steel frames

The shortage of materials and skilled craftsmen that followed the Second World War encouraged local authorities in the UK to develop systems of building employing standardised components that culminated in the Consortium of Local Authorities Special Programme (CLASP) system of building. The early development was carried out by the

Hertfordshire Country Council in 1945, to fulfil their school building programme. A system of prefabricated building components based on a square grid was developed, to utilise light engineering prefabrication techniques, aimed at economy by mass production and the reduction of site labour. Some 10 years later, the Nottinghamshire County Council, faced with a similar problem and in addition, the problem of designing a structure to accommodate subsidence due to mining operations, developed a system of building based on a pin-jointed steel frame, with spring-loaded diagonal braces, and prefabricated components.

To gain the benefits of economy in mass production of component parts, the Nottinghamshire County Council joined with other local authorities to form CLASP, which was able to order, well in advance, considerable quantities of standard components at reasonable cost. The CLASP system of building has since been used for schools, offices, housing and industrial buildings of up to four storeys. The system retained the pin-jointed frame, originally designed for mining subsidence areas, as being the cheapest light structural steel frame. The CLASP system is remarkable in that it was designed by architects for architects and allows a degree of freedom of design, within standard modules and using a variety of standard components, which no other system of prefabrication has yet to achieve. The CLASP building system is illustrated in Figure 6.15.

Wind bracing

The connections of beams to columns in multi-storey skeleton steel frames do not generally provide a sufficiently rigid connection to resist the considerable lateral wind forces that tend to cause the frame to rack. The word 'rack' is used to describe the tendency of a frame to be distorted by lateral forces that cause right-angled connections to close up against the direction of the force (in the same way that books on a shelf will tend to fall over if not firmly packed in place). To resist racking caused by the wind forces acting on the faces of a multi-storey building, it is necessary to include some system of cross bracing between the members of the frame to maintain the right-angled connection of members (see Photograph 6.4). The system of bracing used will depend on the rigidity of the connections, the exposure, height, shape and construction of the building.

The frame for a 'point block' building, where the access and service core is in the centre of the building and the plan is square or near square, is commonly braced against lateral forces by connecting cross braces in the two adjacent sides of the steel frame around the centre core which are not required for access, as illustrated in Figure 6.16. Wind loads are transferred to the braced centre core through solid concrete floors acting as plates or by bracing steel-framed floors.

With the access and service core on one face of a structural frame, as illustrated in Figure 6.17, it may be convenient to provide cross bracing to the opposite sides of the service core, leaving the other two sides free for access and natural lighting for toilets, respectively. The bracing to the service core makes it a vertical cantilever anchored to the ground. To transfer wind forces acting on the four faces of the building, either one or more of the floors or roof are framed with horizontal cross bracing, which is tied to the vertical bracing of the service core. The action of the vertical cross bracing to the service core and the connected horizontal floor cross bracing to a floor, or floor and roof, will generally provide adequate stiffness against wind forces.

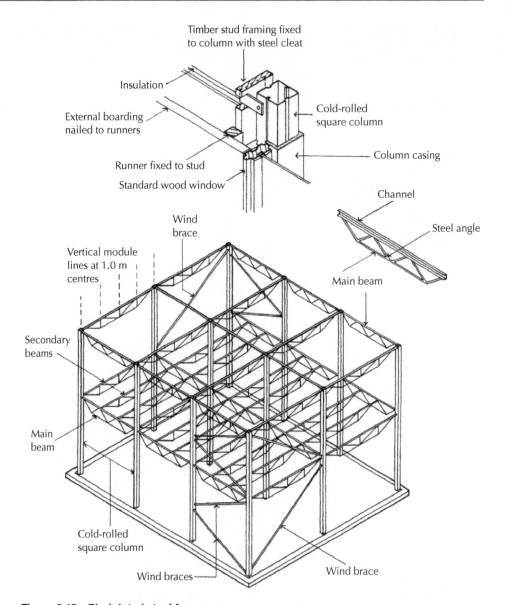

Figure 6.15   Pin-jointed steel frame.

A slab block is a building that is rectangular on plan with two main wall faces much wider than the end walls, as illustrated in Figure 6.18 and Photographs 6.5 and 6.6. With this design, it may be reasonable to accept that the smaller wind forces acting on the end walls will be resisted by the many connections of the two main walls and the horizontal solid plate floors. Here cross bracing to the end walls acting with the horizontal plates of the many solid floors may well provide adequate bracing against wind forces. To provide fire protection to means of escape, service and access cores to multi-storey buildings, it is common to construct a

Photograph 6.4   Wind bracing.

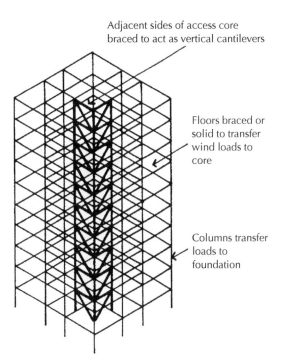

Adjacent sides of access core
braced to act as vertical cantilevers

Floors braced or
solid to transfer
wind loads to
core

Columns transfer
loads to
foundation

Figure 6.16   Wind bracing to central core.

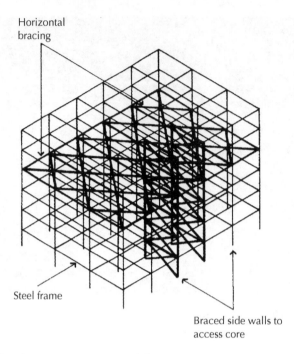

Figure 6.17   **Wind bracing to access core and floors.**

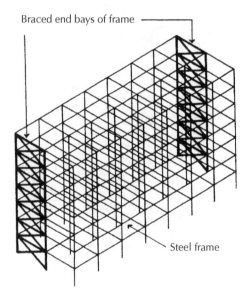

Figure 6.18   **Wind bracing to end walls.**

Photograph 6.5   Wind bracing to the end of a wall.

Photograph 6.6   Wind bracing around stairs.

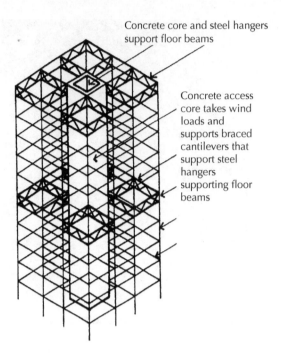

Concrete core and steel hangers support floor beams

Concrete access core takes wind loads and supports braced cantilevers that support steel hangers supporting floor beams

Figure 6.19   Reinforced concrete core supporting cantilever beams and steel hangers.

solid cast in situ reinforced concrete core to contain lifts and escape stairs. A reinforced concrete core by its construction and foundation will act as a very stiff vertical cantilever capable of taking wind forces. To provide wind bracing to a point block, multi-storey structural steel frame with a central reinforced concrete access core, illustrated in Figure 6.19, it is necessary to transfer wind forces, acting on the walls, to the core. The systems of bracing that are used combine bracing through solid concrete floor plates and cross bracing to structural steel floors. The type of cross bracing illustrated in Figure 6.19 takes the form of braced girders hung from the frame to the four corners of the building and carried back, below floor level and tied to the core to act as hung, cantilevered cross wind bracing.

Connections and fasteners

The usual practice is to use long lengths of steel column between which shorter lengths of beam are connected, to minimise the number of column-to-column joints and for the convenience of setting beam ends on shelf angles bolted to columns. In making the connections of four beam ends to a column, it is usual to connect the ends of main beams to the thicker material of column flanges and the secondary, more lightly loaded beam ends to the thinner web material. The ready-cut beams are placed on the shelf or seating angles, which have been shopped or site bolted to columns, as illustrated in Figure 6.20. The beam ends are bolted to the projecting flanges of the shelf angles. Angle side cleats are bolted to the flange of columns and webs to main beams, and angle top cleats to the web of columns and flanges of secondary beams. The side and top cleats serve the purpose of maintaining beams in their correct position. Where convenient, angle cleats are bolted to columns and beams in the fabricator's shop to reduce site connections to a minimum. These simple

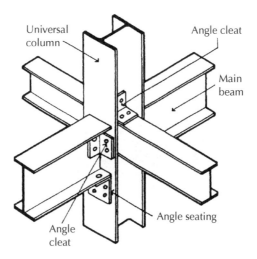

Figure 6.20   Four-beam to column connection.

cleat connections can be accurately and quickly made to provide support and connections between beams and columns.

An alternative to the simple cleat connection is to use end plates welded to the ends of the beams (Figure 6.21 and Photograph 6.7). The end plate is predrilled with holes that

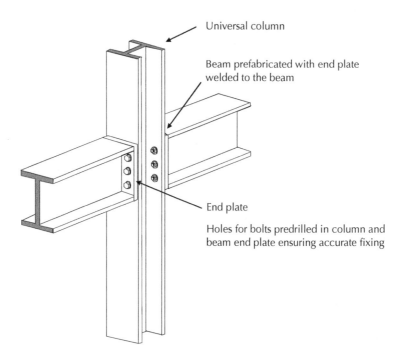

Figure 6.21   Beam connection using end plates welded to the beam.

Photograph 6.7   Column and beam connection using end plates welded to the beam.

are accurately positioned to line in with predrilled holes in the steel column. The plate can then be bolted to the beam to form a more rigid connection, which will transfer some of the bending forces.

Cleat connections of beam to column (Figure 6.20) are generally assumed to provide a simple connection in structural analysis and calculation, as there is little restraint to simple bending by this type of end connection of beams. This simply supported (unrestrained) connection is the usual basis of design calculations for structural steel frames as the simple connection provides little restraint to bending, whereas a welded end connection is rigid and affects beam bending.

Figure 6.22 is an illustration of the connection of a main beam to a column on the external face of a building, with the external face beam connected across the outside flanges of external columns. The internal beam is supported by a bottom seating angle cleat and top angle cleat. The external beam is fixed continuously across the outer face of columns to provide support for external walling or cladding that is built across the face of the frame. This beam is supported on a beam cutting to which a plate has been welded to provide a level seating for the beam and for bolt fixing to the beam. The beam cutting is bolted to the flange of columns. A top-angle cleat is bolted to the beam and column to maintain the beam in its correct upright position.

The connection of long column lengths up the height of a structural steel frame is usually made some little distance above floor beam connections, as illustrated in Figure 6.23. The ends of columns are accurately machined flat and level. Cap plates are welded to the ends of columns in the fabricator's shop and drilled ready for site bolted connections. Columns for the top floors of multi-storey buildings will be less heavily loaded than those to the floors below and it may be possible to use a smaller section of column. The connection of these

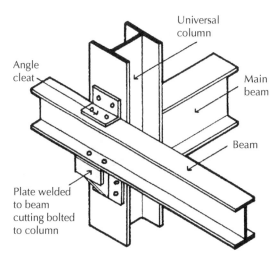

Figure 6.22   External beam to column connection.

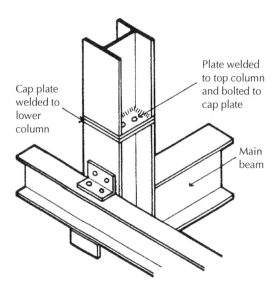

Figure 6.23   Column to column connection.

dissimilar section columns is effected through a thick bearing plate welded to the machined end of the lower column and splice plates welded to the outer flange faces. The thick bearing plate will transfer the load from the smaller section column to that of the larger column. The splice plates provide a means of joining the columns. The upper column is hoisted into position on the bearing plate, and packing plates are fitted into place to make up the difference between the column sections. The connection is then made by bolts through the splice plates, packing pieces, and flanges of the upper column, as illustrated in Figure 6.24. It is also common to splice columns together using a bolted connection, as shown in Figure 6.25 and Photograph 6.8. Where for design purposes a column is required to take its bearing on a main beam, a simple connection will suffice. A bearing plate is welded to the machined end of the column, ready for bolting to the top flange of the main beam, as illustrated in Figure 6.26a. Where a secondary beam is required to take a bearing from a main beam, for

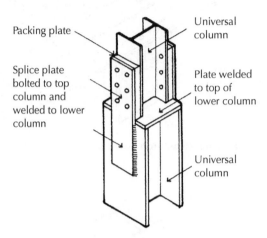

Packing plate

Universal column

Splice plate bolted to top column and welded to lower column

Plate welded to top of lower column

Universal column

Figure 6.24   Small to larger column connection.

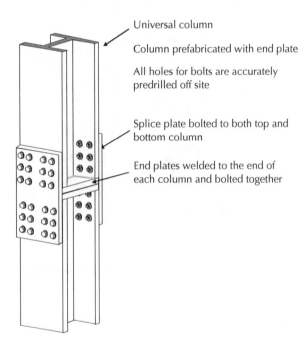

Universal column

Column prefabricated with end plate

All holes for bolts are accurately predrilled off site

Splice plate bolted to both top and bottom column

End plates welded to the end of each column and bolted together

Figure 6.25   Column-to-column spliced-bolted connection.

Photograph 6.8   Spliced column connection.

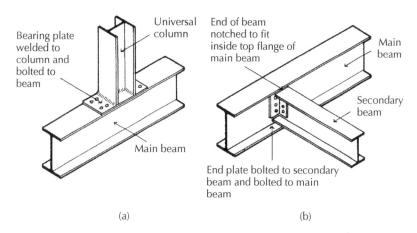

(a)   (b)

Figure 6.26   (a) Column to beam connection. (b) Beam to beam connection.

example, where a floor is trimmed for a stairwell, the end of the secondary beam is notched to fit under and around the top flange of the main beam. The connection is made with angle cleats bolted each side of the web of the secondary beam and to the web of the main beam, as illustrated in Figure 6.26b.

Fasteners

Rivets were used as both shop and field (site) fasteners for structural steelwork up to the early 1950s. Today, rivet fasteners are rarely used and bolts are used as fasteners for site connections, with welding for some shop connections. Site bolting requires less site labour than riveting, requires less skill, is quieter and eliminates fire risk.

Hexagon-headed black bolts

For many years, hexagon-headed black bolts and nuts, illustrated in Figure 6.27, were used for structural steel connections made on site. These bolts were fitted to holes 2 mm larger in diameter than that of the shank of the bolt for ease of fitting. The nut was tightened by hand and the protruding end of the shank of the bolt was burred over the nut by hammering to prevent the nut from working loose. The operation of fitting these bolts, which does not require any great degree of skill, can be quickly completed. The disadvantages of this connection are that the bolts do not make a tight fit into the holes to which they are fitted and there is the possibility of some slight movement in the connection. For this reason, black bolts are presumed to have less strength than fitted bolts and their strength is taken as 80 N/mm². These bolts are little used today for structural steel connections.

Turned and fitted bolts

To obtain more strength from a bolted connection, it may be economical to use steel bolts that have been accurately turned. These bolts are fitted to holes of the same diameter as their shank, and the bolt is driven home by hammering and then secured with a nut. Because of their tight fit, the strength of these bolts is taken as 95 N/mm². These bolts are more expensive than black bolts and have largely been superseded by high-strength friction grip (hsfg) bolts.

High-strength friction grip bolts

Hsfg bolts are made from high-strength steel, which enables them to suffer greater stress due to tightening than ordinary bolts. The combined effect of the greater strength of the bolt itself and the increased friction due to the firm clamping together of the plates being joined makes these bolts capable of taking greater loads than ordinary bolts. Bolts are tightened with a torque wrench, which measures the tightness of the bolt by reference to the torque

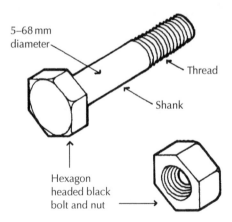

5–68 mm diameter

Thread

Shank

Hexagon headed black bolt and nut

Figure 6.27   Black hexagon bolt.

applied, which in turn gives an accurate indication of the strength of the connection. Hand tightening would give no measure of strength. Though more expensive than ordinary bolts, these bolts and their associated washers are commonly used.

Strength of bolted connections – single shear, double shear
Bolted connections may fail under load for one of two reasons. First, they may fail by the shearing of their shank. Shear is caused by the action of two opposite and equal forces acting on a material. The simplest analogy is the action of the blades of a pair of scissors or shears on a sheet of paper. As the blades close, they exert equal and opposite forces, which tear through the fibres of the paper, forcing one part up and the other down. In the same way, if the two plates move with sufficient force in opposite directions, then the bolt joining them will fail in a single shear, as illustrated in Figure 6.28. The strength of a bolt is determined by its resistance to shear in accordance with the strengths previously noted. Where a bolt joins three plates, it is liable to failure by the movement of adjacent plates in opposite directions, as illustrated in Figure 6.28. Failure is caused by the shank failing in shear at two points simultaneously, hence the term double shear. It is presumed that a bolt is twice as strong in double as in single shear.

Bearing strength
A second type of failure that may occur at a connection is caused by the shank of a bolt bearing so heavily on the metal of the member or members it is joining that the metal becomes crushed, as illustrated in Figure 6.29. The strength of the mild steel used in the majority of steel frames and the connections, in resisting crushing, is taken as $200\,N/mm^2$.

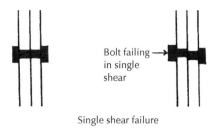

Bolt failing → in single shear

Single shear failure

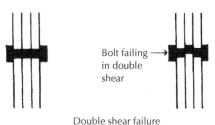

Bolt failing → in double shear

Double shear failure

Figure 6.28   Shear failure.

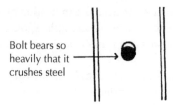

Bolt bears so
heavily that it
crushes steel

Figure 6.29   Bearing failure.

The bearing area of a bolt on the mild steel of a connection is the product of the diameter of
the bolt and the thickness of the thinnest member of the joint. When selecting the diameter
and the number of bolts required for a connection, the shear resistance of the bolts and the
bearing area of the thinnest plate have to be taken into account.

Bolt pitch (spacing)
If bolts are too closely spaced, they may bear so heavily on the section of the members
around them that they tear through the metal, with the result that, instead of the load being
borne by all, it may be transferred to a few bolts, which may then fail in shear. To prevent the
possibility of this type of failure, it is usual to space bolts at least two and a half times their
diameter apart. The distance apart is measured centre to centre. Bolts should be at least one
and three-quarter times their diameter from the edge of the steel member.

6.5 Welding

The term 'welding' describes the operation of running molten weld metal into the heated
junction of steel plates or members so that, when the weld metal has cooled and solidified,
it strongly binds them together. The edges of the members to be joined are cleaned and also
shaped for certain types of weld. For a short period, the weld metal is molten as it runs into
the joint, and for this reason, it is obvious that a weld can be formed more readily with the
operator working above the joint than in any other position. Welding can be carried out
more quickly and accurately in a workshop where the members can be manipulated more
conveniently for welding than they can be on site. Welding is mostly used in the prefabrication of built-up beams, trusses and lattice frames. The use of shop-welded connections for
angle cleats to conventional skeleton frames is less than it was due to the possibility of damage to the protruding cleats during transport, lifting and handling of members.

In the design of welded structures, it is usual practice to prefabricate as far as practical
in the workshop and to make site connections either by bolting or by designing joints that
can readily be welded on site. The advantage of welding as applied to structural steel frames
is that members can be built up to give the required strength for minimum weight of steel,
whereas standard members do not always provide the most economical section. The labour
cost in fabricating welded sections is such that it can only be justified in the main for long-
span and non-traditional frames. The reduction in weight of steel in welded frames may
often justify higher labour costs in large, heavily loaded structures. In buildings, where the
structural frame is partly or wholly exposed, the neat appearance of the welded joints and

connections is an advantage. It is difficult to tell from a visual examination whether a weld has made a secure connection, and X-ray or sonic equipment is the only exact way of testing a weld for adequate bond between weld and parent metal. This equipment is somewhat bulky to use on site, and this is one of the reasons why site welding is not favoured.

Surfaces to be welded must be clean and dry if the weld metal is to be bonded to the parent metal. These conditions are difficult to achieve in the UK's wet climate out on site. The process of welding used in structural steelwork is 'fusion welding', in which the surface of the metal to be joined is raised to a plastic or liquid condition so that the molten welding metal fuses with the plastic or molten parent metal to form a solid weld or join. For fusion welding, the requirements are a heat source, usually electrical, to melt the metal, a consumable electrode to provide the weld metal to fill the gap between the members to be joined, and some form of protection against the entry of atmospheric gases which can adversely affect the strength of the weld. The metal of the members to be joined is described as the parent or base metal and the metal deposited from the consumable electrode, the weld metal. The fusion zone is the area of fusion of weld metal to parent metal.

The method of welding mostly used for structural steelwork is the arc welding process, where an electric current is passed from a consumable electrode to the parent metals and back to the power source. The electric arc from the electrode to the parent metals generates sufficient heat to melt the weld metal and the parent metal to form a fusion weld. The processes of welding mostly used are:

❏   Manual metal arc (MMA) welding
❏   Metal inert gas (MIG) and metal active gas (MAG) welding
❏   Submerged arc (SA) welding

MMA welding

MMA welding, a manually operated process, is the oldest and the most widely used process of arc welding. The equipment for MMA welding is simple and relatively inexpensive, and the process is fully positional in that welding can be carried out vertically and even overhead due to the force with which the arc propels drops of weld metal onto the parent metal. Because of its adaptability, this process is suitable for complex shapes, welds where access is difficult and on site welding. The equipment consists of a power supply and a hand-held, flux-covered, consumable electrode, as illustrated in Figure 6.30. As the electrode is held by hand, the soundness of the weld depends largely on the skill of the operator in controlling the arc length and speed of movement of the electrode. The purpose of the flux coating to the electrode is to stabilise the arc, provide a gas envelope or shield around the weld to inhibit pickup of atmospheric gases, and produce a slag over the weld metal to protect it from the atmosphere. Because this weld process depends on the skill of the operator, there is a high potential for defects.

MIG and MAG welding

The processes of MIG and MAG welding use the same equipment, which is more complicated and expensive than that needed for MMA welding. In this process, the electrode is continuously fed with a bare wire electrode to provide weld metal, and a cylinder to provide

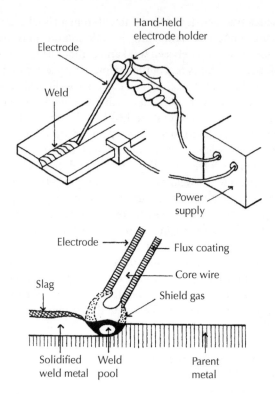

Figure 6.30   MMA welding.

gas through an annulus to the electrode tip to form a gas shield around the weld, as illustrated in Figure 6.31a. The advantage of the continuous electrode wire feed is that there is no break in welding to replace electrodes as there is with MMA welding, which can cause weakness in the weld run, and the continuous gas supply ensures a constant gas shield protection against the entry of atmospheric gases which could weaken the weld. The manually operated electrode of this type of welding equipment can be used by less highly trained welders than the MMA electrode. The bulk of the equipment and the need for shelter to protect the gas envelope limit the use of this process to shop welding.

SA welding

SA welding is a fully automatic bare wire process of welding where the arc is shielded by a blanket of flux that is continuously fed from a hopper around the weld, as illustrated in Figure 6.31b. The equipment is mounted on a gantry that travels over the weld bench to lay down flux over the continuous weld run. The equipment, which is bulky and expensive, is used for long continuous shop weld runs of high quality, requiring less skilled welders.

Types of weld

Two types of weld are used, the fillet weld and the butt weld.

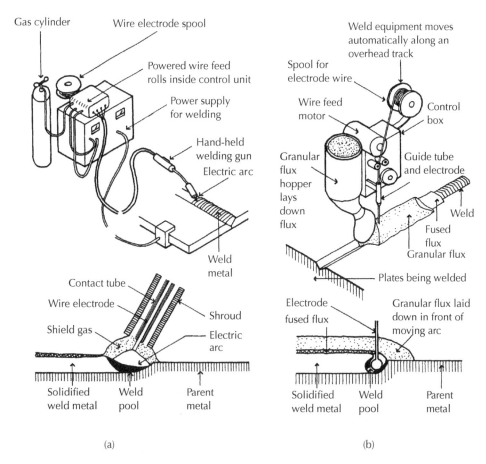

Figure 6.31   (a) MIG welding. (b) Automatic SA welding.

Fillet weld

The fillet weld takes the form of a fillet of weld metal deposited at the junction of two-parent metal membranes to be joined at an angle, the angle usually being a right angle in structural steelwork. The surfaces of the members to be joined are cleaned and the members fixed in position. The parent metals to be joined are connected to one electrode of the supply and the filler rod to the other. When the filler rod electrode is brought up to the join, the resulting arc causes the weld metal to run in to form the typical fillet weld illustrated in Figure 6.32.

The strength of a fillet weld is determined by the throat thickness multiplied by the length of the weld to give the cross-sectional area of the weld, the strength of which is taken as $115\,N/mm^2$. The throat thickness is used to determine the strength of the weld, as it is along a line bisecting the angle of the join that a weld usually fails. The throat thickness does not extend to the convex surface of the weld over the reinforcement weld metal, because this reinforcement metal contains the slag of minerals other than iron that form on the surface of the molten weld metal, which are of uncertain strength. The dotted lines in

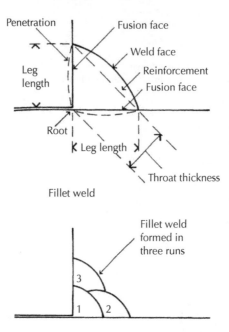

Figure 6.32   Fillet weld.

Figure 6.32 represent the depth of penetration of the weld metal into the parent metal and enclose that part of the parent metal that becomes molten during welding and fuses with the molten weld metal.

The leg lengths of fillet weld used in structural steelwork are 3, 4, 5, 6, 8, 10, 12, 15, 18, 20, 22 and 25 mm (Figure 6.32). Throat thickness is the leg length multiplied by 0.7 mm. Fillet welds of 5–22 mm are those most commonly used in structural steelwork, the larger sizes being used at heavily loaded connections. Fillet welds of up to 10 mm are formed by one run of the filler rod in the arc welding process and the larger welds by two or more runs, as illustrated in Figure 6.32. When filled welds are specified by leg length, the steel fabricator has to calculate the gauge of the filler rod and the current to be used to form the weld. An alternative method is to specify the weld as, for example, a 1–10/225 weld, which signifies that it is a 1-run weld with a 10 gauge filler rod to form 225 mm of weld for each filler rod. As filler rods are of standard length, this specifies the volume of the weld metal used for the specified length of weld and therefore determines the size of the weld. Intermittent fillet welds are generally used in structural steelwork, common lengths being 150, 225 and 300 mm.

Butt welds
Butt welds are used to join plates at their edges. The weld metal fills the gap between them. The section of the butt weld employed depends on the thickness of the plates to be joined and whether welding can be executed from one side only or from both sides. The edges of the plates to be joined are cleaned and shaped as necessary, the plates are then fixed in position and the weld metal run in from the filler rod. Thin plates up to 5 mm thick, require no shaping of their edges and the weld is formed, as illustrated in Figure 6.33. Plates up to 12 mm thick have their edges shaped to form a single V weld, as illustrated in Figure 6.34.

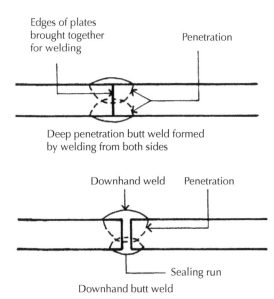

Figure 6.33   Butt welds.

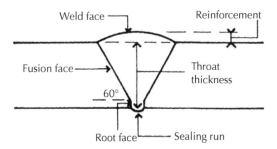

Figure 6.34   Single V butt weld.

The purpose of the V-section is to allow the filler rod to be manipulated inside the V to deposit weld metal throughout the depth of the weld without difficulty. Plates up to 24 mm thick are joined together either with a double V weld, where welding can be carried out from both sides, or by a single U, where welding can only be carried out from one side.

Figure 6.35 is an illustration of a double V and a single U weld. The U-shaped weld section provides room to manipulate the filler rod in the root of the weld but uses less of the expensive weld metal than would a single V weld of similar depth. It is more costly to form the edges of plates to the U-shaped weld than it is to form the V-shaped weld, and the U-shaped weld uses less weld metal than does a V weld of similar depth. Here the designer has to choose the weld that will be the cheapest. Plates over 24 mm thick are joined with a double U weld, as illustrated in Figure 6.35. Butt welds between plates of dissimilar thickness are illustrated in Figure 6.35.

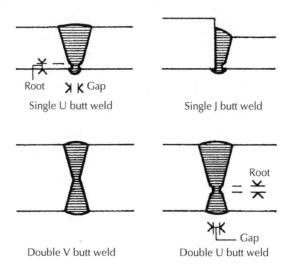

Single U butt weld

Single J butt weld

Double V butt weld

Double U butt weld

Figure 6.35   Butt welds.

The throat thickness of a butt weld is equal to the thickness of the thinnest plate joined by the weld, and the strength of the weld is determined by the throat thickness multiplied by the length of the weld to give the cross-sectional area of throat. The size of a butt weld is specified by the throat thickness; that is, the thickness of the thinnest plate joined by the weld. The shape of the weld may be described in words as, for example, a double V butt weld or by symbols.

Uses of welding in structural frames

Welding can often be used economically in fabricating large-span beams, whereas it is generally cheaper to use standard beam sections for medium and small spans. Figure 6.36 is an illustration of a built-up beam section fabricated from mild steel strip and plates, fillet and butt welded together. It will be seen that the material can be disposed to give maximum thickness of flange plates at mid-span where it is needed. Figure 6.37 illustrates a welded beam end connection, where strength is provided by increasing the size of the plates, which are shaped for welding to the column.

Built-up columns

Columns particularly lend themselves to fabrication by welding, where a fabricated column may be preferable to standard rolled steel sections. The advantages of these fabricated hollow section columns are that the sections may be designed to suit the actual loads, connections for beams and roof frames can be simply made to the square faces, and the appearance of the column may be preferred where there is no necessity for fire-resistant casing. Hollow section columns are fabricated by welding together angle or channel sections or plates, as illustrated in Figure 6.38. The advantage of columns fabricated by welding two angle or channel sections together is the least weld length necessary. The disadvantage is the

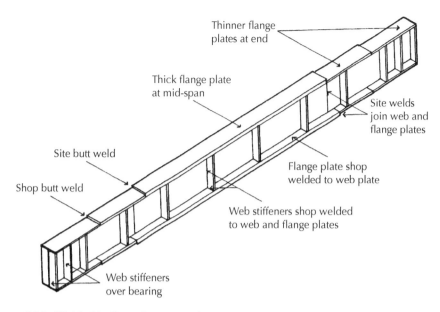

Figure 6.36   Welded built-up long-span beam.

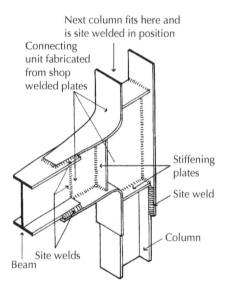

Figure 6.37   Welded beam to column connection.

limited range of sections available. The benefit of welding four plates together is the facility of selecting the precise thickness and width of plate necessary structurally, and the disadvantage is the length of weld necessary. The considerable extra cost of fabricating built-up sections limits their use to one-off special structural designs.

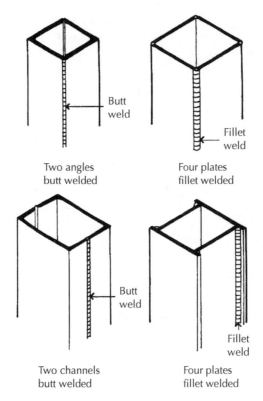

Two angles
butt welded

Four plates
fillet welded

Two channels
butt welded

Four plates
fillet welded

Figure 6.38   Welded built-up columns.

Column bases and foundations

Because of the comparatively small section area of a steel column, it is necessary to weld a steel base plate to it to provide a flat base to bear on the foundation and so spread the load, and to provide a means of fixing with holding-down bolts. The bases of steel columns are accurately machined so that they bear truly on the steel base plates to which they are welded. The three types of steel base plates that are used are the plate base, the gusseted plate base, and the slab or bloom base. For comparatively light loads, it is usual to use a 12 mm thick steel base plate fillet welded to the column. The thin plate is sufficient to spread the light loads over its area without buckling. The plate, illustrated in Figures 6.39 and 6.40, is of sufficient area to provide holes for holding-down bolts.

Prior to the column being positioned, the foundation base is checked for level, and steel shims are used to ensure that the base of the column sits at the required level. The column is hoisted into position over the concrete base so that it is plumb (vertical). The column is then lowered over the bolts. Because the bolts are cast with a void around them, which allows a small amount of moment (lateral tolerance), wedges are used to move the base into the correct position. The base is checked for line and level before being grouted in. A small bund wall of sand is then positioned around the column base and non-shrinking grout is then poured between the base and the foundation. The grout fills all the voids, including those made by the cones, which had allowed movement. Once the grout has been set, the column is securely held in position.

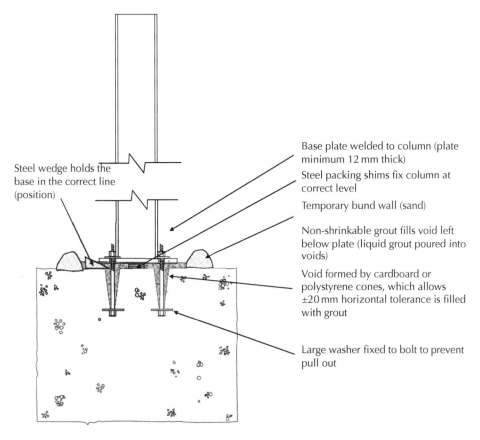

Steel wedge holds the base in the correct line (position)

Base plate welded to column (plate minimum 12 mm thick)

Steel packing shims fix column at correct level

Temporary bund wall (sand)

Non-shrinkable grout fills void left below plate (liquid grout poured into voids)

Void formed by cardboard or polystyrene cones, which allows ±20 mm horizontal tolerance is filled with grout

Large washer fixed to bolt to prevent pull out

Figure 6.39   Column base fixed to holding-down bolts.

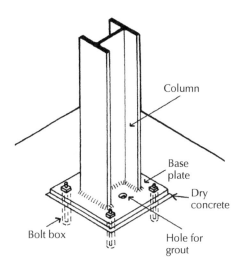

Column

Base plate

Dry concrete

Bolt box

Hole for grout

Figure 6.40   Column base plate.

The purpose of the levelling concrete or grout that is run between the base plate and the concrete is to provide uniform contact between the level underside of the plate and the irregular surface of the concrete foundation. A wet (liquid) mix of self-levelling expanding cement is poured between the column and foundation, and a temporary bund wall is used to ensure that the liquid grout is contained. As the grout dries and hardens, it expands slightly.

A gusseted base plate may also be used to spread the load of the column over a sufficient area of plate. The machined column base is fillet welded to the base plate and four shaped steel gusset plates are welded to the flanges of the column and the base plate, as illustrated in Figure 6.41. The gusset plates effectively spread the loads from the column over the area of the base plate. The column is hoisted into position so that it is plumb, and steel wedges are driven in between the plate and the concrete base ready for levelling concrete or grout. The word 'slab' or 'bloom' is used to describe comparatively thick, flat sections of steel that are produced by the process of hot rolling steel ingots to shape. A slab or bloom is thicker than a plate. Slab or bloom bases for steel columns are used for the benefit of their ability to spread heavy loads over their area without buckling. The machined ends of columns are fillet welded to slab bases and the columns hoisted into position until plumb. Steel wedges are driven in between the base and the concrete foundation ready for holding-down bolts and grouting. Figure 6.42 is an illustration of a slab base.

Steel column bases are secured in position on concrete foundations, with two, four or more holding-down bolts. These bolts are termed 'holding-down' because they hold the columns in position and may hold columns down against uplift that may occur due to the effect of wind pressure on the faces of tall buildings.

Holding-down bolts may be cast into concrete foundations by themselves or inside collars, which make allowance for locating bolts in the correct position. To cast the expanded metal collars in concrete, they are supported by timber templates, which are supported at the sides of bases. Figure 6.43 is an illustration of a timber template supporting collars ready

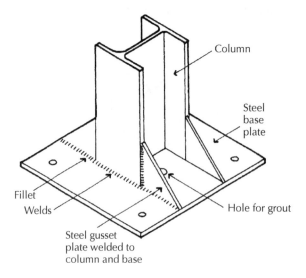

Figure 6.41   Gusseted base plate.

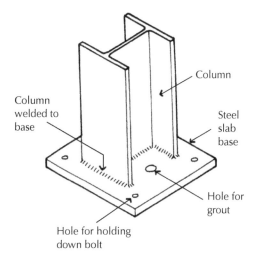

Column welded to base

Column

Steel slab base

Hole for grout

Hole for holding down bolt

Figure 6.42   Slab base plate.

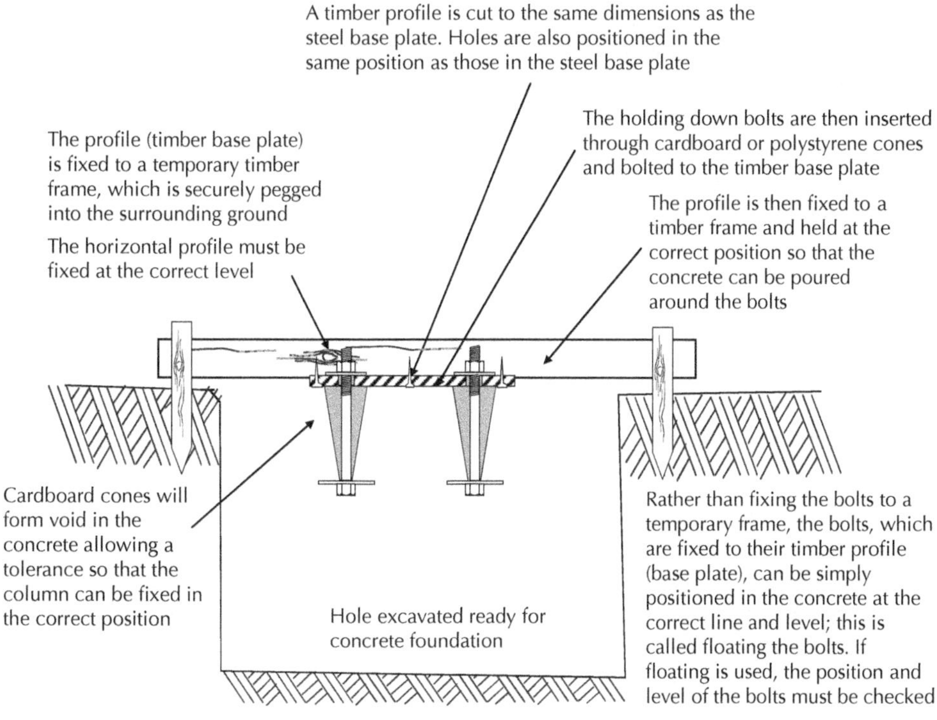

A timber profile is cut to the same dimensions as the steel base plate. Holes are also positioned in the same position as those in the steel base plate

The profile (timber base plate) is fixed to a temporary timber frame, which is securely pegged into the surrounding ground

The horizontal profile must be fixed at the correct level

The holding down bolts are then inserted through cardboard or polystyrene cones and bolted to the timber base plate

The profile is then fixed to a timber frame and held at the correct position so that the concrete can be poured around the bolts

Cardboard cones will form void in the concrete allowing a tolerance so that the column can be fixed in the correct position

Hole excavated ready for concrete foundation

Rather than fixing the bolts to a temporary frame, the bolts, which are fixed to their timber profile (base plate), can be simply positioned in the concrete at the correct line and level; this is called floating the bolts. If floating is used, the position and level of the bolts must be checked

Figure 6.43   Temporary bolt boxes.

for casting into a concrete base. The advantage of these collars is that they allow some little adjustment for aligning the holes in steel bases with the position for casting in holding-down bolts. The boss shown in the diagram at the top of collars is a wood plug to hold the bolts in the correct position. The frame holds the bolts at the correct line and level. Once the column is in position, grout is run to fill the collars around the holding-down bolts, which are fitted with anchor plates to improve contact with the concrete. Another method of fixing holding-down bolts is to drill holes in dry concrete bases and to use expanding bolts in the drilled holes. This necessitates a degree of accuracy in locating the position for drilling holes to coincide with the holes in base plates.

Mass concrete foundation to columns
The base of columns carrying moderate loads of up to, say, 400 kN bearing on soils of good bearing capacity can be formed economically of mass concrete. The size of the base depends on the bearing capacity of the soil and load on the column base, and the depth of the concrete is equal to the projection of the concrete beyond the base plate, assuming an angle of dispersion of load in concrete of 45°.

Reinforced concrete base
The area of the base required to spread the load from heavily loaded columns on subsoils of poor to moderate bearing capacity is such that it is generally more economical to use a reinforced concrete base than a mass concrete one. The steel column base plate is fixed as it is to a mass concrete base. Where column bases are large and closely spaced, it is often economical to combine them in a continuous base or raft. When a heavily reinforced concrete base is used, it may be possible to tie and position the bolt boxes to the reinforcement cage prior to pouring the concrete, rather than erecting a temporary timber frame.

Steel grillage foundation
The steel grillage foundation is a base in which a grillage of steel beams transmits the column load to the subsoil. The base consists of two layers of steel beams, two or three in the top layer under the foot of the column and a lower cross-layer of several beams so that the area covered by the lower layer is sufficient to spread column loads to the requisite area of subsoil. The whole of the steel beam grillage is encased in concrete. This type of base is rarely used today, as a reinforced concrete base is much cheaper.

Hollow rectangular sections

Beam-to-column connections
Bolted connections to closed box section columns may be made with long bolts passing through the section. Long bolts are expensive and difficult to use as they necessitate raising beams on opposite sides of the column at the same time to position the bolts. Beam connections to hollow rectangular and square-section columns may be made through plates, angles or tees welded to the columns. Standard beam sections are bolted to T-section cleats welded to columns and lattice beams by bolting end plates welded to beams to plates welded to columns, as illustrated in Figure 6.44.

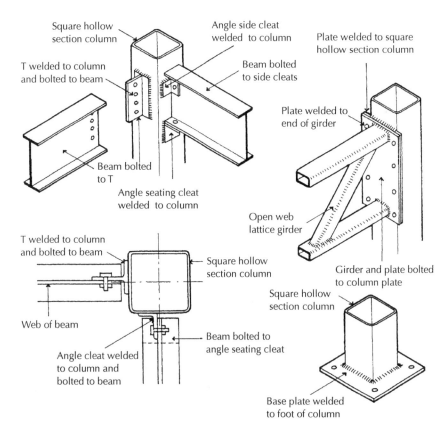

Figure 6.44   Connections to hollow section columns.

Flowdrill jointing

A recent innovation in making joints to hollow rectangular steel (HRS) sections is the use of the flowdrill technique as an alternative to using either long bolting through the hollow sections or welding and site bolting. The flowdrill technique depends on the use of a tungsten carbide bit (drill), which can be used in a conventional power-operated drill. As the tungsten carbide bit rotates at high speed on the surface of the steel, friction generates sufficient heat to soften the steel. As the bit penetrates the now softened wall of the steel section, it redistributes the metal to form an internal bush, as illustrated in Figure 6.45. Once the metal has cooled, the formed internal bush is threaded with a coldform flowtap bit to make a threaded hole ready for a bolt. The beam connection to the hollow steel column is completed by bolting welded on end plates or bolting to web cleats welded to the column through the ready-drilled holes. The execution of this form of connection requires a good deal of skill in setting out centre-punched holes accurately in the face of the hollow section to align exactly the holes to be drilled. Flowdrill jointing is the preferred method of making site connections to hollow sections for the benefit of economy in materials and site labour and the security of the bolted connection.

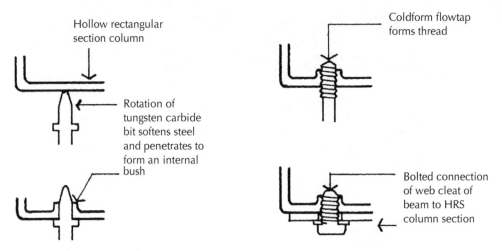

Hollow rectangular
section column

Rotation of
tungsten carbide
bit softens steel
and penetrates to
form an internal
bush

Coldform flowtap
forms thread

Bolted connection
of web cleat of
beam to HRS
column section

Figure 6.45   Flowdrill jointing.

Cold strip sections

Beam-to-column connections

Beam-to-column connections are made by means of protruding studs or T's welded to the columns and bolted to the beams. Studs welded to columns are bolted to small section beams and ties, and larger section beams to T-section cleats welded to columns, as illustrated in Figure 6.46. The T-section cleat is required for larger beams to spread the bearing area over a sufficient area of thin column wall to resist buckling.

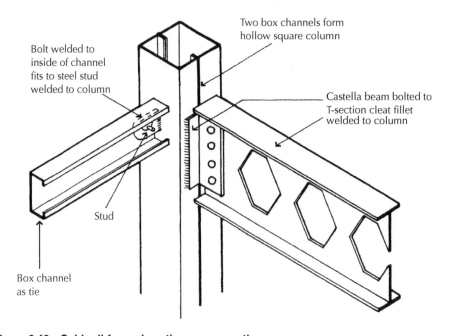

Bolt welded to
inside of channel
fits to steel stud
welded to column

Two box channels form
hollow square column

Castella beam bolted to
T-section cleat fillet
welded to column

Stud

Box channel
as tie

Figure 6.46   Cold roll-formed sections – connections.

6.6 Fire protection of structural steelwork

To limit the growth and spread of fires in buildings, the Regulations classify materials in accordance with the tendency of the materials to support spread of flame over their surface, which is also an indication of the combustibility of the materials. Regulations also impose conditions to contain fires inside compartments to limit the spread of flame. To provide safe means of escape, the Regulations set standards for the containment of fires and the associated smoke and fumes from escape routes for notional periods of time deemed adequate for escape from buildings.

One aspect of fire regulations is to specify notional periods of fire resistance for the load-bearing elements of a building so that they will maintain their strength and stability for a stated period during fires in buildings for the safety of those in the building. Steel, which is non-combustible and makes no contribution to fire, loses so much of its strength at a temperature of 550 °C that a loaded steel member would begin to deform, twist and sag and no longer support its load. Because a temperature of 550 °C may be reached early in the development of fires in buildings, regulations may require a casing to structural steel members to reduce the amount of heat getting to the steel. The larger the section of a structural steel member, the less it will be affected by heat from fires, by absorbing heat before it loses strength. The greater the mass and the smaller the perimeter of a steel section, the longer it will be before it reaches a temperature at which it will fail. This is due to the fact that larger sections will absorb more heat than smaller ones before reaching a critical temperature.

The traditional method of protecting structural steelwork from damage by fire is to cast concrete around beams and columns or to build brick or blockwork around columns with concrete casing to beams. These heavy, bulky and comparatively expensive casings have by and large been replaced by lightweight systems of fire protection employing sprays, boards, preformed casing and intumescent coatings. The materials used for fire protection of structural steelwork may be grouped as:

- ❏ Spray coatings
- ❏ Board casings
- ❏ Preformed casings
- ❏ Plaster and lath
- ❏ Concrete, brick or block casings

Spray coatings

A wide range of products is available for application by spraying on the surface of structural steel sections to provide fire protection. The materials are sprayed onto the surface of the steel sections so that the finished result is a lightweight coating that takes the profile of the coated steel, as illustrated in Figure 6.47. This is one of the cheapest methods of providing a fire protection coating or casing to steel for protection of up to four hours, depending on the thickness of the coating. The finished surface of these materials is generally coarse-textured and because of the lightweight nature of the materials, these coatings are easily damaged by knocks and abrasions. They provide some protection against corrosion of steel and, being lightweight, assist in controlling condensation. These sprayed systems of protection are suitable for use where appearance is not a prime consideration and for beams in floors above suspended ceilings. Being lightweight and porous, spray coatings are not

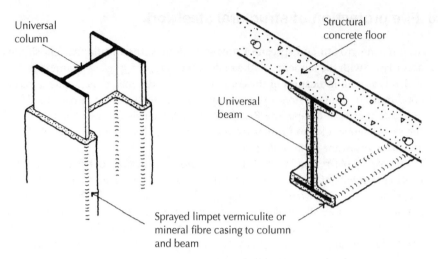

Figure 6.47   Fire protection of structural steelwork by sprayed limpet casing.

generally suited to external use. Spray coatings may be divided into three broad groups as described further.

Mineral fibre coatings

Mineral fibre coatings consist of mineral fibres that are mixed with inorganic binders, the wet mix being sprayed directly onto the clean, dry surface of the steel. The material dries to form a permanent, homogenous insulation that can be applied to any steel profile.

Vermiculite/gypsum/cement coatings

Vermiculite/gypsum/cement coatings consist of mixes of vermiculite or aerated magnesium oxychloride with cement or vermiculite with gypsum plaster. The materials are pre-mixed and water is added on site for spray application directly to the clean, dry surface of steel. The mix dries to a hard, homogenous insulation that can be left roughly textured from spraying or trowelled to a smooth finish. These materials are somewhat more robust than mineral spray coatings but will not withstand knocks. The use of sprayed vermiculite has declined, but still can be found in existing buildings.

Intumescent coatings

Intumescent coatings include mastics and paints that swell when heated to form an insulating protective coat, which acts as a heat shield. The materials are applied by spray or trowel to form a thin coating over the profile of the steel section. They provide a hard finish that can be left textured from spraying or trowelled smooth, and provide protection of up to two hours.

Board casings

There is a wide choice of systems based on the use of various preformed boards that are cut to size and fixed around steel sections as a hollow, insulating fire protection. Board casings may be grouped in relation to the materials that are used in the manufacture of the boards that are used as:

- ❏ Mineral fibreboards or batts
- ❏ Vermiculite/gypsum boards
- ❏ Plasterboard

For these board casings to be effective as fire protection, they must be securely fixed around the steel sections, and joints between boards must be covered, lapped or filled to provide an effective seal to the joints in the board casing. Board casings are only moderately robust and can be easily damaged by moderate impacts; therefore, they are not suitable for external use. Board casings are particularly suitable for use in conjunction with ceiling and wall finishes.

Mineral fibreboards and batts

Mineral fibreboards and batts are made of mineral fibres bound with calcium silicate or cement. The surface of the boards and batts, which is coarse-textured, can be plastered. These comparatively thick boards are screwed to light steel framing around the steel sections (see Photograph 6.9). Mineral fibre batts are semirigid slabs, which are fixed by means of spot-welded pins and lock washers. Mineral fibreboards are moderately robust and are used where appearance is not a prime consideration.

Photograph 6.9   Mineral fibreboard fire protection.

Vermiculite/gypsum boards

Vermiculite/gypsum boards are manufactured from exfoliated vermiculite and gypsum or non-combustible binders (see Photograph 6.10). The boards are cut to size and fixed around steelwork, either to timber noggins wedged inside the webs of beams and columns or screwed together and secured to steel angles or strips, as illustrated in Figure 6.48. The

Photograph 6.10   Gypsum board fire protection.

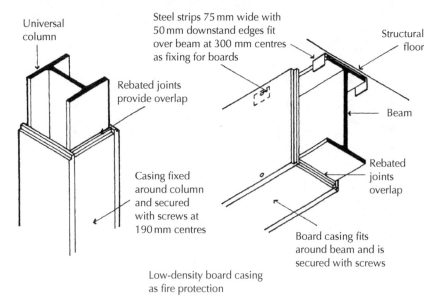

Universal column

Steel strips 75 mm wide with 50 mm downstand edges fit over beam at 300 mm centres as fixing for boards

Structural floor

Rebated joints provide overlap

Beam

Casing fixed around column and secured with screws at 190 mm centres

Rebated joints overlap

Board casing fits around beam and is secured with screws

Low-density board casing as fire protection

Figure 6.48   Vermiculite/gypsum board.

edges of the boards may be square-edged or rebated. The boards, which form a rigid, fairly robust casing to steelwork, can be self-finished or plastered.

Plasterboard casings
Plasterboard casings can be formed from standard-thickness plasterboard or from a board with a gypsum/vermiculite core for improved fire resistance. The boards are cut to size and fixed to metal straps around steel sections. The boards may be self-finished or plastered. This is a moderately robust casing. Figure 6.49 is an illustration of a board casing.

Preformed casings

Preformed casings are made in preformed 'L' or 'U' shapes ready for fixing around the range of standard column or beam sections, respectively. The boards are made of vermiculite and gypsum, or with a sheet steel finish on a fire-resisting lining, as illustrated in Figure 6.50. The vermiculite and gypsum boards are screwed to steel straps fixed around the steel sections and the sheet metal-faced casings by interlocking joints and screws. These preformed casings provide a neat, ready-finished surface, with good resistance to knocks and abrasions in the case of the metal-faced casings.

Plaster and lath casings

Plaster on metal lath casing is one of the traditional methods of fire protection for structural steelwork. Expanded metal lath is stretched and fixed to stainless steel straps fixed around steel sections with metal angle beads at arrises, as illustrated in Figure 6.51. The lath is covered with vermiculite gypsum plaster to provide an insulating fire-protective casing

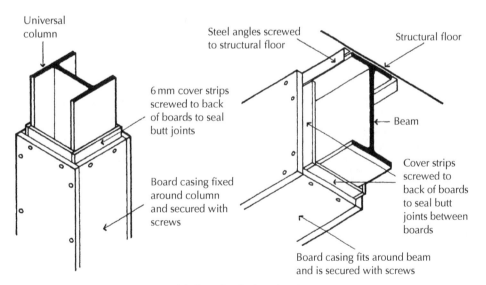

Medium-density board casing

Figure 6.49   Board casing.

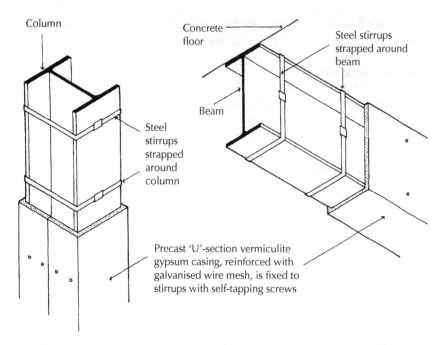

Column

Concrete floor

Steel stirrups strapped around beam

Beam

Steel stirrups strapped around column

Precast 'U'-section vermiculite gypsum casing, reinforced with galvanised wire mesh, is fixed to stirrups with self-tapping screws

Figure 6.50   Preformed casings.

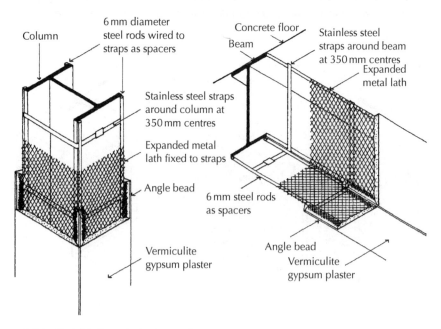

Column

6 mm diameter steel rods wired to straps as spacers

Concrete floor

Beam

Stainless steel straps around beam at 350 mm centres

Expanded metal lath

Stainless steel straps around column at 350 mm centres

Expanded metal lath fixed to straps

Angle bead

6 mm steel rods as spacers

Vermiculite gypsum plaster

Angle bead

Vermiculite gypsum plaster

Figure 6.51   Metal lath and plaster casing.

that is trowelled smooth, ready for decoration. This rigid, robust casing can suffer abrasion and knocks and is particularly suitable for use where a similar finish is used for ceilings and walls.

Concrete, brick or block casing

An in situ cast concrete casing provides fire protection to structural steelwork and protection against corrosion. This solid casing is highly resistant to damage by knocks. To prevent the concrete spalling away from the steelwork during fires, it is lightly reinforced, as illustrated in Figure 6.52. The disadvantages of a concrete casing to steelwork are its mass, which considerably increases the deadweight of the frame, and the cost of on site labour and materials in the formwork and falsework necessary to form and support the wet concrete.

Brick casings to steelwork may be used where brickwork cladding or brick division or compartment walls are a permanent part of the building, or where a brick casing is used for appearance's sake to match surrounding fairface brick. A brick casing is an expensive, labour-intensive operation in the necessary cutting and bonding of brick around columns.

Blockwork may be used as an economic means of casing columns, particularly where blockwork divisions or walls are built up to structural steelwork. The labour in cutting and bonding these larger units is considerably less than with bricks. The blocks encasing steelwork are reinforced in every horizontal joint with steel mesh or expanded metal lath.

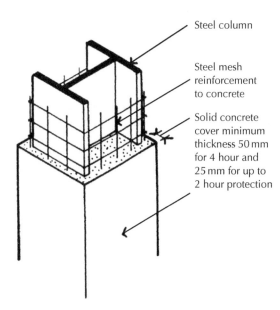

Steel column

Steel mesh
reinforcement
to concrete

Solid concrete
cover minimum
thickness 50 mm
for 4 hour and
25 mm for up to
2 hour protection

Figure 6.52   Concrete fire protection.

6.7 Floor construction for structural steel frames

Functional requirements

The functional requirements of floors are:

- ❏ Strength and stability
- ❏ Durability and freedom from maintenance
- ❏ Fire safety
- ❏ Resistance to the passage of heat (for party floors and exposed floors)
- ❏ Resistance to the passage of sound

Strength and stability

The requirements from the Building Regulations are that buildings be constructed so that the loadbearing elements, foundations, walls, floors and roofs have adequate strength and stability to support the dead loads of the construction and anticipated loads on roofs, floors and walls without such undue deflection or deformation as might adversely affect the strength and stability of parts or the whole of the building. Strength and stability of floors depend on the nature of the materials used in the floor elements, and the section of the materials used in resisting deflection (bending) under the dead and imposed loads. Under load, any horizontal element will deflect (bend) to an extent. Deflection under load is limited to about 1/300 of span to minimise cracking of rigid finishes to floors and ceilings and to limit the sense of insecurity the occupants might have, were the floor to deflect too obviously. In general, the strength and stability of a floor or roof is a product of the depth of the supporting members: the greater the depth, the greater the strength and stability.

Durability and freedom from maintenance

Durability and freedom from maintenance of floors constructed with steel beams, profiled steel decking and reinforced concrete depend on the internal conditions of the building. The majority of multi-storey-framed buildings today are heated and ventilated, so that there is little likelihood of moist internal conditions occurring, such as to cause progressive, destructive corrosion of steel during the useful life of the building.

Fire safety

The practical guidance given in Approved Document B is directed to the safe escape of people from buildings in case of fire rather than the protection of the building and its contents. Insurance companies that provide cover against the risks of damage to the buildings and contents by fire will generally require additional fire protection such as sprinklers and detection equipment.

Internal fire spread

Fire may spread within a building over the surface of materials covering walls and ceilings. Regulations prohibit the use of materials that encourage spread of flame across their surface when subject to intense radiant heat and those which give off appreciable heat when burning. Limits are set on the use of thermoplastic materials used in rooflights and lighting diffusers. As a measure of ability to withstand the effects of fire, the elements of a structure are given notional fire resistance times, in minutes, based on tests. Elements are tested for the ability to withstand the effects of fire in relation to:

❑   Resistance to collapse (loadbearing capacity), which applies to loadbearing elements.
❑   Resistance to fire penetration (integrity), which applies to fire-separating elements (e.g. floors).
❑   Resistance to the transfer of excessive heat (insulation), which applies to fire-separating elements.

The notional fire resistance times, which depend on the size, height and use of the building, are chosen as being sufficient for the escape of occupants in the event of fire. The requirements for the fire resistance of elements of a structure do not apply to:

❑   A structure that supports only a roof, unless:
     (a) the roof acts as a floor (e.g. car parking), or as a means of escape.
     (b) the structure is essential for the stability of an external wall, which needs to have fire resistance.
❑   the lowest floor of the building.

To prevent rapid fire spread that could trap occupants, and to reduce the chances of a fire growing large, it is necessary to subdivide buildings into compartments separated by walls and/or floors of fire-resisting construction. The degree of subdivision into compartments depends on:

❑   The use and fire load (contents) of the building.
❑   The height of the floor of the top storey as a measure of ease of escape and the ability of fire services to be effective.
❑   The availability of a sprinkler system, which can slow the rate of growth of fire.

The necessary compartment walls and/or floors should be of solid construction, sufficient to resist the penetration of fire for the stated notional period of time in minutes. The requirements for compartment walls and floors do not apply to single-storey buildings.

Smoke and flame may spread through concealed spaces, such as voids above suspended ceilings, roof spaces and enclosed ducts and wall cavities in the construction of a building. To restrict the unseen spread of smoke and flames through such spaces, cavity barriers and stops should be fixed as a tight-fitting barrier to the spread of smoke and flames.

Resistance to the passage of heat
The requirements for the conservation of power and fuel by the provision of adequate insulation of floors are described in *Barry's Introduction to Construction of Buildings*.

Resistance to the passage of sound
In multi-storey buildings, the structural frame may provide a ready path for the transmission of impact sound over some considerable distance. The slamming of a door, for example, can cause a sudden disturbing sound clearly heard some distance from the source of the sound by transmission through the frame and floor members. Unexpected sounds are

often more disturbing than continuous background sounds such as external traffic noise. To provide resistance to the passage of sound it is necessary to provide a break in the path between potential sources of impact and continuous solid transmitters. Solid concrete floors tend to be relatively good at not transmitting sound, compared to timber and steel; however, this is influenced by the type of floor finish.

Precast hollow floor beams

The precast, hollow, reinforced concrete floor units illustrated in Figure 6.53 are from 400 to 1200 mm wide, 110 to 300 mm thick for spans of up to 10 m for floors and 13.5 m for the less heavily loaded roofs. The purpose of the voids in the units is to reduce deadweight without affecting strength. The reinforcement is cast into the webs between the hollows. The wide floor units are used where there is powered lifting equipment, which can swing the units into place. These hollow floor units can be used as floor slabs with a non-structural levelling floor screed; alternatively, they may be used with a structural reinforced concrete topping with tie bars over beams for composite action with the concrete casing to beams. Raised floor finishes can also be applied directly to the unfinished surface.

The end bearing of these units is a minimum of 75 mm on steel shelf angles or beams and 100 mm on masonry and brick walls. The ends of these floor units are usually supported by steel shelf angles either welded or bolted to steel beams so that a part of the depth of the beam is inside the depth of the floor, as illustrated in Figure 6.54.

The ends of the floor units may be splayed to fit under the top flange of the beams. A disadvantage of the construction shown in Figure 6.54 is that the deep I-section beam projects some distance below the floor units and increases the overall height of construction for a given minimum clear height between the floor and the underside of the beam. Welded top hat profile beams with the floor units supported by the bottom flange, as illustrated in Figure 6.55, may be used to minimise the overall height of the construction.

The top hat section is preferred because of the difficulty of lowering and manoeuvring the units into the web of broad flange I-section beams. This construction method is

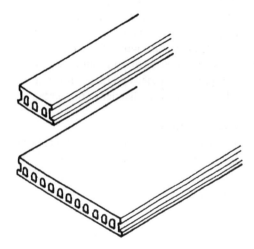

Figure 6.53   Hollow precast floor units.

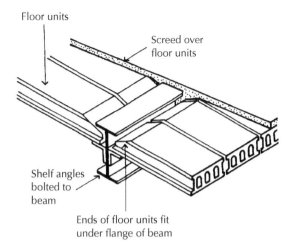

Floor units

Screed over
floor units

Shelf angles
bolted to
beam

Ends of floor units fit
under flange of beam

Figure 6.54   Hollow precast floor units on steel beam.

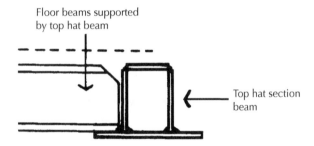

Floor beams supported
by top hat beam

Top hat section
beam

Figure 6.55   Top hat section beam.

particularly suited to multi-storey residential flats, where the comparatively small imposed loads on floors facilitate a combination of overall beam depth and floor units to minimise construction depth. A screed is spread over the floor for lightly loaded floors and roofs, and a reinforced concrete constructional topping for more heavily loaded floors.

Precast prestressed concrete floor units
Precast prestressed concrete floor units are comparatively thin, prestressed solid plank, concrete floor units are designed as permanent centring (shuttering) for composite action with structural reinforced concrete topping, as illustrated in Figure 6.56. The units are 400 and 1200 mm wide, 65, 75 or 100 mm thick and up to 9.5 m long for floors and 10 m for roofs. It may be necessary to provide some temporary propping to the underside of these planks until the concrete topping has gained sufficient strength. A disadvantage of this construction is that as the planks are laid on top of the beams so that the floor spans continuously over beams, there is increase in overall depth of construction from top of floor to underside of beams.

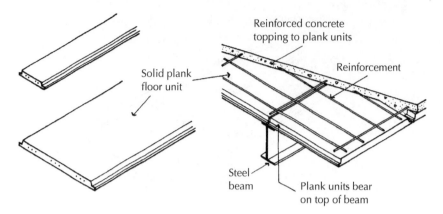

Figure 6.56   Prestressed solid plank floor unit.

Precast beam and filler block floor

The precast beam and filler block floor system of precast reinforced concrete beams or planks to support precast hollow concrete filler blocks is illustrated in Figure 6.57. For use with steel beams, the floor beams are laid between supports such as steel shelf angles fixed to the web of the beams or laid on the top flange of beams, and the filler blocks are then laid between the floor beams. The reinforcement protruding from the top of the planks acts with the concrete topping to form a continuous floor system spanning across the structural beams. These small beams or planks and filler blocks can be positioned without the need for heavy lifting equipment. This type of floor is most used in smaller-scale buildings supporting the lighter imposed floor loads common in residential buildings, for example.

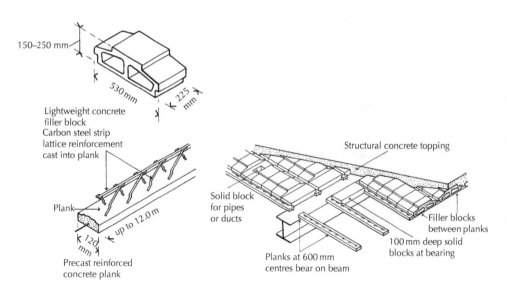

Figure 6.57   Precast beam and filter block floor.

Hollow clay block and concrete floor

The hollow clay block and concrete floor system, illustrated in *Barry's Introduction to Construction of Buildings,* consists of hollow clay blocks and in situ cast concrete reinforced as ribs between the blocks. This floor has to be laid on temporary centring to provide support until the in situ concrete has gained sufficient strength and is labour intensive.

Precast concrete T-beams

Precast concrete T-beam floors are mostly used for long-span floors and particularly roofs of such buildings as stores, supermarkets, swimming pools and multi-storey car parks, where there is a need for wide-span floors and roofs, and the depth of the floor is no disadvantage. The floor units are cast in the form of a double T, as illustrated in Figure 6.58. The strength of these units is in the depth of the tail of the T, which supports and acts with the comparatively thin top web. A structurally reinforced concrete topping is cast on top of the floor units.

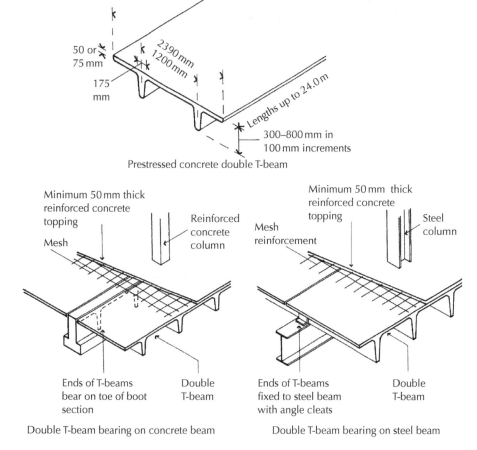

Prestressed concrete double T-beam

Double T-beam bearing on concrete beam   Double T-beam bearing on steel beam

Figure 6.58   Prestressed concrete double T-beam.

Cold-rolled steel deck and concrete floor

The traditional concrete floor to a structural steel frame consisted of reinforced concrete, cast in situ with the concrete casing to beams, cast on timber centring and falsework supported at each floor level until the concrete had sufficient strength to be self-supporting. The considerable material and labour costs in erecting and striking the support for the concrete floor led to the adoption of the precast concrete self-centring systems such as the hollow beam and plank, and beam and infill block floors. The term 'self-centring' derives from the word 'centring' used to describe the temporary platform of wood or steel on which in situ cast concrete is formed. The precast concrete beam, plank and beam, and block floors do not require temporary support, hence the term self-centring. A disadvantage of the precast concrete beam and plank floors for use with a structural steel frame is that it is usual practice to erect the steel frame in one operation. Raising the heavy, long precast concrete floor units and moving them into position is, to an extent, impeded by the skeleton steel frame.

Profiled cold-rolled steel decking

Profiled cold-rolled steel decking, as permanent formwork, acting as the whole or a part of the reinforcement to concrete has become the principal floor system for structural steel frames (see Photographs 6.11–6.14). The profiled steel deck is easily handled and fixed in place as formwork (centring) for concrete. The profiled cold roll-formed, steel sheet

Photograph 6.11   Steel deck floor.

Photograph 6.12   Composite construction – studs welded to the steel frame embedded in the concrete.

Photograph 6.13   Boxing out to leave a service void in the concrete floor.

Photograph 6.14 Permanent steel deck floor – formwork for the floor and edge of the composite floor.

decking, illustrated in Figure 6.59, is galvanised on both sides as a protection against corrosion. The profile is shaped to bond to the concrete, using projections that taper in from the top of the deck. Another profile is of trapezoidal section with chevron embossing for key to concrete. The steel deck may be laid on the top flange of beams, as illustrated in Figure 6.59, or supported by shelf angles bolted to the web of the beam to reduce overall height and fixed in position on the steelwork with shot-fired pins, self-tapping screws or by welding, with two fixings to each sheet. Side laps of deck are fixed at intervals of not more than 1 m with self-tapping screws or welding.

For medium spans between structural steel beams, the profiled steel deck acts as both permanent formwork and as reinforcement for the concrete slab that is cast in situ on the deck.

A mesh of anti-crack reinforcement is cast into the upper section of the slab, as illustrated in Figure 6.59. For long spans and heavy loads, the steel deck can be used with additional reinforcement cast into the bottom of the concrete between the upstanding profiles and, for composite action between the floor and the beams, shear studs are welded to the beams and cast into the concrete. The steel mesh reinforcement cast into the concrete slab floor is sufficient to provide protection against damage by fire in most situations. For high fire rating, the underside of the deck can be coated with sprayed-on protection or an intumescent coating. Where there is to be a flush ceiling for appearance and as a housing for services, a suspended ceiling is hung from hangers slotted into the profile or hangers bolted to the underside of the deck.

For particularly large spans or where cuts are made through the profile metal sheet for services, some temporary propping is required until the concrete has reached its seven-day maturity.

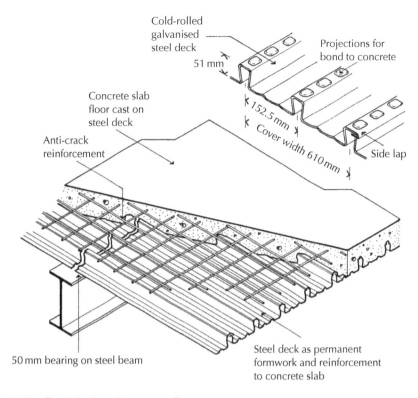

Cold-rolled
galvanised
steel deck

51 mm

Projections for
bond to concrete

152.5 mm

Cover width 610 mm

Side lap

Concrete slab
floor cast on
steel deck

Anti-crack
reinforcement

50 mm bearing on steel beam

Steel deck as permanent
formwork and reinforcement
to concrete slab

Figure 6.59   Steel deck and concrete floor.

Slimfloor floor construction

'Slimfloor' is the name adopted by British Steel (now Corus) for a form of floor construction
for skeleton steel-framed buildings. This form of construction is an adaptation of a form
of construction developed in Sweden, where restrictions on the overall height of buildings
dictated the development of a floor system with the least depth of floor construction to
gain the maximum number of storeys within the height limitations. Slimfloor construction comprises beams fabricated from universal column sections to which flange plates
are welded, as illustrated in Figure 6.60. The flange plates, which are wider overall than the
flanges of the beams, provide support to profiled steel decking that acts in part as reinforcement and provides support for the reinforced concrete constructional topping. The galvanised, profiled steel deck units are 210 mm deep with ribs at 600 mm centres. The ribs and
the top of the decking are ribbed to stiffen the plates and to provide some bond to concrete.
To seal the ends of the ribs in the decking, to contain the concrete that will be cast around
beams, sheet steel stop ends are fixed through the decking to the flange plates, as illustrated
in Figure 6.61. Constructional concrete topping is spread over the decking and into the ribs
around reinforcement in the base of the ribs and anti-crack reinforcement in the floor slab.

The galvanised pressed steel deck units are designed for spans of 6 m for use with the
typical grid of 9 m beam spans at 6 m centres. For spans of over 6 m and up to 7.5 m, the
decking will need temporary propping at mid-span until the concrete has developed

Cross section

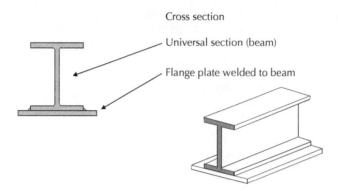

Universal section (beam)

Flange plate welded to beam

Figure 6.60   Slimfloor beam.

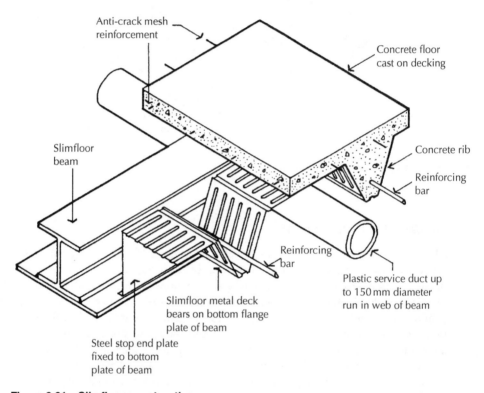

Anti-crack mesh reinforcement

Concrete floor cast on decking

Slimfloor beam

Concrete rib

Reinforcing bar

Reinforcing bar

Plastic service duct up to 150 mm diameter run in web of beam

Slimfloor metal deck bears on bottom flange plate of beam

Steel stop end plate fixed to bottom plate of beam

Figure 6.61   Slimfloor construction.

adequate strength. The slimfloor may be designed as a non-composite form of construction where the floor is assumed to have no composite action with the beams, as illustrated in Figure 6.61. This non-composite type of floor construction is usual where the imposed floor loads are low, as in residential buildings, and the floor does not act as a form of bracing to the structural frame. A particular advantage of the slimfloor is that all or some of the various services, common to some modern buildings, may be accommodated within the deck depth rather than being slung below the structural floor over a false ceiling. Calculations and tests have shown that 150 mm diameter holes may be cut centrally through the web of the beams at 600 mm spacing along the middle third of the length of the beam without significantly affecting the load-carrying capacity of the beam. Figure 6.61 is an illustration of the floor system showing a plastic tube sleeve run through the web of a beam for service pipes and cables.

The ceiling finish may be fixed to the underside of the decking or hung from the decking to provide space for services such as ducting. Because of the concrete encasement to the beams, most slimfloor constructions achieve one hour's fire resistance rating without the need for applied fire protection to the underside of the beam. Where fire resistance requirement is over 60 minutes, it is necessary to apply fire protection to the underside of the bottom flange plate.

The advantages of the slimfloor construction are:

❑   Speed of construction through ease of manhandling and ease of fixing the lightweight deck units, which provide a safe working platform.
❑   Pumping of concrete obviates the need for mechanical lifting equipment.
❑   The floor slab is lightweight as compared with in situ or precast concrete floors.
❑   The deck profile provides space for both horizontal services in the depth of the floor and vertical services through the wide top flange of the profile.
❑   Least overall depth of floor to provide minimum constructional depth consistent with robustness requirements dictated by design codes.

Composite construction

Composite construction is the name given to structural systems in which the separate structural characteristics and advantages of structural steel sections and reinforced concrete are combined as, for example, in the T-beam system. A steel frame, cased in concrete and designed to allow for the strength of the concrete in addition to that of the steel, is a form of composite construction. Where concrete encases steel sections, it is accepted that the stiffening and strengthening effect of the concrete on the steel can be allowed for in engineer's calculations. By reinforcing the concrete casing and allowing for its composite effect with the steel frame, a saving in steel and a reduction in the overall size of members can be achieved.

Shear stud connectors

A concrete floor slab bearing on a steel beam may be considered to act with the beam and serve as the beam's compressive flange, as a form of composite construction. This composite construction effect will work only if there is a sufficiently strong bond between the concrete and the steel, to make them act together in resisting shear stresses developed under load. The adhesion bond between the concrete and the top flange of the beam is not generally sufficient, and it is usually necessary to fix shear studs or connectors to the top flange of the beam, which are then cast in the floor slab. The purpose of these studs and connectors is to provide a positive resistance to shear. Figure 6.62 is an illustration of typical shear stud connectors and Figure 6.63 is an illustration of composite floor and beam construction.

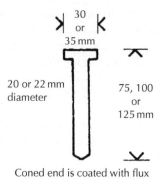

Figure 6.62   Shear stud connector.

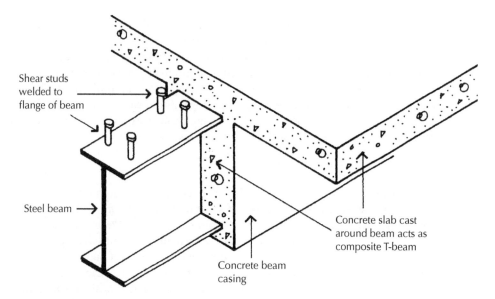

Figure 6.63   Composite construction.

Inverted T-beam composite construction

The composite beam and floor construction described earlier employ the standard I-section beams. The top flange of the beam is not a necessary part of the construction, as the concrete floor slab can be designed to carry the whole of the compressive stress so that the steel in the top flange of the beam is wastefully deployed. By using an inverted T-section member, steel is placed in the tension area and concrete in the compression area, where their characteristics are most useful. A cage of mild steel binders, cast into the beam casing and linked to the reinforcement in the floor slab, serves to make the slab and beam act as a form of composite construction by the adhesion bond of the concrete to the whole of the T-section.

Preflex beams

The use of high-tensile steel sections for long-span beams has been limited owing to the deflection of the beams under load, which causes cracking of concrete casing, and possible damage to partitions and finishes. Preflex beams are made by applying and maintaining loads, which are greater than working loads, to pairs of steel beams. In this deflected position, reinforced concrete is cast around the tension flanges of the beams. When the concrete has developed sufficient strength, the load is released and the beams tend to return to their former position. In so doing, the beams induce a compressive stress in the concrete around the tension flange, which prevents the beams from wholly regaining their original shape. The beams now have a slight upward camber. Under loads, the deflection of these beams will be resisted by the compressive stress in the concrete around the bottom flange, which will also prevent cracking of concrete. Further stiffening of the beam to reduce deflection is gained by the concrete casing to the web of the beam. By linking the reinforcement in the concrete web casing to the floor slab, the concrete and steel can be made to act as a composite form of construction. These beams may be connected to steel columns, with end plates welded to beam ends and bolted to column flanges, or may be cast into reinforced concrete columns.

Preflex beams are considerably more expensive than standard mild steel beams and are designed, in the main, for use in long-span heavily loaded floors. The slimfloor beam may be used in composite construction. For composite action, shear studs are welded to the top flange of a universal column section to which a wide bottom plate has been welded. This bottom plate serves as a bearing for hollow, precast reinforced concrete floor units. Structural concrete topping is spread and consolidated around the beam and as reinforced structural topping around transverse reinforcement, as illustrated in Figure 6.64. The result is a reinforced concrete floor acting compositely with the steel beam, the concrete casing tying to the beam and shear studs. The advantage of this construction is the least depth of floor of uniform depth. This type of floor is more expensive than a straightforward beam and slab floor.

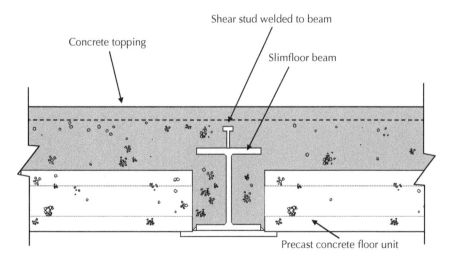

Figure 6.64   Composite floor construction.

Further reading

See, for example www.worldsteel.org and www.designingbuildings.co.uk.

Reflective exercises

Why would you choose a structural steel frame over a structural concrete or structural timber frame? What factors do you need to consider and how would you justify your decision to your client?

Your client is concerned about the amount of energy used to produce steel and has asked you to justify its use for a large development designed with a structural steel frame and lightweight steel modules. Your client is keen to use local materials and labour and aims to have zero-carbon building developments by 2030.

❏   How would you advise your client and how would you justify and support your reasoning?

Structural Concrete Frames: Chapter 7 AT A GLANCE

What? Structural concrete frames are constructed from reinforced concrete that is cast in situ, assembled from precast reinforced concrete components, or erected using a combination of in situ and precast reinforced concrete. The structural concrete frame is designed and constructed to produce a skeleton frame of columns, beams and floors onto which the walls and roof are fixed. Efforts are ongoing to reduce the embodied carbon of concrete and reduce carbon dioxide emissions associated with the production of cement. Concrete is a durable material and over the life of the building and the material (concrete and steel reinforcement) can be recovered and recycled at the end of the building's service life. Depending on the detailing and fixing, it may be possible to recover precast beams and columns to be reused.

Why? Reinforced structural concrete frames are inherently fire-resistant and offer good sound resistance and thermal resistance due to their mass. The framed structure allows freedom for positioning of internal partition walls and reasonable clear spans. Prestressed concrete beams provide the option for longer clear spans. Structural concrete frames are a popular choice for a variety of framed buildings, ranging from low-rise to high-rise construction. Framed construction allows considerable design freedom for the building envelope. External walls and glazing may be positioned between the structural concrete frame to express the structure. Alternatively, the external envelope may be positioned to completely cover the frame.

When? The structural concrete frame will be erected after the foundations are complete. The construction sequence provides a structural framework onto which the external envelope and roof are fixed. This provides a weatherproof envelope within which internal works can proceed unhindered by wind and rain.

How? The reinforced concrete frame will be constructed off the concrete foundations. Columns are erected on starter bars and joined together using concrete beams to form a skeleton framework onto which the floor is placed. This allows a safe working platform from which the next storey is constructed, allowing the structure to be constructed one storey at a time. Depending on the sequence and type of construction, it may be possible to fix the external wall cladding/envelope as each storey is completed. Alternatively, it may be more economical to install the building envelope once the structural frame is complete. Precast concrete components may be delivered to site as individual units (e.g. beams and columns), or they may be delivered as large wall and floor elements. The size of the pre-assembled elements is limited by transportation restrictions and physical access to the construction site.

7 Structural Concrete Frames

Reinforced concrete is one of the primary structural materials used in engineering and building works. Concrete has been used for a long time, but it was the pioneering work of French engineer Joseph Monier that led to a patent in 1867 for the process of strengthening concrete by embedding steel in it. Since the early days of reinforcing concrete with steel to enhance its strength, reinforced concrete has become a common material on construction sites, used extensively for civil engineering works and widely applied in the construction of buildings. Concrete is the most widely used construction material in the world given its versatility and practical applications. It is, however, responsible for approximately 8% of global CO_2 emissions and researchers and manufacturers are exploring ways to reduce the carbon footprint of cement and concrete. There may be alternatives to concrete that have a comparatively better carbon footprint, but given the versatility and durability of concrete it is not always possible or desirable to change to other materials. Concerns over the carbon footprint of the cement and concrete industry have led to innovations in production and materials science to reduce CO_2 emissions. Reinforced concrete is either placed in situ or manufactured as precast units offsite in factories and then delivered and installed on site. Reinforced concrete structures can be recycled at the end of the building's life: precast units can sometimes be removed with little damage occurring to the units and be reused, whereas aggregate and steel reinforcement can be recovered from the demolition of the in situ reinforced concrete.

7.1 Concrete

The materials used in the production of concrete are cement, aggregate and water.

Cement

The cement used today was first developed by Joseph Aspdin, a Leeds builder, who took out a patent in 1824 for the manufacture of Portland cement. He developed the material for the production of artificial stone and named it Portland cement because, in its hardened state, it resembled natural Portland limestone in texture and colour. The materials of Aspdin's cement, limestone and clay, were later burnt at a high temperature by Isaac Johnson in 1845, to produce a clinker which, ground to a fine powder, is what we now term Portland cement. The characteristics of cement depend on the proportions of the compounds of the raw

Barry's Advanced Construction of Buildings, Fifth Edition. Stephen Emmitt.
© 2023 John Wiley & Sons Ltd. Published 2023 by John Wiley & Sons Ltd.

materials used and the fineness of the grinding of the clinker, produced by burning the raw materials, as described later. Cement accounts for around 7% of global CO_2 emissions and the production of cement is responsible for around 95% of the carbon footprint of concrete, and alternatives to Portland cement are now available.

Ordinary Portland cement

Ordinary Portland cement is the cheapest and most commonly used cement, accounting for about 90% of all cement production. It is made by heating limestone and clay to a temperature of about 1300 °C to form a clinker, rich in calcium silicates. The clinker is ground to a fine powder with a small proportion of gypsum, which regulates the rate of setting when the cement is mixed with water. This type of cement is affected by sulphates such as those present in groundwater in some clay soils. The sulphates have a disintegrating effect on ordinary Portland cement. For this reason, sulphate-resisting cements are produced for use in concrete in sulphate-bearing soils, marine works, sewage installations and manufacturing processes where soluble salts are present.

Rapid hardening Portland cement

Rapid hardening Portland cement is similar to ordinary Portland, except that the cement powder is more finely ground. The effect of the finer grinding is that the constituents of the cement powder react more quickly with water, and the cement develops strength more rapidly. Rapid hardening cement develops in three days, a strength that is similar to that developed by ordinary Portland in seven days. With the advantage of the cement's early strength development, it is possible to speed up construction. With rapid hardening cement, the initial set is much shorter and formwork systems can be removed earlier. Although rapid hardening is more expensive than ordinary Portland cement, it is often used because of its early strength advantage. Rapid-hardening Portland cement is not a quick-setting cement. Several months after mixing there is little difference in the characteristics of ordinary and rapid hardening cement.

Sulphate-resisting Portland cement

The effect of sulphates on ordinary cement is to combine with the constituents of the cement. As the sulphates react there is an increase in volume on crystallisation, which causes the concrete to disintegrate. Disintegration is severe where the concrete is alternately wet and dry, as in marine works. To counteract this, the aluminates within the cement, which are affected by sulphates, are reduced to provide increased resistance to the effect of sulphates. Because it is necessary to carefully control the composition of the raw materials of this cement, it is more expensive than ordinary cement. High-alumina cement, described later, is also a sulphate-resisting cement.

White Portland cement

White Portland cement is manufactured from China clay and pure chalk or limestone and is used to produce white concrete finishes. Both the raw materials and the manufacturing process are comparatively expensive; therefore, the cement is mainly used for the surface of exposed concrete and for cement renderings. Pigments may be added to the cement to produce pastel colours.

Low-heat Portland cement

Low-heat Portland cement is used mainly for mass concrete works in dams and other constructions, where the heat developed by hydration of other cements would cause serious shrinkage cracking. The heat developed by the hydration of cement in concrete in construction works is dissipated to the surrounding air, whereas in large mass concrete works it dissipates slowly. Control of the constituents of low-heat Portland causes it to harden more slowly and therefore develop less rapidly than other cements. The slow rate of hardening does not affect the ultimate strength of the cement, yet allows the low heat of hydration to dissipate through the mass of concrete to the surrounding air.

Portland blast furnace cement

Portland blast furnace cement is manufactured by grinding Portland cement clinker with blast furnace slag, the proportion of slag being up to 65% by weight and the percentage of cement clinker no less than 35%. This cement develops heat more slowly than ordinary cement and is used in mass concrete works as a low-heat cement. It has good resistance to the destructive effects of sulphates and is commonly used in marine works.

Water-repellent cement

Water-repellent cement is made by mixing a metallic soap with ordinary or white Portland cement. Concrete made with this cement is more water repellent and therefore absorbs less rainwater than concrete made with other cements and is thus less liable to dirt staining. This cement is used for cast concrete and cast stone for its water-repellent property.

High-alumina (aluminous) cement

High-alumina (aluminous) cement is not one of the Portland cements. It is manufactured from bauxite and limestone or chalk in equal proportions. Bauxite is a mineral containing a higher proportion of alumina (aluminium oxide) than the clays used in the manufacture of Portland cements, hence the name given to this cement. The disadvantages of this cement are that there is a serious falling off in strength in hot moist atmospheres, and it is attacked by alkalis. This cement is little used for concrete in the UK.

Blended cement (low-carbon cement)

The terms 'blended cement' or 'low-carbon cement' are used to describe a combination of Portland cement and cement replacing material comprising industrial waste. This helps to reduce the carbon footprint of the Portland cement and also deal with an industrial by-product.

Aggregates

Concrete is a mix of particles of hard material, the aggregate, bound with a paste of cement and water, with at least three-quarters of the volume of concrete being occupied by aggregate. Aggregate for concrete should be hard, durable and contain no materials that are likely to decompose or change in volume or affect reinforcement. Clay, coal or pyrites in aggregate may soften, swell, decompose and cause stains in concrete. The aggregate should be clean and free from organic impurities and coatings of dust or clay, which would prevent the particles of aggregate from being adequately coated with cement and so lower the strength of the concrete.

Volume for volume, cement is generally more expensive than aggregate and it is advantageous, therefore, to use as little cement as necessary to produce a dense, durable concrete.

There is a direct relation between the density and strength of finished concrete and the ease with which concrete can be compacted. The characteristics of the aggregate play a considerable part in the ease with which concrete can be compacted. The measure of the ease with which concrete can be compacted is described as the workability of the mix. Workability is affected by the characteristics of the particles of aggregate such as size and shape so that, for a given mix, workability can be improved by careful selection of aggregate.

The grading of the size and the shape of the particles of aggregate affects the amount of cement and water required to produce a mix of concrete that is sufficiently workable to be compacted to a dense mass. The more cement and water that are needed for the sake of workability, the greater the drying shrinkage there will be by loss of water as the concrete dries and hardens.

Natural aggregates

Sand and gravel are the cheapest and most commonly used aggregate in the UK and consist of particles of broken stone deposited by the action of rivers and streams or from glacial action. Sand and gravel deposited by rivers and streams are generally more satisfactory than glacial deposits because the former comprise rounded particles in a wide range of sizes and weaker materials have been eroded by the washing and abrasive action of moving water. Glacial deposits tend to have angular particles of a wide variety of sizes, poorly graded, which adversely affect the workability of a concrete in which they are used.

Crushed rock aggregates are generally more expensive than sand and gravel, owing to the cost of quarrying and crushing the stone. Provided the stone is hard, inert and well-graded, it serves as an admirable aggregate for concrete. The term 'granite aggregate' is used commercially to describe a wide range of crushed natural stones, some of which are not true igneous rocks. Natural granite is hard and dense and serves as an excellent aggregate. Hard sandstone and close-grained crystalline limestone, when crushed and graded, are commonly used as aggregate in areas where sand and gravel are not readily available.

Because of the depletion of inland deposits of sand and gravel, marine aggregates are used. They are obtained by dredging deposits of broken stone from the bed of the sea. Most of these deposits contain shells and salt. Though not normally harmful in reinforced concrete, limits should be set to the proportion of shells and salt in marine aggregates used for concrete. One of the disadvantages of marine fine aggregate is that it has a preponderance of one size of particle, which can make design mix difficult. Sand from the beach is often of mainly single-sized particles and contains an accumulation of salts. Beach sands to be used as fine aggregate in concrete should be carefully washed to reduce the concentration of salts.

Artificial aggregates

Blast furnace slag is the by-product of the conversion of iron ore to pig iron and consists of the non-ferrous constituents of iron ore. The molten slag is tapped from the blast furnace and is cooled and crushed. In areas where there is a plentiful supply of blast furnace slag, it is an economical and satisfactory aggregate for concrete.

Clean broken brick is used as an aggregate for concrete required to have a good resistance to damage by fire. The strength of the concrete produced with this aggregate depends on the strength and density of the bricks from which the aggregate is produced. Crushed engineering brick aggregate will produce a concrete of medium crushing strength. Porous brick aggregate should not be used for reinforced concrete work in exposed positions, as the aggregate will absorb moisture and encourage the corrosion of the reinforcement.

Fine and coarse aggregate

'Fine aggregate' is the term used to describe natural sand, crushed rock and gravel, most of which passes through a number 5 British Standard (BS) sieve. 'Coarse aggregate' is the term used to describe natural gravel, crushed gravel or crushed rock, most of which is retained on a 5 BS sieve. The differentiation of fine and coarse aggregate is made because in practice the fine and coarse aggregate are ordered separately for mixing to produce a determined mix for particular uses and strengths of concrete.

Grading of aggregate

The word 'grading' is used to describe the percentage of particles of a particular range of sizes in a given aggregate, from fines (sand) to the largest particle size. A sound concrete is produced from a mix that can be readily placed and compacted in position; that is, a mix that has good workability and after compaction is reasonably free of voids. This is affected by the grading of the aggregate and the water/cement ratio.

The grading of aggregate is usually given by the percentage by weight passing the various sieves used for grading. Continuously graded aggregate should contain particles graded in size from the largest to the smallest to produce a dense concrete. Sieve sizes from 75 to 5 mm (from 3 to $^3/_{16}$ inch) are used for coarse aggregate. An aggregate containing a large proportion of large particles is referred to as being 'coarsely' graded and one having a large proportion of small particles as 'finely' graded.

Particle shape and surface texture

The shape and surface texture of the particles of an aggregate affect the workability of a concrete mix. An aggregate with angular edges and a rough surface, such as crushed stone, requires more water in the mix to act as a lubricant to facilitate compaction than does one with rounded smooth faces to produce a concrete of the same workability. It is often necessary to increase the cement content of a mix made with crushed aggregate or irregularly shaped gravels to provide the optimum water/cement ratio to produce concrete of the necessary strength. This additional water, on evaporation, tends to leave small void spaces in the concrete, which will be less dense than concrete made with rounded particle aggregate. The addition of extra water, beyond that required in the chemical reaction (hydration), will weaken the concrete. Water that does not take part in the chemical reaction leaves voids as it evaporates out of the concrete.

The nature of the surface of the particles of an aggregate will affect workability. Gravel dredged from a river will have smooth surfaced particles, which will afford little frictional resistance to the arrangement of particles that takes place during compaction of concrete. A crushed granite aggregate will have coarse surfaced particles that will offer some resistance during compaction. The shape of particles of aggregate is measured by an angularity index and the surface by a surface coefficient. Engineers use these to determine the true workability of a concrete mix, which cannot be judged solely from the grading of particles.

Water

Water for concrete should be reasonably free from such impurities as suspended solids, organic matter and dissolved salts, which may adversely affect the properties of concrete. Water that is fit for drinking is accepted as being satisfactory for mixing water for concrete.

Low-carbon concrete

The term 'low-carbon concrete' is used to describe concrete mixes that contain a very small amount of cement, and hence have a lower carbon footprint compared to traditional concrete. Although manufacturers have taken measures to reduce the amount of energy used and the amount of CO_2 produced in the production of cement, it still remains a concern for many, prompting the drive to reduce the amount of cement contained in concrete mixes. The cement is replaced, or partially replaced, with low-carbon binders and/or recycled materials such as industrial waste. It is claimed that the carbon footprint is reduced by up to 50% compared to traditional concrete.

7.2 Concrete mixes

The strength and durability of concrete are affected by the voids in the concrete caused by poor grading of aggregate, incomplete compaction or excessive water in the mix.

Water/cement ratio

Workability

The materials used in concrete are mixed with water for two reasons: first, to enable the reaction with the cement, which causes setting and hardening to take place; and second, to act as a lubricant to render the mix sufficiently plastic for placing and compaction.

About a quarter part by weight of water to one part by weight of cement is required for the completion of the setting and hardening process. This proportion of water to cement will result in a concrete mix far too stiff (dry) to be adequately placed and compacted. About a half by weight of water to one part by weight of cement is required to make a concrete mix workable. The greater the proportion of water to cement used in a concrete mix, the weaker the ultimate strength of the concrete. The principal reason for this is that the water, in excess of the amount required to complete the hardening of the cement, evaporates and leaves voids in the concrete, which reduces its strength. It is usual practice, therefore, to define a ratio of water to cement in concrete mixes to achieve a dense concrete. The water/cement ratio is expressed as the ratio of water to cement by weight and the limits of this ratio for most concrete lie between 0.4 and 0.65. Outside these limits, there is a great loss of workability below the lower figure and a loss of strength of concrete above the upper figure.

Water-reducing admixtures

The addition of 0.2% by weight of calcium lignosulphonate, commonly known as 'lignin', to cement will reduce the amount of water required in concrete by 10% without loss of workability. This allows the cement content of a concrete mix to be reduced for a given water/cement ratio. Calcium lignosulphonate acts as a surface-active additive that disperses the cement particles, which then need less water to lubricate and disperse them in concrete.

Water-reducing admixtures such as lignin are promoted by suppliers as densifiers, hardeners, water proofers and plasticisers, on the basis that the reduction of water content leads to a denser concrete due to there being fewer voids after the evaporation of water. To ensure that the use of these admixtures does not adversely affect the durability of a concrete, it is usual practice to specify a minimum content of cement.

Nominal mixes

Volume batching

The constituents of concrete may be measured by volume in batch boxes, in which a nominal volume of aggregate and a nominal volume of cement are measured for a nominal mix, as for example, in a mix of 1 : 2 : 4 of cement : fine : coarse aggregate. A batch box usually takes the form of an open-top wooden box in which volumes of cement, fine and coarse aggregate are measured separately for the selected nominal volume mix. For a mix such as 1 : 2 : 4, one batch box will suffice, the mix proportions being gauged by the number of fillings of the box with each of the constituents of the mix.

Measuring the materials of concrete by volume is not an accurate way of proportioning and cannot be relied on to produce concrete with a uniformly high strength. Cement powder cannot be accurately proportioned by volume because, while it may be poured into and fill a box, it can be readily compressed to occupy considerably less space. Proportioning aggregates by volume takes no account of the amount of water retained in the aggregate, which may affect the water/cement ratio of the mix and affect the proportioning because wet sand occupies a greater volume than does the same amount of sand when dry. Volume batch mixing is mostly used for the concrete for the foundations and oversite concrete of small buildings such as houses. In these cases, the concrete is not required to suffer any large amounts of stress, and the strength and uniformity of the mix is relatively unimportant. The scale of the building operation does not justify more exact methods of batching.

Weight batching

A more accurate method of proportioning the materials of concrete is by weight batching, by proportioning the fine and coarse aggregate by weight by reference to the weight of a standard bag of cement. Where nominal mixes are weight batched, it is best to take samples of the aggregate and dry them to ascertain the weight of water retained in the aggregate and so adjust the proportion of water added to the mix to allow for the water retained in the aggregate. Water can be proportioned by volume or by weight.

Designed mixes

Designed mixes of concrete are those where strength is the main criterion of the specified mix, which is judged on the basis of strength tests. The position in which concrete is to be placed, the means used and the ease of compacting it, the nature of the aggregate and the water/cement ratio all affect the ultimate strength of concrete. A designed concrete mix is one where the variable factors are adjusted (selected) by the engineer to produce a concrete with the desired minimum compressive strength at the lowest possible cost. If, for example, the cheapest available local aggregate in a particular district will not produce a very workable mix, it would be necessary to use a wet mix to facilitate placing and compaction, and this in turn would necessitate the use of a cement-rich mix to maintain a reasonable water/cement ratio. In this example, it might be cheaper to import a different aggregate, more expensive than the local one, which would produce a comparatively dry but workable mix requiring less cement. These are the considerations the engineer and the contractor have in designing a concrete mix.

Prescribed mixes and standard mixes

Prescribed mixes and standard mixes are mixes of concrete where the constituents are of fixed proportion by weight, to produce a 'grade' of concrete with minimum characteristics strength.

Mixing, placing and compacting concrete

Mixing concrete

Concrete may be mixed by hand when the volume to be used does not warrant the use of a mechanical mixing plant. The materials are measured out by volume in timber gauge boxes, turned over on a clean surface several times dry and then water is added. The mix is turned over again until it has a suitable consistency and uniform colour. It is obviously difficult to produce mixes of uniform quality by hand mixing. A small hand-tilting mixer is often used. The mixing drum is rotated by a petrol or electric motor, the drum being tilted by hand to fill and empty it. This type of mixer takes over a deal of the backbreaking work of mixing but does not control the quality of mixes as materials are measured by volume.

A concrete batch mixer mechanically feeds the materials into the drum where they are mixed and from which the wet concrete is poured. The materials are batched by either weight or volume. For extensive works, plant is installed on site that stores cement (delivered in bulk), measures the materials by weight and mechanically mixes them. Concrete for high-strength reinforced concrete work can only be produced from batches (mixes) of uniform quality. Such mixes are produced by plants capable of accurately measuring and thoroughly mixing the materials.

Ready-mixed concrete

Ready-mixed concrete is extensively used today. It is prepared in mechanical, concrete mixing depots where the materials are stored, weight-batched and mixed, and the wet concrete is transported to site in rotating drums mounted on lorries (cement mixers). The action of the rotating drum prevents aggregates from segregating and the concrete from setting and hardening for an hour or more. Once delivered it must be placed and compacted quickly as it rapidly hardens.

Placing and compacting concrete

The initial set of Portland cement takes place from half an hour to one hour after it is mixed with water. If a concrete mix is disturbed after the initial set has occurred, the strength of the concrete may be adversely affected. It is usual to specify that concrete be placed as soon after mixing as possible and not more than half an hour after mixing.

A concrete mix consists of particles varying in size from powder to coarse aggregate graded to, say, 40 mm. If a wet mix of concrete is poured from some height and allowed to fall freely, the larger particles tend to separate from the smaller. This action is termed 'segregation of particles'. Concrete should not, therefore, be tipped or poured into place from too great a height. It is usual to specify that concrete is to be placed from a height not greater than 1 m.

Once in place, concrete should be thoroughly consolidated or compacted. The purpose of compaction is to cause entrapped bubbles of air to rise to the surface to produce as dense and void-free concrete as possible. Compaction may be effected by agitating the mix with a spade or heavy iron bar. If the mix is dry and stiff, this is a very laborious process and not very effective. A more satisfactory method is to employ a pneumatically operated poker vibrator, which is inserted into the concrete and, by vibration, liberates air bubbles and compacts the concrete. As an alternative, the formwork of reinforced concrete may be vibrated by means of a motor attached to it.

Construction joints

Because it is not possible to place concrete continuously (on the vast majority of construction sites), it is necessary to form construction joints. A construction joint is the junction of freshly placed concrete with concrete that has been placed and set, for example, concrete poured on the previous day. These construction joints are a potential source of weakness, because there may not be a good bond between the two placings of concrete. When forming a construction joint, the previously placed concrete needs to be clean, with a sound surface exposed. The top surface of the concrete is usually broken away by means of a mechanical scabbier. This hammers the surface of the set concrete breaking away the loose surface, leaving a clean, surface for the new concrete to form a mechanical and chemical bond. There should be as few construction joints as practical and joints should be either vertical or horizontal. Joints in columns are made as near as possible to beam haunching, and those in beams at the centre or within the middle third of the span. Vertical joints are formed against a strip board. Water bars are fixed across or cast into construction joints where there is a need to provide a barrier to the movement of water through the joint (see Chapter 3).

Curing concrete

Concrete gradually hardens and gains strength after its initial set. For this hardening process to proceed and the concrete to develop its maximum strength, there must be water present in the mix. If during the early days after the initial set, there is too rapid a loss of water, the concrete will not develop its maximum strength. The process of preventing a rapid loss of water is termed 'curing concrete'. Large exposed areas of concrete such as road surfaces are cured by covering the surface for at least a week after placing, with building paper, plastic sheets or wet sacks to retard evaporation of water. In very dry weather, the surface of concrete may have to be sprayed with water in addition to covering it. The formwork around reinforced concrete is often kept in position for some days after the concrete is placed, to give support until the concrete has gained sufficient strength to be self-supporting. This formwork also serves to prevent too rapid a loss of water and so helps to cure the concrete. In very dry weather, it may be necessary to spray the formwork to compensate for too rapid a loss of water.

Specially designed curing agents can also be used. These are chemical liquids that are sprayed over the concrete to produce a thin film that effectively seals the water, needed for hydration, within the concrete.

Self-compacting concrete (SCC)

Self-compacting concrete (SCC) is a concrete that does not require vibration for placing and compaction. SCC was first developed in Japan in the late 1980s and has since become popular because it offers a rapid rate of concrete placement and hence faster construction times. Vibration equipment is not required, helping to reduce the noise and vibration suffered by construction workers. SCC also has the benefit of being easier to place around closely spaced reinforcement compared with normal concrete, due to its ease of flow. The engineering properties of SCC are similar to concrete for the same specification, although the surface finish is usually of a higher quality.

Additions are used to improve and maintain the cohesion and segregation resistance of SCC. The additions range from inert mineral fillers (e.g. limestone), to pozzolanic (fly ash, silica fume) and hydraulic fillers (ground granulated blast furnace slag). Admixtures such as superplasticisers are an essential component of SCC, helping to bring about the water reduction and improve fluidity. Polymer fibres may be added to improve the stability of the SCC.

SCC is delivered ready mixed direct to site by the manufacturer and pumped to its required position. SCC must be placed in one pour without a break in placing to maintain its integrity. Because of the special qualities of SCC, it is essential that site personnel are trained in the specific requirements of placing SCC and that adequate supervision is in place to monitor the pour. Guidance is provided by manufacturers into the rate of placing and measures to take if a problem occurs while the SCC is being placed, such as a stoppage in the flow of the SCC.

SCC can also be used in the manufacture of precast concrete products, especially when a high-quality surface appearance is required.

Deformation of concrete

Hardened concrete will suffer deformation due to:

❑ Elastic deformation that occurs instantaneously and is dependent on applied stress.
❑ Drying shrinkage that occurs over a long period and is independent of the stress in concrete.
❑ Creep, which occurs over a long period and is dependent on stress in concrete.
❑ Expansion and contraction due to changes in temperature and moisture.
❑ Alkali–silica reaction (ASR)

Elastic deformation
Under the stress of dead and applied loads of a building, hardened concrete deforms elastically. Vertical elements such as columns and walls are compressed and shortened in height, and horizontal elements such as beams and floors lengthen due to bending. These comparatively small deformations, which are related to the strength of the concrete, are predictable and allowance is made in design.

Drying shrinkage
The drying shrinkage of concrete is affected principally by the amount of water in the concrete at the time of mixing and to a lesser extent by the cement content of the concrete. It can also be affected by a porous aggregate losing water. Drying shrinkage is restrained by the amount of reinforcement in concrete. The rate of shrinkage is affected by the humidity and temperature of the surrounding air, the rate of airflow over the surface and the proportion of surface area to volume of concrete. Where concrete dries in the open air in summer,

small masses of concrete will suffer about a half of the total drying shrinkage a month after placing, and large masses about a half of the total shrinkage a year after placing. Shrinkage will not generally affect the strength or stability of a concrete structure but is sufficient to require the need for movement joints where solid materials such as brick and block are built up to the concrete frame.

Creep

Under sustained load, concrete deforms as a result of the mobility of absorbed water within the cement gel under the action of sustained stress. From the point of view of design, creep may be considered as an irrecoverable deformation that occurs with time at an ever decreasing rate under the action of sustained load. Creep deformation continues over very long periods of time to the extent that measurable deformation can occur 30 years after the concrete has been placed. The factors that affect creep of concrete are the concrete mix, relative humidity and temperature, size of member and applied stress.

Most aggregates used in dense concrete are inert and do not suffer creep deformation under load. The hardened cement water paste surrounding the particles of aggregate is subject to creep deformation under stress due to movements of absorbed water. The relative volume of cement gel to aggregate, therefore, affects deformation due to creep. Changing from a 1 : 1 : 2 to a 1 : 2 : 4 cement, fine and coarse aggregate mix increases the volume of aggregate from 60% to 75%, yet causes a reduction in creep by as much as 50%.

Temperature, relative humidity and the size of members have an effect on the hydration of cement and migration of water around the cement gel towards the surface of concrete. In general, creep is greater the lower the humidity and increases with a rise in temperature. Small section members of concrete will lose water more rapidly than large members and will suffer greater creep deformation during the period of initial drying. The effect of creep deformation has the most serious effect through stress loss in prestressed concrete, deflection increase in large-span beams, buckling of slender columns and buckling of cladding in tall buildings.

Alkali–silica reaction

The chemical reaction of high silica-content aggregate with alkaline cement causes a gel to form, which expands and causes concrete to crack. The expansion, cracking and damage to concrete is often most severe where there is an external source of water in large quantities. Foundations, motorway bridges and concrete subject to heavy condensation have suffered severe damage through ASR. The expansion caused by the gel formed by the reaction is not uniform in time or location. The reaction may develop slowly in some structures yet very rapidly in others and may affect one part of a structure but not another. Changes in the method of manufacture of cement, which has produced a cement with higher alkalinity, are thought to be one of the causes of some noted failures. To minimise the effect of ASR, it is recommended that cement-rich mixes and high silica-content aggregates be avoided.

Self-healing concrete

'Self-healing concrete' and 'bio-concrete' are terms used to describe concrete that has the capacity to repair its own cracks, thus helping to prevent deterioration. All concrete will deteriorate over time and their susceptibility to cracking leads to water entering the structure and leads to corrosion of the reinforcement and then spalling of the concrete. This is expensive to repair and or replace. The principle of self-healing concrete is similar to that of a human body healing a fracture in a bone through mineralisation. Research into the effectiveness and practicalities of self-healing concrete is ongoing, although at present there are four approaches that use bacteria, mineral-based agents, shape-memory polymers and pumped healing agents.

❏ Concrete is mixed with bacteria that biologically interact with a nutrient (starch) to produce calcite that helps to close up small cracks in the concrete. Alternatively, the bacteria, calcium lactate, nitrogen and phosphorus are sealed into biodegradable clay capsules. These capsules are mixed into the concrete and the bacteria remain dormant until they are activated by water as it enters cracks in the concrete. As the bacteria from the genus bacillus feed on their food source, they excrete calcite that helps to heal the cracks in the concrete structure.
❏ Mineral-based agents contained in microcapsules are mixed into the concrete. Minerals are released after contact with water to help seal tiny cracks in the concrete, similar to the biological approach.
❏ *Shape-memory polymers.* Shape-shifting materials are embedded into the concrete mix. When repair is required the polymers are heated by passing a small electrical current through the structure and the material transforms into a memorised shape to fill the cracks.
❏ *Pumped healing agents.* This technique relies on pumping healing agents through a flow network, a network of thin channels within the concrete.

The advantage of self-healing concrete is primarily that it helps to reduce the frequency and cost of maintenance and repair. The disadvantages primarily relate to cost and the technical challenges or, for example, embedding sufficient clay capsules to be effective without compromising the strength of the concrete. Research is ongoing into self-healing mortars that can be applied to the face of the concrete structure once deterioration has been detected.

7.3 Reinforcement

Concrete is strong in resisting compressive stress but comparatively weak in resisting tensile stress. The tensile strength of concrete is between one-tenth and one-twentieth of its compressive strength. Steel, which has good tensile strength, is cast into reinforced concrete members in the position or positions where maximum tensile stress occurs. To determine where tensile and compressive stresses occur in a structural member, it is convenient to consider the behaviour of an elastic material under stress. A bar of rubber laid across (not fixed) two supports will bend under load. The top surface will shorten and become compressed under stress, while the bottom surface becomes stretched under tensile stress, as illustrated in Figure 7.1.

A member that is supported so that the supports do not restrain bending while under load is termed 'simply supported'. From Figure 7.1 it will be seen that maximum stretching due to tension occurs at the outwardly curved underside of the rubber bar. If the bar were

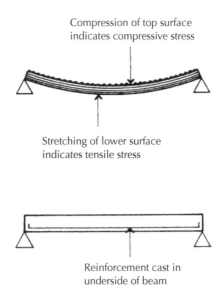

Figure 7.1   Simply supported beam.

of concrete, it would seem logical to cast steel reinforcement on the underside of the bar. In that position, the steel would be exposed to the surrounding air and it would rust and gradually lose strength. Furthermore, if a fire occurred in the building, near to the beam, the steel might lose so much strength as to impair its reinforcing effect and the beam would collapse. It is usual practice, therefore, to cast the steel reinforcement into concrete so that there is at least 15 mm of concrete cover between the reinforcement and the surface of the concrete.

Concrete cover

For internal concrete structures, a 15 mm cover of concrete is considered sufficient to protect steel reinforcement from corrosion. External members need considerably more cover; in areas up to where reinforced concrete is exposed to seawater and abrasion, the concrete cover to the reinforcement should be 60 mm. From laboratory tests and experience of damage caused by fires in buildings, it has been established that various thicknesses of concrete cover will prevent an excessive loss of strength in steel reinforcement for particular periods of time. The presumption is that the concrete cover will protect the reinforcement for a period of time for the occupants to escape from the particular building during a fire. The statutory period of time for the concrete cover to provide protection against damage by fire varies with the size and type of building, from half an hour to four hours.

Bond and anchorage of reinforcement

The cement in concrete cast around steel reinforcement adheres to the steel just as it does to the particles of the aggregate, and this adhesion plays its part in the transfer of tensile stress from the concrete to the steel. It is important, therefore, that the steel reinforcement be clean and free from scale, rust and oily or greasy coatings. Under load, tensile stress tends

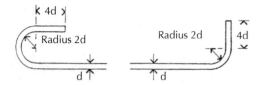

Figure 7.2 Hooked ends for reinforcing bars.

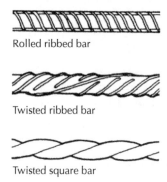

Rolled ribbed bar

Twisted ribbed bar

Twisted square bar

Figure 7.3 Deformed reinforcing bars.

to cause the reinforcement to slip out of bond with the surrounding concrete due to the elongation of the member. This slip is resisted partly by the adhesion of the cement to the steel and by the frictional resistance between steel and concrete. To secure a firm anchorage of reinforcement to concrete and to prevent slip, it is usual practice to hook or bend the ends of bars, as illustrated in Figure 7.2.

As an alternative to hooked or bent ends of reinforcing bars, deformed bars may be used. The simplest form of bar is the twisted, square bar illustrated in Figure 7.3, which through its twisted surface presents some resistance to slip. The round section, ribbed bar and the twisted, ribbed bars provide resistance to slip in concrete by the many projecting ribs illustrated in Figure 7.3. Deformed bars, which are more expensive than plain round bars, are used for heavily loaded structural concrete.

Shear

Beams are subjected to shear stresses due to the shearing action of the supports, and the self-weight and imposed loads of beams. Shear stress is greatest at the points of support and zero at mid-span in uniformly loaded beams. Shear failure occurs at an angle of 45°, as illustrated in Figure 7.4. Due to its poor tensile strength, concrete does not have great shear resistance and it is usual to introduce steel shear reinforcement in most beams of over, say, 2.5 m span.

To maintain the main reinforcing bars in place while concrete is being placed and until it has hardened, it is usual practice to use a system of stirrups or links, which are formed from light section reinforcing bars. These rectangular stirrups are attached by binding wire to the main reinforcement. To provide shear reinforcement at points of support in beams, the stirrups are more closely spaced, as illustrated in Figure 7.4.

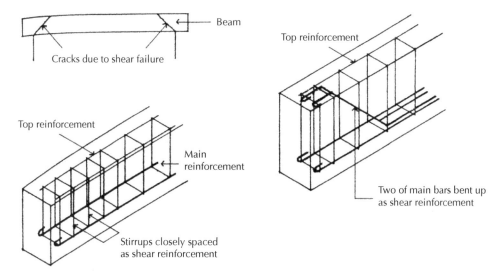

Figure 7.4   Shear reinforcement.

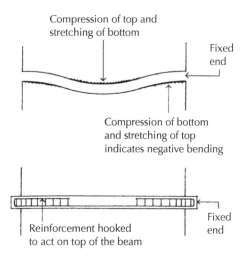

Figure 7.5   Beams with fixed ends (top reinforcement omitted for clarity).

Fixed end support

A beam with pin-jointed end support will suffer simple bending under load, whereas a beam with fixed end support is restrained from simple bending by the fixed ends, as illustrated in Figure 7.5. Because of the upward, negative bending close to the fixed ends, the top of the beam is in tension and the underside at mid-span is in tension due to positive bending. In a concrete beam with fixed ends, it is not sufficient to cast reinforcement into the lower face of the beam only, as the concrete will not have sufficient tensile strength to resist tensile stresses in the top of the beam near points of support. Both top and bottom reinforcement are necessary, as illustrated in Figure 7.5, where the main bottom reinforcement

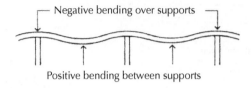

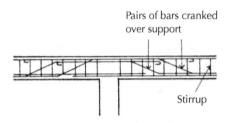

Figure 7.6   Reinforced concrete beam to span continuously over supports.

is bent up at the support ends and continued as top reinforcement along the length of the main tensile end support. Top reinforcement has been omitted for the sake of clarity.

Because of the fixed end support, the upward negative bending at supports will cause some appreciable deformation bending of columns around the connection of beam to column. Where a beam is designed to span continuously over several supports, as illustrated in Figure 7.6, it will suffer negative, upward bending over the supports. At these points, the top of the beam will suffer tensile stress and additional top reinforcement will be necessary. Here additional top reinforcement is used against tensile stress, and the bottom reinforcement is cranked up over the support to provide shear reinforcement, as illustrated in Figure 7.6.

Cantilever beams

A cantilever beam projects from a wall or structural frame. The cantilever illustrated in Figure 7.7 may take the form of a reinforced concrete slab projecting from a building as a balcony or as several projecting cantilever beams projecting from a structural frame to

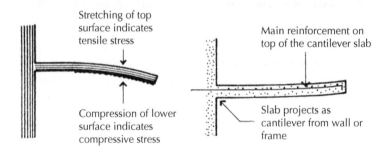

Figure 7.7   Cantilever slab.

support a reinforced concrete slab. As a simple explanation of the stress in a fixed-end cantilever, assume it is made of rubber. Under load, the rubber will bend, as illustrated in Figure 7.7. The top surface of the rubber will stretch and the bottom surface will be compressed, indicating tensile stress in the top and compressive stress in the bottom. Under load, a reinforced concrete cantilever will suffer similar but less obvious bending with the main reinforcement in its top, as illustrated in Figure 7.7, with the necessary cover of reinforcement. Under appreciable load, shear reinforcement will also be necessary close to the point of support.

Reinforcement to concrete columns

Columns are designed to support the loads of roofs, floors and walls. If all these loads acted concentrically on the section of the column, then it would suffer only compressive stress and it would be sufficient to construct the column of either concrete by itself or of reinforced concrete to reduce the required section area. In practice, the loads of floor and roof beams, and walls and wind pressure, act eccentrically; that is, off the centre of the section of columns, and so cause some bending and tensile stress in columns. The steel reinforcement in columns is designed primarily to sustain compressive stress to reinforce the compressive strength of concrete, but also to reinforce the poor tensile strength of concrete against tensile stress due to bending from fixed end beams, eccentric loading and wind pressure.

Mild steel reinforcement

The cheapest and most commonly used reinforcement is round section mild steel rods of diameter from 6 to 40 mm. These rods are manufactured in long lengths and can be quickly cut and easily bent without damage. The disadvantages of ordinary mild steel reinforcement are that if the steel is stressed up to its yield point, it suffers permanent elongation; if exposed to moisture, it progressively corrodes; and on exposure to the heat generated by fires, it loses strength.

In tension, mild steel suffers elastic elongation, which is proportional to stress up to the yield stress, and it returns to its former length once stress is removed. At yield stress point, mild steel suffers permanent elongation and then, with further increase in stress again, suffers elastic elongation. If the permanent elongation of mild steel, which occurs at yield stress, was to occur in reinforcement in reinforced concrete, the loss of bond between the steel and the concrete and consequent cracking of concrete around reinforcement would be so pronounced as to seriously affect the strength of the member. For this reason, maximum likely stresses in mild steel reinforcement are kept to a figure some two-thirds below yield stress. In consequence, the mild steel reinforcement is working at stresses well below its ultimate strength.

Cold-worked steel reinforcement

If mild steel bars are stressed up to yield point and permanent plastic elongation takes place and the stress is then released, subsequent stressing up to and beyond the former yield stress will not cause a repetition of the initial permanent elongation at yield stress. This change of behaviour is said to be due to a reorientation of the steel crystals during the initial stress at yield point. In the design of reinforced concrete members, using this type of reinforcement, maximum stress need not be limited to a figure below yield stress, to avoid loss of bond between concrete and reinforcement, and the calculated design stresses may be considerably higher than with ordinary mild steel. In practice, it is convenient to simultaneously stress cold-drawn steel bars up to yield point and to twist them axially to produce cold-worked deformed bars with improved bond to concrete.

Deformed bars

To limit the cracks that may develop in reinforced concrete around mild steel bars, due to the stretching of the bars and some loss of bond under load, it is common to use deformed bars that have projecting ribs or are twisted to improve the bond to concrete. The types of deformed reinforcing bars generally used are ribbed bars that are rolled from mild steel and ribbed along their length, ribbed mild steel bars that are cold drawn as high-yield ribbed bars, ribbed cold drawn and twisted bars, high-tensile steel bars that are rolled with projecting ribs, and cold twisted square bars. Figure 7.3 is an illustration of some typical deformed bars.

Galvanised steel reinforcing bars

Where reinforced concrete is exposed externally or is exposed to corrosive industrial atmospheres, it is sound practice to use galvanised reinforcement as a protection against corrosion of the steel to prevent rust staining of fairface finishes and inhibit rusting of reinforcement that might weaken the structure. The steel reinforcing bars are cut to length, bent and then coated with zinc by the hot dip galvanising process.

Stainless steel reinforcement

Stainless steel is an alloy of iron, chromium and nickel on which an invisible corrosion-resistant film forms on exposure to air. Stainless steel is about 10 times the cost of ordinary mild steel. It is used for reinforcing bars in concrete where the cover of concrete for corrosion protection would be much greater than that required for fire protection and the least section of reinforced concrete is a critical consideration.

Assembling and fixing reinforcement

Reinforcement for structural beams and columns is usually assembled in the form of a cage within temporary or permanent formwork, with the main and secondary reinforcement being fixed to links or stirrups that hold it in position. The principal purpose of these links is to secure the longitudinal reinforcing bars in position when concrete is being placed and compacted. They also serve to some extent in anchoring reinforcement in concrete and in addition provide some resistance to shear, with closely spaced links at points of support in beams. Links are formed from small section reinforcing bars that are cut and bent to contain the longitudinal reinforcement. Stirrups or links are usually cold bent to contain top and bottom longitudinal reinforcement to beams and the main reinforcement to columns with the ends of each link overlapping, as illustrated in Figure 7.8. As an alternative, links may be formed from two lengths of bar, the main U-shaped part of the link and a top section, as illustrated in Figure 7.8. The advantage of this arrangement of links is that where there are several longitudinal reinforcing bars in a cage, they can be dropped in from the top of the links rather than being threaded through the links as the cage is wired up, thus

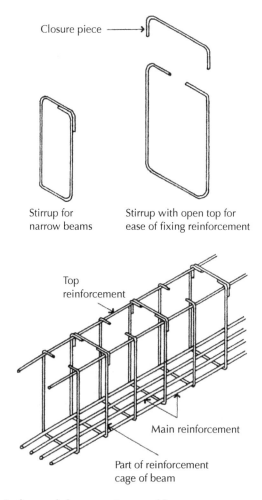

Closure piece ⟶

Stirrup for
narrow beams

Stirrup with open top for
ease of fixing reinforcement

Top
reinforcement

Main reinforcement

Part of reinforcement
cage of beam

Figure 7.8   Stirrups to form reinforcement cage of beams.

saving time. Figure 7.8 is an illustration of part of a reinforcement cage for a reinforced concrete beam.

The separate cages of reinforcement for individual beams and column lengths are usually made up on site with the longitudinal reinforcement wired to the links with 1.6 mm soft iron binding wire that is cut to short lengths, bent in the form of a hairpin, and looped and twisted around all intersections to secure reinforcing bars to links. The ends of binding wire must be flattened so that they do not protrude into the cover of concrete, where they might cause rust staining. Considerable skill, care and labour are required in accurately making up the reinforcing cages and assembling them in the formwork. This is one of the disadvantages of reinforced concrete where unit labour costs are high. At the junction of beams and columns, there is a considerable confusion of reinforcement, compounded by large bars to

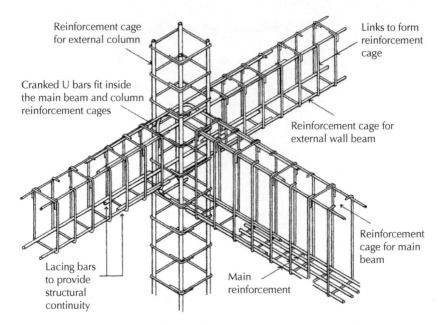

Figure 7.9   Reinforcement cages for reinforced concrete beam and column.

provide structural continuity at the points of support and cranked bars for shear resistance. An alternative is to assemble reinforcement cages offsite in a factory and deliver the prefabricated units to site. These will then be positioned and joined to one another to form the reinforcement for beams and columns. The advantage of using preformed reinforcement is that it reduces the amount of time on site and also contributes to the wellbeing of workers, as there is much less bending and stretching involved when the units are assembled in a factory at a controlled working height.

Figure 7.9 is an illustration of the junction of the reinforcement for a main beam with an external beam and an external column. The longitudinal bars for the beams finish just short of the column reinforcement for ease of positioning the beam cages. Continuity bars are fixed through the column and wired to beam reinforcement. U bars fixed inside the column reinforcement and wired to the main beam serve to anchor the beam to the column against lateral forces.

Figure 7.10a is an illustration of the reinforcement for the junction of four beams with a column. It can be seen that the reinforcement for intersecting beams is arranged to cross over at the intersection inside the columns. Figure 7.10b shows the next stage where the column, floor and beam have been cast and encased in concrete; starter bars are left protruding so that the lower column reinforcement cage can be tied to the next column cage. The correct term for linking the column starter bars to the next column cage is a 'column splice' Figure 7.10b is an illustration of a column splice made in vertical cages for convenience in erecting formwork floor by floor and handling cages. In the reinforcement illustrated in Figure 7.9 and Figure 7.10, the reinforcing bars are deformed to improve anchorage and to obviate the necessity for hooked or bent ends of bars that considerably increase the labour of assembling reinforcing cages.

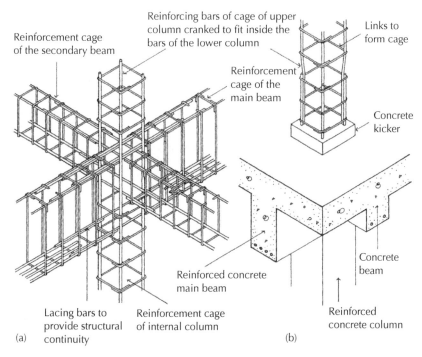

Reinforcement cage
of the secondary beam

Reinforcing bars of cage of upper
column cranked to fit inside the
bars of the lower column

Links to
form cage

Reinforcement
cage of the
main beam

Concrete
kicker

Concrete
beam

Reinforced concrete
main beam

Reinforced
concrete column

(a)   Lacing bars to   Reinforcement cage
provide structural   of internal column
continuity

Reinforcement cage
of internal column

(b)

Figure 7.10   (a) Reinforcement cages of internal columns and beams. (b) Upper column cage spliced to lower cage.

Spacers for reinforcement

To ensure that there is the correct cover of concrete around reinforcement to protect the steel from corrosion and to provide adequate fire protection, it is necessary to fix spacers to reinforcing bars between the bars and the formwork. The spacers hold the reinforcement a set distance away from the face of the formwork, which will also be the face of the concrete. These spacers must be securely fixed so that they are not displaced during placing and compacting of concrete, and are strong enough to maintain the required cover of concrete. Spacer blocks can be made from plastic, concrete or steel. Concrete spacer blocks are cast to the thickness of the required cover; they can be cast on site from sand and cement with a loop of binding wire protruding for binding to reinforcement, or ready-prepared concrete spacers illustrated in Figure 7.11 may be used. The holes in the spacers are for binding wire. Plastic spacers are preferred to made-up cement and sand spacers for ease of use and security of fixing reinforcement in position.

Plastic wheel spacers, as illustrated in Figure 7.12, are used with reinforcing bars to columns and to reinforcement in beams, with the spacers bearing on the inside face of formwork to provide the necessary cover for concrete around steel. The reinforcing bars clip into and are held firmly in place through the wheel spacers. The plastic pylon spacers illustrated in Figure 7.12 are designed to provide support and fixing to the bottom main reinforcement of beams. The reinforcement slips into and is held firmly in the spacer, which bears on the inside face of the formwork to provide the necessary cover of concrete.

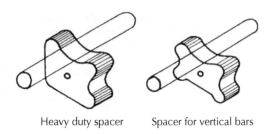

Heavy duty spacer   Spacer for vertical bars

Figure 7.11   Concrete spacers for reinforcement.

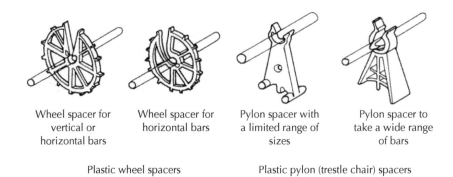

| Wheel spacer for vertical or horizontal bars | Wheel spacer for horizontal bars | Pylon spacer with a limited range of sizes | Pylon spacer to take a wide range of bars |

Plastic wheel spacers   Plastic pylon (trestle chair) spacers

Figure 7.12   Plastic wheel and pylon spacers.

These plastic spacers, which are not affected by concrete, are sufficiently rigid to provide accurate spacing and will not cause surface staining of concrete. They are commonly used in reinforced concrete.

To provide adequate support for top reinforcement, which is cast into reinforced concrete floors, a system of chairs is used. The steel chairs are fabricated from round section mild steel rods to form a system of inverted U's which are linked by rods welded to them. Chairs are galvanised after fabrication. Chairs are either selected from a range of ready-made depths or purpose made to order. The steel chairs are placed on top of the main bottom reinforcement, which is supported by pylon spacers. The top bar of the chairs supports the top reinforcement, which is secured in place with binding wire. The chairs, illustrated in Figures 7.13 and 7.14, must be substantial enough to support the weight of those spreading and compacting the concrete. Reinforcement should be securely tied together; care should be taken to ensure that protruding ends of the tie wire do not intrude into the concrete cover.

Fibre-reinforced concrete (FRC)

It is also possible to reinforce concrete with fibres, to produce what is known as fibre reinforced concrete (FRC). The fibres used include steel, macro-synthetic fibres (polypropylene and nylon fibres) and glass fibres, as well as fibres made from natural materials and recycled products/materials such as rubber tyres. Some of these fibres are used as a substitute to traditional steel reinforcement, especially in specialised applications such as thin precast elements. Fibres are used to increase the workability of the concrete and to help reduce cracking through plastic and drying shrinkage.

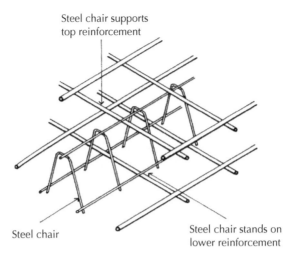

Steel chair supports
top reinforcement

Steel chair

Steel chair stands on
lower reinforcement

Figure 7.13   Steel chair.

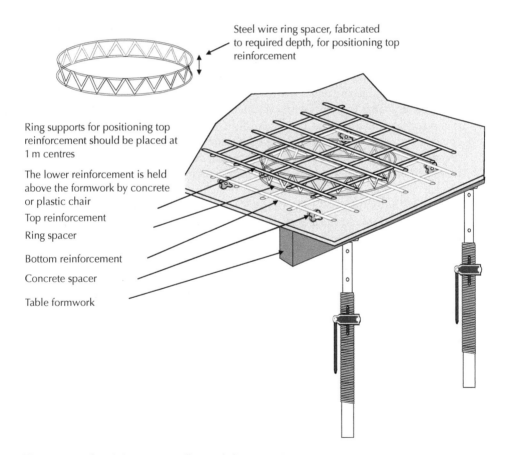

Steel wire ring spacer, fabricated
to required depth, for positioning top
reinforcement

Ring supports for positioning top
reinforcement should be placed at
1 m centres

The lower reinforcement is held
above the formwork by concrete
or plastic chair

Top reinforcement

Ring spacer

Bottom reinforcement

Concrete spacer

Table formwork

Figure 7.14   Steel ring spacer: floor reinforcement cage.

7.4 Formwork and falsework

Formwork is the term used for the temporary timber, plywood, metal or other material, such as a fabric that is used to contain, support and form wet concrete until it has gained sufficient strength to be self-supporting. Falsework is the term used to describe the temporary system or systems of support to the formwork. Formwork and falsework should be strong enough to support the weight of wet concrete and the pressure from placing and compacting the concrete inside the forms. Formwork should be sufficiently rigid to prevent any undue deflection of the forms out of true line and level and be sufficiently tight to prevent excessive loss of water and mortar from the concrete. The size and arrangement of the units of formwork should permit ease of handling, erection and striking. 'Striking' is the term used for dismantling formwork once the concrete is sufficiently hard.

The traditional material for formwork was sawn timber square-edged boarding that can be readily cut to size, fixed, struck and subsequently reused. This approach is still used where the designers require a fair face surface finish that expresses the grain of the timber formwork. The timber boarding can also be finished with an angle to produce a 'V' shape within the concrete and hence a distinctive surface appearance. Marine plywood provides a more watertight lining than sawn boards and a smoother surface finish. Joints between plywood are sealed with foamed plastic strips. Other materials used for formwork are steel sheet, glass-reinforced plastics (GRPs) and hardboard. Where concrete is to be left exposed, a variety of surface linings, such as steel, rubber, thermoplastics or other material, may be used to provide a finished textured surface to concrete.

Fabric formwork is a structural membrane used to form the concrete mould. The material is highly flexible and deflects under the pressure of the concrete to create organic and geometrically complex shapes with a distinctive surface pattern. Research and development is ongoing into the structural and architectural potential of fabric formwork.

Insulated concrete formwork (ICF) is used mainly for self-build and smaller-scale construction projects. The formwork comprises lightweight hollow blocks that are stacked together to create a formwork, into which concrete is then poured. Some of the proprietary systems also provide reinforcing spacers within the blocks to help retain structural integrity during the concrete pour. The formwork is light to assemble by hand and is held together once the concrete has been poured into the formwork. Unlike traditional formwork, this formwork is designed to stay in place once the concrete has hardened and provides thermal insulation to the walls.

Honeycombing and leaks

Formwork should be reasonably watertight to prevent small leaks causing unsightly stains on exposed concrete surfaces and large leaks causing honeycombing. Honeycombing is caused by the loss of water, fine aggregate and cement from concrete through large cracks, which results in a very coarse textured concrete finish, which will reduce bond and encourage corrosion of reinforcement. To control leaks from formwork, it is common to use foamed plastic strips in joints.

Release agents

To facilitate the removal of formwork and avoid damage to concrete as forms are struck, the surface of forms in contact with concrete should be coated with a release agent that prevents wet concrete adhering strongly to the forms. The more commonly used release agents are neat oils with surfactants, mould cream emulsions and chemical release agents that are applied as a thin film to the inside faces of formwork before it is fixed in position.

Formwork support

The support for formwork is usually of timber in the form of bearers, ledgers, soldiers and struts (see Photograph 7.1). For beams, formwork usually comprises bearers at fairly close centres, with soldiers and struts to the sides, and falsework ledgers and adjustable steel props, as illustrated in Figure 7.15. Formwork for columns is formed with plywood facings, vertical backing members and adjustable steel clamps, as illustrated in Figures 7.16 and 7.17. Falsework consists of adjustable steel props fixed as struts to the sides. Temporary falsework and formwork are struck and removed once the concrete they support and contain has developed sufficient strength to be self-supporting. In normal weather conditions, the minimum period after placing ordinary Portland cement concrete that formwork can be struck is from 9 to 12 hours for columns, walls and sides of large beams, 11–14 days for the soffit of slabs and 15–21 days for the soffit of beams.

Column formwork

Column formwork can be constructed from short wall panels, with the ends of the panel overlapping (Figures 7.15–7.17), or it can be constructed using specially designed and fabricated column formwork (Figures 7.18 and 7.19 and Photograph 7.2). Column formwork is made of timber, prefabricated steel units (Figure 7.18), or a combination of timber and steel. GRP, expanded polystyrene, hardboard and plastic formers are also becoming more popular, as is cardboard. Where a circular column is required, it is possible to use disposable formwork, such as single-use cardboard formers. These are lightweight, easy to handle and can be quickly positioned. The cardboard tubes used to form the concrete column must be firmly held in position, adequately propped and clamped in place, before the concrete is poured. Extra care should be taken when pouring the concrete into lightweight formers because they can easily be knocked out of position.

With the sectional column formwork shown in Figure 7.16, the panels are erected around a concrete kicker. The concrete kicker is a small (40–50 mm) upstand cast in the concrete floor, which provides a firm object around which the column formwork can be erected.

Prior to the formwork being erected, the concrete kicker is accurately cast onto the concrete floor. The small amount of formwork used to cast the kicker must be exactly the same cross-sectional dimensions as the column. Once the column is firmly clamped around the kicker, the formwork is checked for line and plumb. Adjustable props and clamps hold the columns firmly in position (Photograph 7.1). Having poured and vibrated the concrete, the columns should be checked to ensure that they are still line and plumb. If slight movement has occurred, it is possible to adjust the props and bring the column back in line. The column should be cast 20 mm above the soffit of the underside of the next floor slab. This will provide a solid surface, which the table formwork for the floor slab can butt up against.

Horizontal formwork

It is common to use patented formwork systems to cast large concrete floor slabs. These usually take the form of plywood decks, fixed to steel or aluminium bearers, which are held in position by interlinked and braced adjustable props (Photograph 7.3). Figures 7.20 provide a sequence of events for the erection of prop, beam and panel formwork. Alternatively,

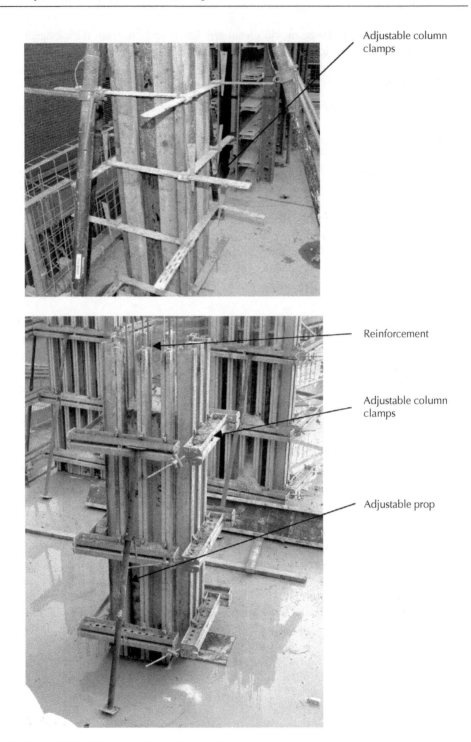

Adjustable column clamps

Reinforcement

Adjustable column clamps

Adjustable prop

Photograph 7.1   Square column formwork systems.

table formwork can be used. The advantage of table formwork is that the supporting structure is already partly assembled as a braced table. The table can be hoisted by crane into the building and wheeled into position, before being properly secured. Where the concrete floor differs in soffit level, due to dropdown beams, additional props can be used to the formwork pieces that make up the beam (Figure 7.20).

To reduce the time that the main floor formwork is needed and the subsequent cost of hiring the formwork, intermediate props are used between the main formwork. The system of table and props is positioned, fixed in place and the concrete is then poured. Once the concrete has reached sufficient maturity, the majority of the horizontal support is removed. The intermediate props remain in position, continuing to provide support to the concrete

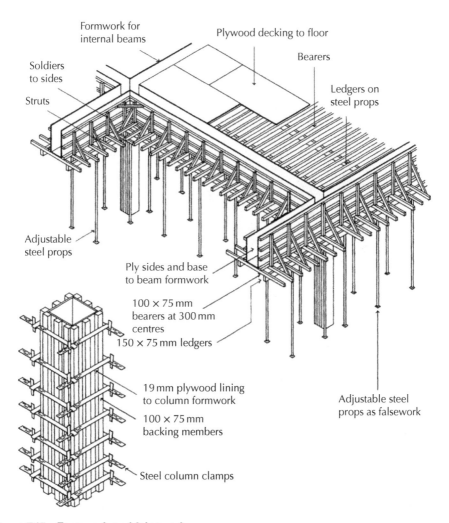

Figure 7.15   Formwork and falsework.

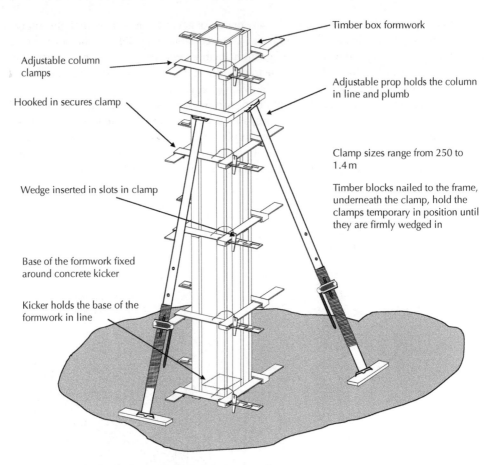

Timber box formwork

Adjustable column clamps

Hooked in secures clamp

Adjustable prop holds the column in line and plumb

Clamp sizes range from 250 to 1.4 m

Wedge inserted in slots in clamp

Timber blocks nailed to the frame, underneath the clamp, hold the clamps temporary in position until they are firmly wedged in

Base of the formwork fixed around concrete kicker

Kicker holds the base of the formwork in line

Figure 7.16   Column clamps and timber formwork.

slab. The table formwork or the system of beams can be removed early, cleaned and quickly used on the next floor.

The props that are used to provide the longer-term support are usually designed to support both the concrete slab and the beams that support the main horizontal formwork. To ensure that the beams and the main horizontal formwork can be released early, the support mechanism is in two parts (Figure 7.21). At the head of the prop, a collar or outer tube supports the horizontal beams and the main shaft of the prop is in direct contact with the concrete floor. Because the collar, sleeve or head that holds the horizontal bearers can be lowered and released independently of the main head, the prop is often called a drop head prop (Figure 7.21 and Photograph 7.4). Depending on the depth of the concrete floor and the properties of the concrete, using drop head props can allow the majority of the horizontal support to be struck after one day. This is a considerable saving in time, compared with the four to seven days' maturity normally required before removing the horizontal support.

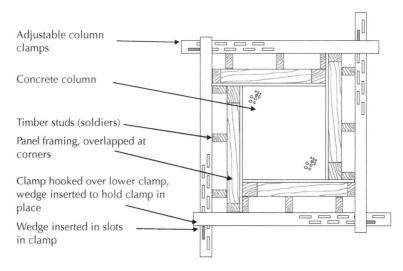

Adjustable column clamps

Concrete column

Timber studs (soldiers)

Panel framing, overlapped at corners

Clamp hooked over lower clamp, wedge inserted to hold clamp in place

Wedge inserted in slots in clamp

Figure 7.17   Plan of column clamps and timber formwork.

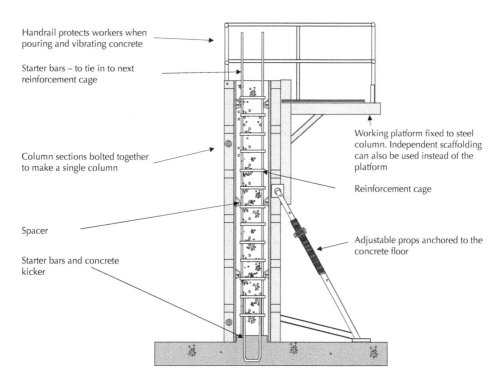

Handrail protects workers when pouring and vibrating concrete

Starter bars – to tie in to next reinforcement cage

Column sections bolted together to make a single column

Spacer

Starter bars and concrete kicker

Working platform fixed to steel column. Independent scaffolding can also be used instead of the platform

Reinforcement cage

Adjustable props anchored to the concrete floor

Figure 7.18   Proprietary steel column formwork. (*Source*: adapted from http://www.peri. ltd.uk).

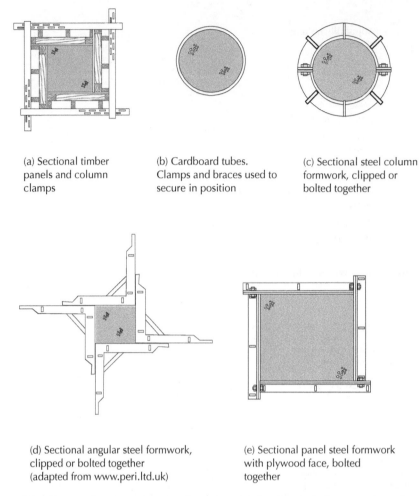

(a) Sectional timber panels and column clamps

(b) Cardboard tubes. Clamps and braces used to secure in position

(c) Sectional steel column formwork, clipped or bolted together

(d) Sectional angular steel formwork, clipped or bolted together (adapted from www.peri.ltd.uk)

(e) Sectional panel steel formwork with plywood face, bolted together

Figure 7.19 Types of column formwork – plan. (a) Sectional timber panels and column clamps. (b) Cardboard tubes. Clamps and braces used to secure in position. (c) Sectional steel column formwork, clipped or bolted together. (d) Sectional angular steel formwork, clipped or bolted together (*Source*: adapted from www.peri.ltd.uk). (e) Sectional panel steel formwork with plywood face, bolted together.

Because it is easy to assemble and can be quickly manoeuvred into position, table formwork is used to construct horizontal concrete surfaces. The tables are designed so that they can accommodate columns that would penetrate the decking (Figure 7.22). Table forms can be bolted or clipped together to provide the formwork for large horizontal surfaces (Figure 7.23 and Photograph 7.5).

Columns and other vertical concrete structures should be cast 20 mm above the soffit of the underside of the next floor slab. This provides a solid surface for the horizontal formwork to butt up against. The joint between vertical concrete surfaces (i.e. columns and

Concrete kicker formwork

Concrete kicker formed around
the starter bars

Integral working
platform

Spacer

Reinforcement cage with
concrete spacer blocks to ensure
concrete cover is maintained

Steel column formwork erected
around concrete kicker and
reinforcement cage

Photograph 7.2   Circular steel column formwork.

walls) and horizontal formwork should be tight and sealed. Foam plastic sealing strips or
gunned silicone rubber helps to seal the joint between the concrete and timber.

Prior to the concrete being poured, it is vital that the surface of the formwork is cleaned.
All joints should be taped and sealed; all loose debris and wire must be cleared away. A
compressed air hose is generally used to blow away debris and dust (the reinforcement usually prevents brushes from being used effectively). To make striking the formwork easier
and to reduce blowholes, the surface of the form should be coated with a release agent.

Photograph 7.3   Prop and beam formwork.

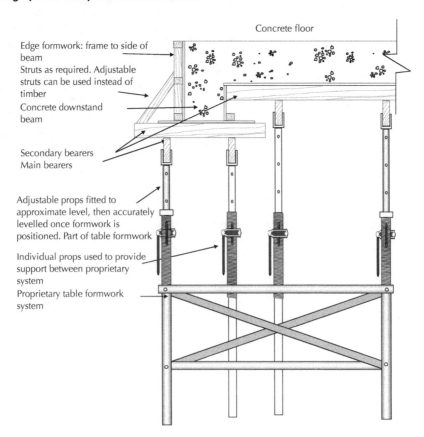

Figure 7.20   Beam and slab formwork.

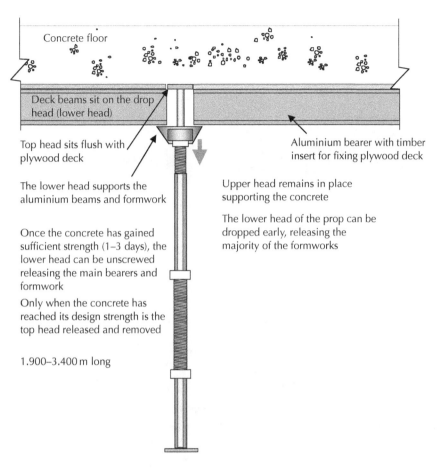

Concrete floor

Deck beams sit on the drop
head (lower head)

Top head sits flush with
plywood deck

Aluminium bearer with timber
insert for fixing plywood deck

The lower head supports the
aluminium beams and formwork

Upper head remains in place
supporting the concrete

The lower head of the prop can be
dropped early, releasing the
majority of the formworks

Once the concrete has gained
sufficient strength (1–3 days), the
lower head can be unscrewed
releasing the main bearers and
formwork

Only when the concrete has
reached its design strength is the
top head released and removed

1.900–3.400 m long

Figure 7.21   Double headed/drop head props.

Release agents are either chemical-based, emulsion (mould cream) or neat oil with sur-
factant. Mould cream should not be used on steel forms, and oil can lead to staining of the
concrete surface; therefore, a chemical release agent is used. Chemical agents help to produce high-quality finishes, are easily applied and are the least messy of the release agents.

Flying formwork

Table formwork comes in a range of sizes. It is not uncommon to see tables in excess of
5 m long. Large systems of prefabricated formwork reduce the assembly time. Where the
building structure is designed in regular grid patterns, large tables prove economical. The
large formwork systems are manoeuvred on trolleys, which wheel the forms to the edge
of the building where they are lifted from the building using 'C hooks' fitted to cranes
(Figure 7.24). Because the large frames are easily manoeuvred from the edge of the building and 'slung' through the air to their next position, they tend to be termed 'flying forms'.
Using such large tables considerably reduces the cycle time to complete the formwork for
the next floor. Large flying forms can halve the time required to assemble smaller tables
or beam and prop systems. Horizontal formwork must be struck gradually to avoid shock

Main beams and
secondary beams
carried by drop head
prop

Adjustable prop

Concrete floor at partial
maturity

Drop head (outer
sleeve) can be released
and the main and
secondary beams
removed

Main head of the prop
remains in contact with
the concrete floor at all
times, continuing to
provide support until the
concrete reaches full
maturity

Photograph 7.4   Drop head prop.

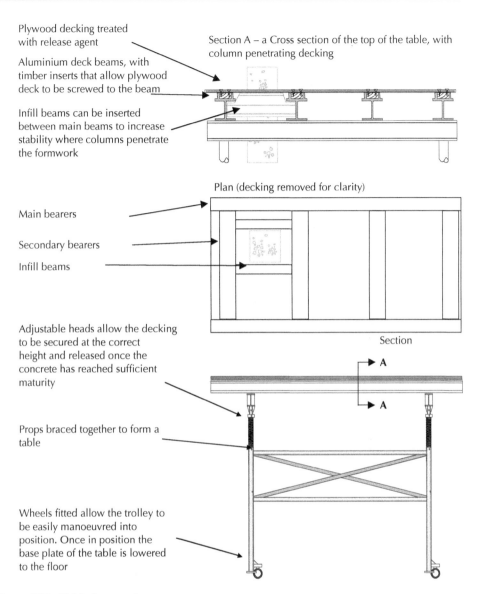

Plywood decking treated with release agent

Aluminium deck beams, with timber inserts that allow plywood deck to be screwed to the beam

Infill beams can be inserted between main beams to increase stability where columns penetrate the formwork

Section A – a Cross section of the top of the table, with column penetrating decking

Plan (decking removed for clarity)

Main bearers

Secondary bearers

Infill beams

Adjustable heads allow the decking to be secured at the correct height and released once the concrete has reached sufficient maturity

Props braced together to form a table

Wheels fitted allow the trolley to be easily manoeuvred into position. Once in position the base plate of the table is lowered to the floor

Section

A

A

Figure 7.22   Table formwork.

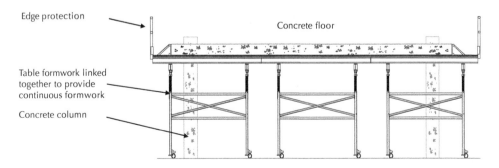

Edge protection

Concrete floor

Table formwork linked together to provide continuous formwork

Concrete column

Figure 7.23   Table formwork system.

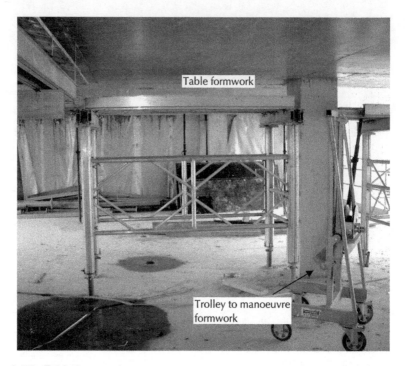

Photograph 7.5   Table formwork.

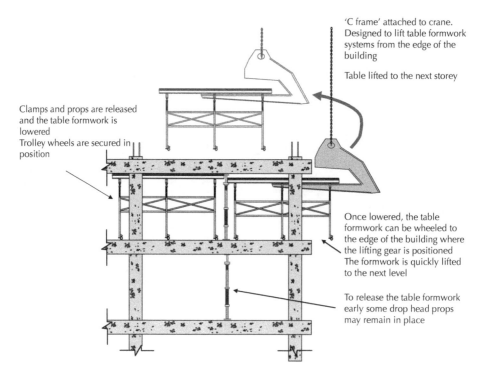

Figure 7.24   Flying formwork.

overloading in the concrete slab, which could cause the slab to fail. To eliminate any unnecessary vibration and shock, large table forms must be released slowly and the jacks slightly lowered first.

Wall formwork

It is common to see prefabricated steel or aluminium patient systems used on large concrete structures. Figure 7.25 shows a typical steel wall system. Photographs 7.6 and 7.7 show the formwork and steel reinforcement within the wall being assembled. As the working platform is already an integral part of the formwork, extra scaffolding around the formwork is not required. The formwork is manoeuvred into position by crane and quickly bolted together and propped (Photograph 7.7). Threaded steel ties are used to hold the two faces of formwork together. A steel or plastic sleeve, through which the tie runs, acts as a spacer holding the faces of the formwork the required distance apart. The sleeve is cut to the width of the wall. Various proprietary ties are available and not all require the use of a sleeve.

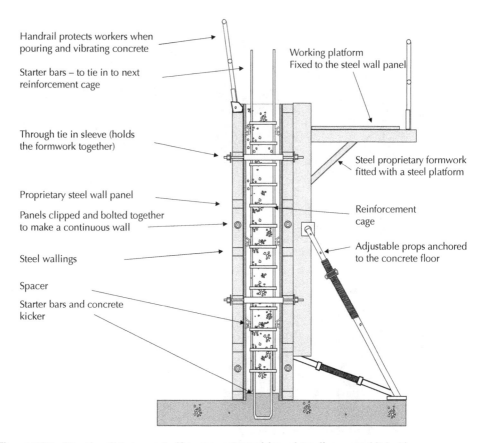

Figure 7.25   Panel wall formwork. (*Source*: adapted from http://www.peri.ltd.uk).

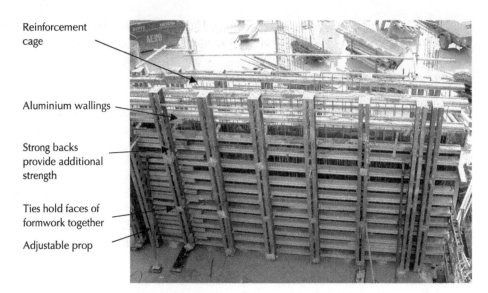

Reinforcement cage

Aluminium wallings

Strong backs provide additional strength

Ties hold faces of formwork together

Adjustable prop

Photograph 7.6 Wall formwork.

One side of the wall formwork is erected against the concrete kicker

The reinforcement is then tied to the starter bars

Void formers are installed to allow for door openings

Finally the other side of the wall formwork is placed in position

Ties hold faces of formwork together

Tower scaffolding erected to pour concrete

Photograph 7.7 Wall formwork erection sequence.

7.5 Prestressed concrete

Because concrete has poor tensile strength, a large part of the area of an ordinary reinforced concrete beam plays little part in the flexural strength of the beam under load. In
the calculation of stresses in a simply supported beam, the strength of the concrete in the
lower part of the beam is usually ignored. When reinforcement is stretched before or after
the concrete is cast and the stretched reinforcement is anchored to the concrete, it causes
a compressive prestress in the concrete as it resists the tendency of the reinforcement to
return to its original length. This compressive prestress makes more economical use of the
concrete by allowing all of the section of concrete to play some part in supporting load. In
prestressed concrete, the whole or part of the concrete section is compressed before the
load is applied, so that when the load is applied, the compressive prestress is reduced by
flexural tension.

In ordinary reinforced concrete, the concrete around reinforcement is bonded to it and
must, therefore, take some part in resisting tensile stress. Because the tensile strength of
concrete is low, it will crack around the reinforcement under load, and hair cracks on the
surface of concrete are not only unsightly, they also reduce the protection against fire and
corrosion. The effectiveness of the concrete cover will be reduced when cracking occurs.
In designing reinforced concrete members, it is usual to limit the anticipated tensile stress
to limit deflection and the extent of cracking of concrete around reinforcement. This is
a serious limitation in the most efficient use of reinforced concrete, particularly in long-
span beams.

When reinforcement is stretched and put under tensile stress and then fixed in the
concrete, once the prestress is released, the tendency of the reinforcement to return to its
original length induces a compressive stress in concrete. The stretching of reinforcement
before it is cast into concrete is described as pre-tensioning and stretching reinforcement
after the concrete has been cast as post-tensioning. The advantage of the induced compressive prestress, caused either by pre- or post-tensioning, is that under load the tensile stress
developed by bending is acting against the compressive stress induced in the concrete, and
in consequence cracking is minimised. If cracking of the concrete surface does occur and
the load is reduced or removed, then the cracks close up due to the compressive prestress.
Another advantage of the prestress is that the compressive strength of the whole of the section of concrete is utilised and the resistance to shear is considerably improved, so obviating
the necessity for shear reinforcement.

For the prestress to be maintained, the steel reinforcement must not suffer permanent
elongation or creep under load. High-tensile wire is used in prestressed concrete to maintain the prestress under load. Under load, a prestressed concrete member will bend or
deflect, and compressive and tensile stresses will be developed in opposite faces, as previously explained. Concrete in parts of the member will therefore have to resist compressive
stress induced by the prestress as well as compressive stress developed during bending.
For this reason, high compressive strength concrete is used in prestressed work to gain
the maximum advantage of the prestress. A consequence of the need to use high-strength
concrete is that prestressed members are generally smaller in section than comparable reinforced concrete ones.

Pre-tensioning

High-tensile steel reinforcing wires are stretched between anchorages at each end of a casting bed and concrete is cast around the wires inside timber or steel moulds. The tension

in the wires is maintained until the concrete around them has attained sufficient strength to take up the prestress caused by releasing the wires from the anchorages. The bond between the stretched wires and the concrete is maintained by the adhesion of the cement to the wires, by frictional resistance and the tendency of the wires to shorten on release and wedge into the concrete. To improve frictional resistance, the wires may be crimped or indented, as illustrated in Figure 7.26. When stressing wires are cut and released from the anchorages in the stressing frame, the wires tend to shorten, and this shortening is accompanied by an increase in diameter of the wires, which wedge into the ends of the member, as illustrated in Figure 7.27. Pre-tensioning of concrete is mainly confined to the manufacture of precast large-span members such as floor beams, slabs and piles. The stressing beds required for this work are too bulky for use on site.

Post-tensioning

After the concrete has been cast inside moulds or formwork and has developed sufficient strength to resist the stress, stressing wires are threaded through ducts or sheaths cast in along the length of the member. These prestressing wires are anchored at one end of the member and are then stretched and anchored at the opposite end to induce the compressive stress. The advantage of post-tensioning is that the stressing wires or rods are stressed against the concrete and there is no loss of stress as there is in pre-tensioning due to the shortening of the wires when they are cut from the stressing bed. The major part of the drying shrinkage of concrete will have taken place before it is post-tensioned and this minimises loss of stress due to shrinkage of concrete. The systems of post-tensioning used are Freyssinet, Gifford–Udall-CCL, Lee–McCall, Magnel–Blaton and the PSC one-wire system.

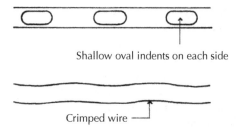

Shallow oval indents on each side

Crimped wire

Figure 7.26   Prestressing wires.

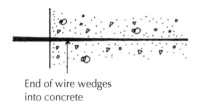

End of wire wedges into concrete

Figure 7.27   Anchorage to stressing wire.

The Freyssinet system

In the Freyssinet system, a duct is formed along the length of the concrete beam as it is cast. The duct is formed by casting concrete around an inflatable tube, which is withdrawn when the concrete has hardened. At each end of the beam, a high-tension, concrete anchor cylinder is cast into the beam. The purpose of the anchor cylinder is to provide a firm base for the head of the jack used to tension wires. A 7 mm diameter cable of high-tensile wires arranged around a core of fine-coiled wire is threaded through the duct. The wires are held at one end between the cast-in concrete anchor cylinder and a loose concrete anchor cone that is hammered in tightly to secure the wires. A hydraulically operated ram is anchored to the stressing wires, and the ram of the jack is positioned to bear on a loose ring that bears on the concrete anchor cone around the wires. A piston on the ram applies stress to the wires, which are anchored by the ram, forcing the anchor cone into the anchor cylinder, as illustrated in Figure 7.28. The stressing wires are released from the jack and the protruding ends of wire are cut off. A grout of cement and water is forced under pressure into the cable duct to protect the wires from corrosion (Photograph 7.8a and b).

The Gifford–Udall–CCL system

In the Gifford–Udall–CLL system, a duct is formed along the length of a concrete beam around an inflatable former, which is withdrawn when the concrete has hardened. Steel thrust rings are fixed to the ends of the beam, through which the stressing wires are threaded

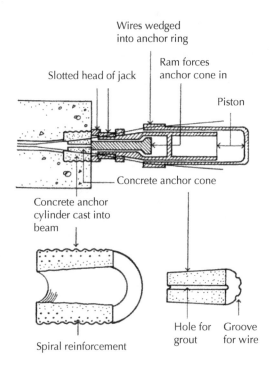

Figure 7.28   Freyssinet system.

Post-tension cables and ducts are positioned within the reinforcement

The tension cables are simply anchored into the concrete at one end and tension exerted at the other

(a)

Reinforcement spacer chair

Grouting tube, used once tension is exerted

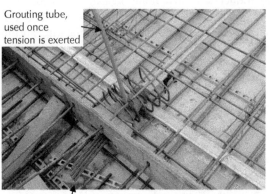

Steel cables are left protruding at the access end

Once the concrete is poured tension is exerted and a report is prepared on each tension cable reporting the level of force exerted

(b)

Photograph 7.8   (a) and (b) Post-tensioned concrete floor.

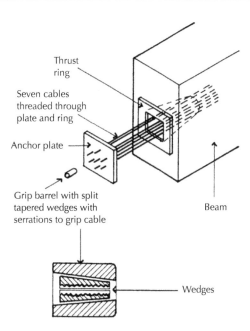

Thrust ring

Seven cables threaded through plate and ring

Anchor plate

Grip barrel with split tapered wedges with serrations to grip cable

Beam

Wedges

Figure 7.29   CCL system.

the length of the beam. The individual stressing wires are threaded through holes in a steel anchor plate at one end of the beam and firmly secured with steel grip barrels and wedges, as illustrated in Figure 7.29. The stressing wires are threaded through a steel anchor plate at the other end of the beam. Each wire is separately stressed by a jack and secured by ramming in split tapered wedges into a grip barrel. When the stressing operation is complete, the duct for the wires is filled under pressure with cement grout. The advantage of this system is that the precise stress in each wire is controlled, whereas in the Freyssinet system all wires are jacked together and, if one wire were to break, the remaining wires would take up their share of the total stress and might be overstressed.

The Lee–McCall system

In the Lee–McCall system comprises an alloy bar that is threaded through a duct in the concrete member and stressed by locking a nut to one end and stressing the rod to the other end with a jack and anchoring it with a nut. The simplicity of this system is self-evident.

The Magnel–Blaton system

In the Magnel–Blaton system, high-tensile wires are arranged in layers of four wires each and are held in position by metal spacers. The layers of wire are threaded through a duct in the concrete member. One end of the wires is fixed in metal sandwich plates against an anchor plate cast into the concrete. Pairs of wires are stressed in turn and wedged in position. The stressed wires are grouted in position in the duct by introducing cement grout through a hole in the top of the member leading to the duct.

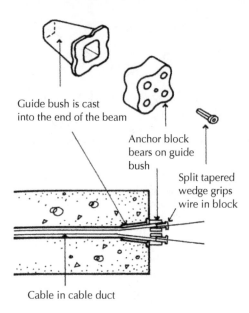

Guide bush is cast
into the end of the beam

Anchor block
bears on guide
bush

Split tapered
wedge grips
wire in block

Cable in cable duct

Figure 7.30   PSC one-wire system.

The PSC one-wire system

In the PSC one-wire system, a duct is formed along the length of a concrete beam by an inflatable former. Guide bushes are cast into the ends of the beam, as illustrated in Figure 7.30. One, two or four high-tensile wires are threaded through holes in anchor blocks at each end of the beam. The wires at one end are secured in the anchor block by ramming in split, tapered wedges around the wires in the holes in the anchor block. At the other end of the beam, each wire is separately stressed by a jack and then secured by ramming in split tapered wedges. The cable duct is then filled with cement grout through the centre hole in the anchor blocks.

The advantage of the Gifford–Udall–CCL, the Lee–McCall, the Magnel–Blaton and the PSC systems over the Freyssinet system is that each wire or pair of wires is stressed individually so that the stress can be controlled and measured, whereas with the Freyssinet system, there is no such control.

7.6 Lightweight concrete

It may be advantageous to employ lightweight concrete such as no-fines concrete for the monolithic loadbearing walls of buildings and aerated concrete for structural members such as roof slabs supporting comparatively light loads, to combine the advantage of reduced deadweight and improved thermal insulation. The various methods of producing lightweight concrete depend on:

❏   The presence of voids in the aggregate.
❏   Air voids in the concrete.
❏   Omitting fine aggregate.
❏   The formation of air voids by the addition of a foaming agent to the concrete mix.

The aggregates described for use in lightweight concrete building blocks are also used for mass concrete or reinforced concrete structural members, where improved thermal insulation is necessary and where the members, such as roof slabs, do not sustain large loads.

No-fines concrete

No-fines concrete consists of concrete made from a mix containing only coarse aggregate, cement and water. The coarse aggregate may be gravel, crushed brick or one of the lightweight aggregates. The coarse aggregate used in no-fines concrete should be as near one size as practicable to produce a uniform distribution of voids throughout the concrete. To ensure a uniform coating of the aggregate particles with cement/water paste, it is important that the aggregate be wetted before mixing and the maximum possible water/cement ratio, consistent with strength, be used to prevent separation of the aggregate and cement paste.

Construction joints should be as few as possible and vertical construction joints are to be avoided if practicable, because successive placings of no-fines concrete do not bond together firmly as do those of ordinary concrete. Because of the porous nature of this concrete, it must be rendered externally or covered with some protective coating or cladding material, and the no-fines concrete plastered or covered internally. A no-fines concrete wall provides similar insulation to a sealed brick cavity wall of similar thickness.

Aerated and foamed concretes

An addition of one part of powdered zinc or aluminium to every thousand parts of cement causes hydrogen to evolve when mixed with water. As the cement hardens, a great number of small sealed voids form in the cement to produce aerated concrete, which usually consists of a mix of sand, cement and water. Foamed concrete is produced by adding a foaming agent, such as resin soap, to the concrete mix. The foam is produced by mixing in a high-speed mixer or by passing compressed air through the mix to encourage foaming. As the concrete hardens, many sealed voids are entrained. Aerated and foamed concretes are used for building blocks and lightweight roofing slabs.

Surface finishes of concrete

A wide variety of surface finishes of concrete are available, the choice of one over another largely dictated by architectural fashion and the preference of the client.

Plain concrete finishes

On drying, concrete shrinks and fine irregular shrinkage cracks appear in the surface in addition to the cracks and variations in colour and texture due to successive placings. One school of thought is to accept the cracks and variations in texture and colour as a fundamental of the material and to make no attempt to control or mask them. Another school of thought is at pains to mask cracks and variations by means of designed joints and profiles on the surface.

Board-marked concrete finishes are produced by compacting concrete by vibration against the surface of the timber formwork so that the finish is a mirror of the grain of the timber boards and the joints between them. This type of finish varies from the regular shallow profile of planed boards to the irregular marks of rough-sawn boards and the deeper profile of boards that have been sand blasted to pronounce the grain of the wood.

A necessary requirement of this type of finish is that the formwork be absolutely rigid to allow dense compaction of concrete to it and that the boards be non-absorbent.

One method of masking construction joints is to form a horizontal indentation or protrusion in the surface of the concrete where construction joints occur, by nailing a fillet of wood to the inside face of the timber forms or by making a groove in the boards so that the groove or protrusion in the concrete masks the construction joint. Various plain concrete finishes can be produced by casting against plywood, hardboard or sheet metal to produce a flat finish or against corrugated sheets or crepe rubber to produce a profiled finish.

Tooled surface finishes

One way of masking construction joints, surface crazing of concrete and variations in colour is to tool the surface with hand or power-operated tools. The action of tooling the surface is to break up the fine particles of cement and fine aggregate that find their way to the surface when wet concrete is compacted inside formwork, and also to expose the coarse texture of aggregate.

Bush hammering

In bush hammering, a round-headed hammer with several hammer points on it is vibrated by a power-driven tool, which is held against the surface and moved successively over small areas of the surface of the concrete. The hammer crushes and breaks off the smooth cement film to expose a coarse surface. This coarse texture effectively masks the less obvious construction joints and shrinkage cracks.

Point tooling

In point tooling, a sharp pointed power vibrated tool is held on the surface, causing irregular indentations and at the same time spalls off the fine cement paste finish. By moving the tool over the surface, a coarse pitted finish is obtained, the depth of pitting and the pattern of the pits being controlled by the pressure exerted and the movement of the tool over the surface. For best effect with this finish, as large an aggregate size as possible should be used to maintain an adequate cover of concrete to reinforcement. The depth of the pitting should be allowed for in determining the cover required.

Dragged finish

In dragged finishing, a series of parallel furrows is tooled across the surface by means of a power-operated chisel pointed tool. The depth and spacing of the furrows depend on the type of aggregate used in the concrete and the size of the member to be treated. This highly skilled operation should be performed by an experienced mason.

Margins to tooled finishes

Bush-hammered and point-tooled finishes should not extend to the edges or arrises of members, as the hammering operation required would cause irregular and unsightly spalling at angles. A margin of at least 50 mm should be left untreated at all angles. As an alternative, a dragged finish margin may be used with the furrows of the dragging at right angles to the angle.

Exposed aggregate finish

An exposed aggregate finish is produced by exposing the aggregate used in the concrete or a specially selected aggregate applied to the face or faces of the concrete. To expose the aggregate, it is necessary either to wash or brush away the cement paste on the face of the concrete or to ensure that the cement paste does not find its way to the face of the aggregate to be exposed. Because of the difficulties of achieving this with in situ cast concrete, exposed aggregate finishes are confined in the main to precast concrete members and cladding panels. One method of exposing the aggregate in concrete is to spray the surface with water, while the concrete is still green, to remove cement paste on the surface. The same effect can be achieved by brushing and washing the surface of green concrete. The pattern and disposition of the aggregate exposed this way is dictated by the proportioning of the mix and placing and compaction of the concrete, and the finish cannot be closely controlled.

To produce a distinct pattern or texture of exposed aggregate particles, it is necessary to select and place the particles of aggregate in the bed of a mould or alternatively to press them into the surface of green concrete. This is carried out by precasting concrete. Members cast face down are prepared by covering the bed of the mould with selected aggregate placed at random or in some pattern. Concrete is then carefully cast and compacted on top of the aggregate so as not to disturb the face aggregate in the bed of the mould. If the aggregate is to be exposed in some definite pattern, it is necessary to bed it in water-soluble glue in the bed of the mould on sheets of brown paper that are washed off later. Once the concrete member has gained sufficient strength, it is lifted from the mould and the face is washed to remove cement paste. Large aggregate particles that are to be exposed are pressed into a bed of sand in the bed of the mould and the concrete is then cast on the large aggregate. When the member is removed from the mould after curing, the sand around the exposed aggregate is washed off. Alternatively, large particles may be pressed into the surface of green concrete and rolled, to bed them firmly and evenly.

7.7 Concrete structural frames

François Hennebique was chiefly responsible for the development of reinforced concrete for use in buildings, first as reinforced concrete piles and later as reinforced concrete beams and columns. In 1930, Freyssinet began development work that led to the use of prestressed concrete in building. The first reinforced concrete-framed building to be built in the UK was the General Post Office building in London, which was completed in 1910. Subsequently, comparatively little use was made of reinforced concrete in the UK until the end of the Second World War (1945). The great shortage of steel that followed the end of the Second World War prompted engineers to use reinforced concrete as a substitute for steel in structural building frames. The shortage of steel continued for some years after the end of the war. Up to the early 1980s, the majority of framed buildings in the UK were constructed with reinforced concrete frames. More recently steel has become a more economic alternative for some building types, such as multi-storey-framed structures and wide-span single-storey shed buildings, described in Chapters 4 and 5, respectively. Composite structures, which make use of the various structural qualities of both steel and concrete, are also widely used.

The members of a reinforced concrete frame can be moulded to any required shape so that they can be designed to use concrete where compressive strength is required and steel reinforcement where tensile strength is required. The members do not need to be of uniform section along their length or height. The singular characteristics of concrete are that it is initially a wet plastic material that can be formed to any shape inside formwork, for economy in section as a structural material or for reasons of appearance, and when it is cast in situ, it will act monolithically as a rigid structure. A monolithically cast reinforced concrete frame has advantageous rigidity of connections in a frame and in a solid wall or shell structure, but this rigidity is a disadvantage in that it is less able to accommodate movements due to settlement, wind pressure, and temperature and moisture changes than is a more flexible structure.

Unlimited choice of shape is an advantage structurally and aesthetically but may well be a disadvantage economically in the complication of formwork and falsework necessary to form irregular shapes. The cost of formwork for concrete can be considerably reduced by repetitive casting in the same mould in the production of precast concrete cladding and structural frames, and the rigidity of the concrete frame can be of advantage on subsoils of poor or irregular bearing capacity and where severe earth movements occur as in areas subject to earthquakes.

In situ cast frames

The in situ cast reinforced concrete frame is extensively used for both single- and multi-storey buildings.

Structural frame construction

The principal use of reinforced in situ cast concrete as a structural material for building is as a skeleton frame of columns and beams with reinforced concrete floors and roof. In this use, reinforced concrete differs little from structural steel skeleton frames cased in concrete. In those countries, where unit labour costs are low and structural steel is comparatively expensive, a reinforced concrete frame is widely used as a frame for both single- and multi-storey buildings such as the small framed building, with solid end walls and projecting balconies with upstands, as illustrated in Figure 7.31.

The in situ cast, reinforced concrete structural frame is much used for multi-storey buildings such as flats and offices. Repetitive floor plans can be formed inside a skeleton frame of continuous columns and floors. To use the same formwork and falsework, floor by floor, variations in the reinforcement and/or mix of concrete in columns, to support variations in loads, can provide a uniform column section. The uniformity of column section and formwork makes for a speedily erected and economic structural frame. An advantage of the reinforced concrete structural frame is that the columns, beams and floor slabs provide a level, solid surface on which walls and partitions can be built and between which walls, partitions and framing may be built and secured by bolting directly to a solid concrete backing.

A reinforced concrete structural frame with one-way spanning floors is generally designed on a rectangular grid for economy in the use of materials in the same way as a structural steel frame. Where floors are cast monolithically with a reinforced concrete frame, the tie beams that are a necessary part of a structural steel frame may be omitted as the monolithically cast floors will act as ties. The in situ reinforced concrete floors, illustrated in Figure 7.32, span one way between the upstand beams in external walls and the

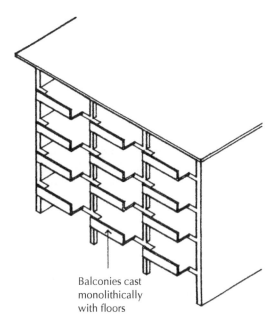

Balconies cast
monolithically
with floors

Figure 7.31   In situ cast concrete frames.

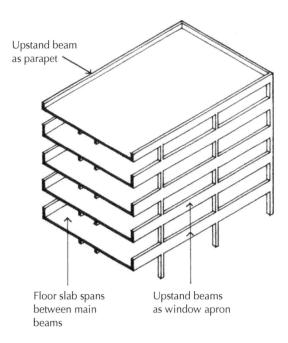

Upstand beam
as parapet

Floor slab spans
between main
beams

Upstand beams
as window apron

Figure 7.32   In situ cast frames.

pair of internal beams supported by internal columns. This arrangement provides open-plan floor areas each side of a central access corridor. An advantage of the upstand beams in the external walls is that the head of windows may be level with or just below the soffit of the floor above for the maximum penetration of daylight.

Cross-wall and box frame construction

Multi-storey structures, such as blocks of flats and hotels with identical compartments planned on successive floors one above the other, require permanent, solid, vertical divisions between compartments for privacy, and sound and fire resistance. In this type of building, it is illogical to construct a frame and then build solid heavy walls within the frame to provide vertical separation, with the walls taking no part in loadbearing. A system of reinforced concrete cross walls provides sound and fire separation and acts as a structural frame supporting floors, as illustrated in Figure 7.33.

Between the internal cross walls, reinforced concrete beam and slab or plate floors may be used. Where flats are planned on two floors as maisonettes, the intermediate floor of the maisonette may be of timber joist and concrete beam construction to reduce cost and deadweight. The intermediate timber floor inside maisonettes is possible where Building Regulations require vertical and horizontal separation between adjacent maisonettes.

A system of box frame, in situ cast external and internal walls and floors may be used where identical floor plans are used for a multi-storey building without columns or beams. The inherent strength and stability of the rigidly connected walls and floors is used to advantage, with both internal and external walls perforated for door and window openings as required, as illustrated in Figure 7.34. This does not necessarily result in the most economical form of building, because of the considerable labour cost in the extensive form-work and falsework needed. A straightforward system of skeleton frame with external cladding and solid internal division walls may often be cheaper.

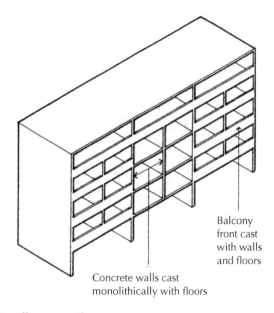

Balcony
front cast
with walls
and floors

Concrete walls cast
monolithically with floors

Figure 7.33   Cross-wall construction.

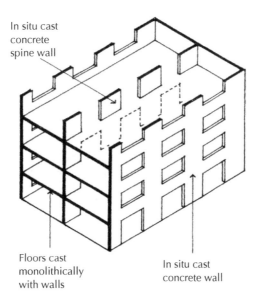

In situ cast
concrete
spine wall

Floors cast
monolithically
with walls

In situ cast
concrete wall

Figure 7.34   Box frame construction.

Wind bracing

In multi-storey reinforced concrete-framed buildings, it is usual to contain the lifts, stairs
and lavatories within a service core, contained in reinforced concrete walls, as part of
the frame. The hollow reinforced concrete column containing the services and stairs is
immensely stiff and will strengthen the attached skeleton frame against wind pressure. In
addition to stiffening the whole building, such a service core may also carry a considerable
part of floor loads by cantilevering floors from the core and using props in the form of slender columns on the face of the building. Similarly, monolithically cast reinforced concrete
flank end walls of slab blocks may be used to stiffen a skeleton frame structure against wind
pressure on its long façade.

Floor construction

In situ cast concrete floors

The principal types of reinforced in situ cast concrete floor construction are:

- ❏ Beam and slab.
- ❏ Waffle grid slab.
- ❏ Drop beam and slab.
- ❏ Flat slab.

Beam and slab floor

A beam and slab floor is generally the most economic and therefore most usual form of
floor construction for reinforced concrete frames. When a reinforced concrete frame is
cast monolithically with reinforced concrete floors, it is logical to design the floor slabs to
span in both directions, so that all the beams around a floor slab can bear part of the load.
This two-way span of floor slabs affects some reduction in the overall depth of floors as

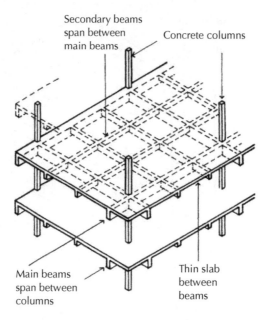

Figure 7.35   Square grid beam and slab floor.

compared with a one-way spanning floor slab construction. Since the most economical shape for a two-way spanning slab is square, the best column grid for a reinforced concrete frame with monolithically cast floors is a square one, as illustrated in Figure 7.35.

The in situ cast reinforced concrete floor illustrated in Figure 7.35 combines main and secondary beams as a grid to provide the least thickness of slab for economy in the mass of concrete in construction, and comparatively widely spaced columns. This square grid results in the minimum thickness of floor slab and minimum depth of beams, and therefore the minimum deadweight of construction. Departure from the square column grid, because of user requirements and circulation needs in a building, will increase the overall depth, weight and therefore cost of construction of a reinforced concrete frame.

The rectangular column grid, illustrated in Figure 7.36, supports main beams between columns that support one-way spanning floors with the beams between columns. The floor slab can be cast in situ on centring and falsework, or precast concrete floor beams or planks may be used. This arrangement involves closely spaced columns and the least mass of concrete in floors.

In a steel frame, the skeleton of columns and beams is designed to carry the total weight of the building. The floors, which span between beams, act independently of the frame. With an in situ cast reinforced concrete frame and floor construction, columns, beams and floors are cast and act monolithically. The floor construction, therefore, acts with and affects the frame and should be considered as part of it.

Waffle grid slab floor
If the column grid is increased from about 6.0 to about 12.0 m^2, or near square, it becomes economical to use a floor with intermediate cross beams supporting thin floor slabs, as illustrated in Figure 7.37. The intermediate cross beams are cast on a regular square grid

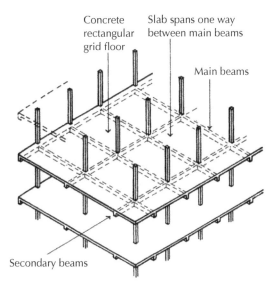

Figure 7.36   Rectangular grid beam and slab floor.

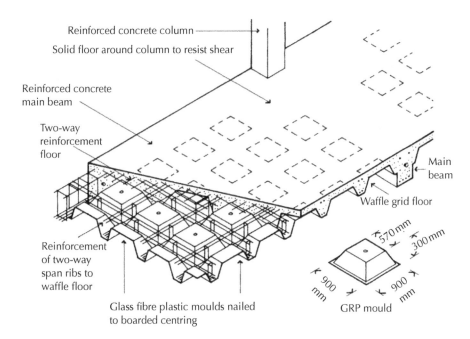

Figure 7.37   Waffle grid in situ cast reinforced concrete floor.

that gives the underside of the floor the appearance of a waffle, hence the name. The advantage of the intermediate beams of the waffle is that they support a thin floor slab and so reduce the deadweight of the floor as compared to a flush slab of similar span. This type of floor is used where a widely spaced square column grid is necessary and floors support comparatively heavy loads. The economic span of floor slabs between intermediate beams

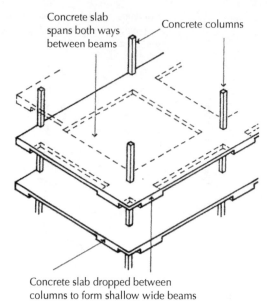

Concrete slab
spans both ways
between beams

Concrete columns

Concrete slab dropped between
columns to form shallow wide beams

Figure 7.38   Drop slab floor.

lies between 900 mm and 3.5 m. The waffle grid form of the floor may be cast around plastic or metal formers (as illustrated in Figure 7.37) laid on timber centring, so that the smooth finish of the soffit may be left exposed.

Drop slab floor
The drop slab floor construction consists of a floor slab, which is thickened between columns in the form of a shallow but wide beam, as illustrated in Figure 7.38. A drop slab floor is of about the same deadweight as a comparable slab and beam floor and will have up to half the depth of floor construction from top of slab to soffit of beams. On a 12 m² column grid, the overall depth of a slab and beam floor would be about 1200 mm, whereas the depth of a drop slab floor would be about 600 mm. This difference would cause a significant reduction in overall height of construction of a multi-storey building. This form of construction is best suited to a square grid of comparatively widely spaced columns selected for large, unobstructed areas of floor. Because of the additional reinforcement required for shallow depth, wide-span beams, this type of floor is more expensive than a traditional rectangular grid beam and slab floor.

Flat slab (plate) floor
In flat slab (plate) floor construction, the slab is of uniform thickness throughout, without downstand beams and with the reinforcement more closely spaced between the points of support from columns. To provide sufficient resistance to shear at the junction of columns and floor, haunched or square-headed columns are often formed. Figure 7.39 is an illustration of this floor. The deadweight of this floor and its cost are greater than for the floor systems previously described, but its depth is less and this latter advantage provides the least

Concrete flat slab floor
heavily reinforced in wide
bands between columns

Concrete columns

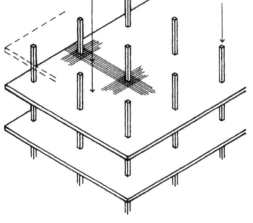

Figure 7.39   Flat slab (plate) floor.

overall depth of construction in multi-storey buildings. The floor slabs in the floor systems described earlier may be of solid reinforced construction or constructed with one of the hollow, or beam or plank floor systems.

In modern buildings, it is common to run air conditioning, heating, lighting and fire fighting services on the soffit of floors above a false ceiling, and these services occupy some depth below which minimum floor heights have to be provided. Even though the beam and slab or waffle grid floors are the most economic forms of construction in themselves, they may well not be the most advantageous where the services have to be fixed below and so increase the overall depth of the floor from the top of the slab to the soffit of the false ceiling below, because the services will have to be run below beams and so increase the depth between false ceiling and soffit of slab. Here it may be economic to bear the cost of a flat slab or drop slab floor to achieve the least overall height of construction and its attendant saving in cost. Up to about a third of the cost of an in situ cast reinforced concrete frame goes to providing, erecting and striking the formwork and falsework for the frame and the centring for the floors. It is important, therefore, to maintain a uniform section of column up the height of the building and repetitive floor and beam design as far as possible, so that the same formwork may be used at each succeeding floor. Alteration of floor design and column section involves extravagant use of formwork. Uniformity of column section is maintained by using high-strength concrete with a comparatively large percentage of reinforcement in the lower, more heavily loaded storey heights of the columns, and progressively less strong concrete and less reinforcement up the height of the building.

Precast reinforced concrete floor systems
Precast reinforced concrete floor beams, planks, T-beams or beam and infill blocks that require little or no temporary support and on which a screed or structural concrete topping

is spread, are commonly used with structural steel frames and may be used for in situ cast concrete frames instead of in situ cast floors. Precast beams and plank floors that require no temporary support in the form of centring are sometimes referred to as self-centring floors. The use of these floor systems with skeleton frame reinforced concrete multi-storey buildings is limited by the difficulty of hoisting and placing them in position and the degree to which the operation would interrupt the normal floor by floor casting of slabs and columns.

Precast hollow floor units

Precast hollow floor units are large precast reinforced concrete, hollow floor units, usually 400 or 1200 mm wide, 110, 150, 200, 250 or 300 mm thick and up to 10 m long for floors, and 13.5 m long for roofs. The purpose of the voids or hollows in the floor units is to reduce deadweight without affecting strength. The reinforcement is cast into the webs between hollows. Hollow precast reinforced concrete floor units can be used by themselves as floor slab with a levelling floor screed or they may be used with a structural reinforced concrete topping with tie bars over beams for composite action with the beams. When used for composite action, it is usual to fix the reinforcing tie bars into slots in the ends of units. These tie bars are wired to loops of reinforcement cast in and protruding from the top of beams for the purpose of continuity of structural action. End bearing of these units should be a minimum of 75 mm on steel and concrete beams, and 100 mm on masonry and brick walls. Figure 7.40 is an illustration of precast hollow floor units bearing on an in situ cast beam.

Precast concrete plank floor units

Precast concrete plank floor units are comparatively thin, prestressed solid plank, concrete floor units, which are little used with skeleton frame concrete structures, are designed as permanent centring and for composite action with reinforced concrete topping. The units are 400 or 1200 mm wide, 65, 75 or 100 mm thick and up to 9.5 m long for floors, and 10 m

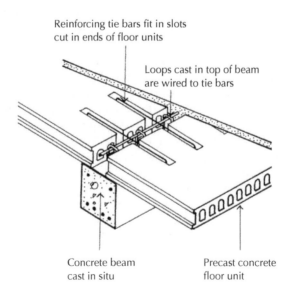

Reinforcing tie bars fit in slots
cut in ends of floor units

Loops cast in top of beam
are wired to tie bars

Concrete beam
cast in situ

Precast concrete
floor unit

Figure 7.40   Precast concrete floor units.

long for roofs. It may be necessary to provide some temporary propping to the underside of these planks until the concrete topping has gained sufficient strength.

Precast concrete T-beams

Precast prestressed concrete T-beam floors are mostly used for long-span floors in such buildings as stores, supermarkets, swimming pools and multi-storey car parks, where there is a need for wide-span floors and the depth of this type of floor is not a disadvantage. The floor units are cast in the form of a double T. The strength of these units is in the depth of the ribs that support and act with the comparatively thin top web. A structural reinforced concrete topping is cast on top of the floor units, which bear on the toe of a boot section concrete beam.

Precast beam and filler block floor

The precast beam and filler block floor system consists of precast reinforced concrete planks or beams that support precast hollow concrete filler blocks, as illustrated in Figure 7.41. The planks or beams are laid between supports with the filler blocks between them, and a concrete topping is spread over the planks and filler blocks. The reinforcement protruding from the top of the plank acts with the concrete topping to form a reinforced concrete

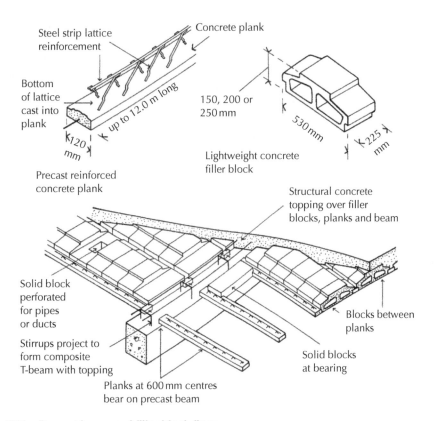

Figure 7.41   Precast beam and filler block floor.

beam. The advantage of this system is that the lightweight planks or beams and filler blocks can be lifted and placed in position much more easily than the much larger hollow concrete floor units.

Hollow clay block and concrete floor
A floor system of hollow clay blocks and in situ cast reinforced concrete beams between the blocks and concrete topping, cast on centring and falsework, was for many years extensively used for the fire-resisting properties of the blocks. This floor system is not much used nowadays because of the considerable labour in laying the floor.

7.8 Precast reinforced concrete frames

Precast concrete has been established as a sound, durable material for framing and cladding buildings, where repetitive casting of units is an acceptable and economic form of construction. Precast concrete elements are manufactured in highly controlled factory conditions, which enable high degrees of accuracy and finishing to be achieved consistently and economically. Most manufacturers stock a range of standard items and can also produce highly unusual and complex architectural forms simply by adjusting the design of the moulds used. The shape or profile of a precast unit can be straight or curved, and the level of detail on the surface finish (known as the degree of 'articulation') can be highly complex if required. The range of colours is also extensive, with different aggregates and additions used to provide the required colour. Precast units, such as lintels and floor units, can be reclaimed and reused when buildings are deconstructed, or the units can be crushed and the aggregates and steel reinforcement recycled and incorporated into new recycled content products.

The chief challenge with precast concrete framework is joining the members on site, particularly if the frame is to be exposed, to provide a solid, rigid bearing in column joints and a strong, rigid bearing of beams to columns that adequately ties beams to columns for structural rigidity. Where the frame is made up of separate precast column and beam units, there is a proliferation of joints. The number of site joints is reduced by the use of precast units that combine two or more column lengths with beams, as illustrated in Figure 7.42. The number of columns and beams that can be combined in one precast unit depends on the particular design of the building and the facilities for casting, transporting, hoisting and fixing units on site. Precast companies usually work with specialist fixing teams to help ensure that the units are installed correctly. Barcoding strips or e-tags can be embedded in the units to assist with timely distribution and delivery to site and the accurate identification and installation of units on site.

The general arrangement of precast structural units is as separate columns, often two-storey height and as cruciform H or M frames. The H frame unit is often combined with under window walling, as illustrated in Figure 7.42.

The two basic systems of jointing used for connections of column to column are by direct end bearing or by connection to a bearing plate welded to protruding studs. Direct bearing of ends is effected through a locating dowel, which can also be used as a post-tensioning connection, as illustrated in Figure 7.43. A coupling plate connection is made by welding

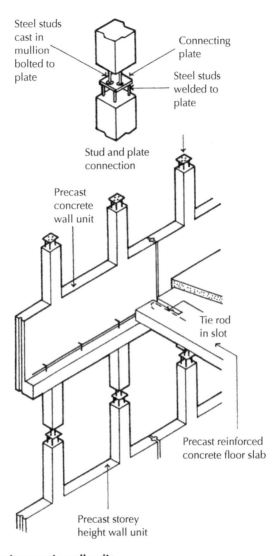

Steel studs cast in mullion bolted to plate

Connecting plate

Steel studs welded to plate

Stud and plate connection

Precast concrete wall unit

Tie rod in slot

Precast reinforced concrete floor slab

Precast storey height wall unit

Figure 7.42   Precast concrete wall units.

a plate to studs protruding from the end of one column and bolting studs protruding from the other to the plate, as illustrated in Figure 7.43. The studs and plate must be accurately located or there will be an excessive amount of site labour in making this connection. The completed joint is usually finished by casting concrete around the joint. Alternatively, the joint may be made with bronze studs and plate and left exposed as a feature of an externally exposed frame.

One method of joining beams to columns is by bearing on a haunch cast in the columns and by connecting a steel box, cast in the end of beams, to an angle or plate set in a housing in columns, as illustrated in Figure 7.44.

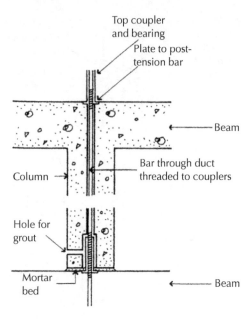

Figure 7.43 Precast concrete frame to frame joint.

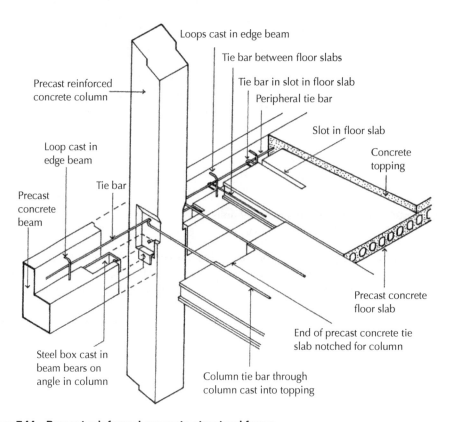

Figure 7.44 Precast reinforced concrete structural frame.

A steel box is cast into the protruding ends of beams, which bear on to a steel angle plate cast into a housing in the column. A bolt is threaded through a hole in the end of the steel box end of one beam, a hole in the column and a hole in the box in the end of the next beam, and secured with a nut. This firmly clamps the beams to the columns.

The precast, hollow floor units bear on the rebate in beams. Slots cast in the floor units accommodate steel tie bars, which are hooked over peripheral tie bars. The tie bars are wired to loops, which are cast into beams. Structural concrete topping is spread, compacted and levelled over the floor slabs and into the space between the ends of slabs and beams to form an in situ reinforced concrete floor. Precast floor units bear either directly on concrete beams or, more usually, on supporting nibs cast for the purpose. Ends of floor units are tied to beams through protruding studs or in situ cast reinforcement so that the floor units serve to transfer wind pressure back to an in situ cast service and access core.

The precast reinforced concrete wall frames, illustrated in Figure 7.45, that combine four columns with a beam, were used with drop-in beams as the structural wall frame system for a 22-storey block of flats. The precast framework is tied to the central core through the precast concrete floor units at each floor level, which are dowel fixed to the precast frame and tied with reinforcement to the in situ core; the precast framework is vertically tensioned by couplers through columns, as illustrated in Figure 7.46, so that column ends are compressed to the dry mortar bed. Storey height frames are linked by short lengths of beam, which are dropped in and tied to the frames. The precast frame-work was designed for rapid assembly through precasting and direct bearing of beams on columns and end bearing of columns, to avoid the use of in situ cast joints that are laborious to make and which necessitate support of beams while the in situ concrete hardens. The top hung, exposed aggregate, precast concrete cladding panels have deep rebate horizontal joints and open vertical joints with mastic seals to columns, as illustrated in Figures 7.45 and 7.46.

Precast concrete wall frames

Precast concrete wall frames were used extensively in Russia and northern European countries in the construction of multi-storey housing, where repetitive units of accommodation were framed and enclosed by large precast reinforced concrete wall panels that served as both external and internal walls and as a structural frame. The advantages of this system of building are that large, standard, precast concrete wall units can be cast offsite and rapidly assembled on site largely independently of weather conditions, a prime consideration in countries where temperatures are below freezing for many months of the year.

Reinforced concrete wall frames can support the loads of a multi-storey building, can be given an external finish of exposed aggregate or textured finish that requires no maintenance, can incorporate insulation either as a sandwich or lining and have an internal finish ready for decoration. Window and door openings are incorporated in the panels so that the panels can be delivered to site with windows and doors fixed in position. In this system of construction, the prime consideration is the mass production of complete wall units off the site, under cover, by unskilled or semi-skilled labour assisted by mechanisation as far as practical towards the most efficient and speedy erection of a building. The appearance of the building is a consequence of the chosen system of production and erection.

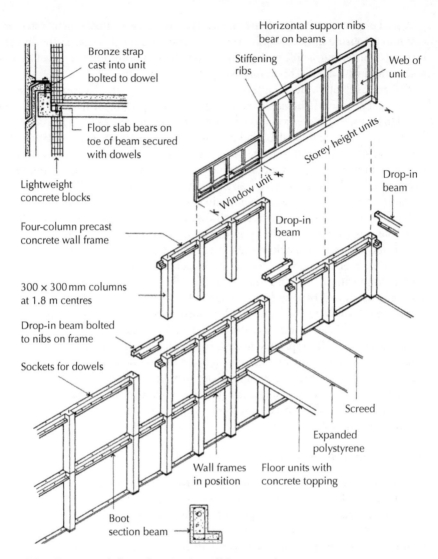

Figure 7.45 **Precast reinforced concrete wall frames.**

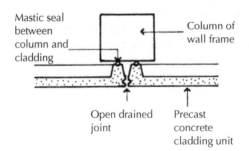

Figure 7.46 **Vertical joint to cladding units.**

The concrete wall units will give adequate protection against wind and rain by the use of rebated horizontal joints and open-drained vertical joints with back-up air seals similar to the joints used with precast concrete cladding panels. Some systems of wall frame incorporate a sandwich of insulation in the thickness of the panel, with the two skins of concrete tied together across the insulation with non-ferrous ties. This is not a very satisfactory method of providing insulation as a sufficient thickness of insulation, for present-day standards will require substantial ties between the two concrete skins, and the insulation may well absorb water from drying out of concrete and rain penetration, and so be less effective as an insulant. For the best effect, the insulation should be applied to the inside face of the wall as an inner lining to panels, or as a site fixed or built inner lining or skin.

The wall frame system of construction depends, for the structural stability of the building, on the solid, secure bearing of frames on each other, and the firm bearing and anchorage of floor units to the wall frames and back to some rigid component of the structure, such as in situ cast service and access cores. Figure 7.47 is an illustration of a typical precast concrete wall frame.

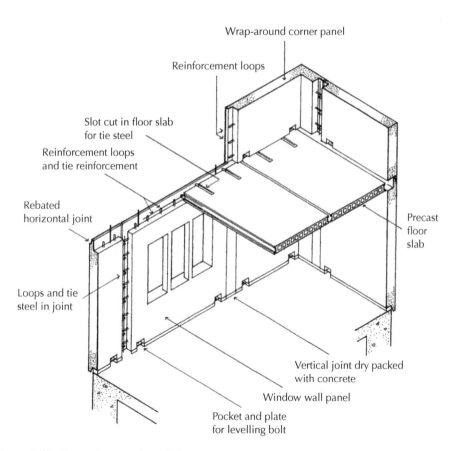

Figure 7.47   Precast concrete wall frame.

7.9 Lift slab construction

In the lift slab system of construction, the flat roof and floor slabs are cast one on the other at ground level around columns or in situ cast service, stair and lift cores. Jacks operating from the columns or cores pull the roof and floor slabs up into position. This system of construction was first employed in America in 1950. Since then, many buildings in America, Europe and Australia have been constructed by this method. The advantage of the system is that the only formwork required is to the edges of the slabs, and no centring whatever is required to the soffit of roof or floors. The slabs are cast monolithically and can be designed to span continuously between and across points of support, and so employ the least thickness of slab. Where it is convenient to cantilever slabs beyond the edge columns and where cantilevers for balconies, for example, are required, they can, without difficulty, be arranged as part of the slab. The advantages of this system are employed most fully in simple, isolated point block buildings of up to five storeys, where the floor plans are the same throughout the height of the building and a flush slab floor may be an advantage. The system can be employed for beam and slab, and waffle grid floors, but the forms necessary between the floors to give the required soffit take most of advantage of simplicity of casting on the ground.

Steel or concrete columns are first fixed in position and rigidly connected to the foundation, and the ground floor slab is then cast. When it has matured, it is sprayed with two or three coats of a separating medium consisting of wax dissolved in a volatile spirit. As an alternative, polythene sheet or building paper may be used as a separating medium. The first floor slab is cast inside the edge formwork on top of the ground floor slab, and when it is mature it is in turn coated or covered with the separating medium and the next floor slab is cast on top of it. The casting of successive slabs continues until all the floors and roof have been cast one on the other on the ground. Lifting collars are cast into each slab around each column. The slabs are lifted by jacks, operating on the top of each column, which lift a pair of steel rods attached to each lifting collar in the slab being raised. A central control synchronises the operation of the hydraulically operated reciprocating ram-type jacks to ensure a uniform and regular lift.

The sequence of lifting the slabs depends on the height of the building, the weight of the slabs and extension columns, the lifting capacity of the jacks and the cross-sectional area of the columns during the initial lifting. The bases of the columns are rigidly fixed to the foundations so that when lifting commences the columns act as vertical cantilevers. The load that the columns can safely support at the beginning of the lift limits the length of the lower column height and the number of slabs that can be raised at one time. As the slabs are raised, they serve as horizontal props to the vertical cantilever of the columns and so increasingly stiffen the columns, the higher the slabs are raised. The sequence of lifting illustrated in Figure 7.48 is adopted so that the roof slab, which is raised first, stiffens the columns, which are then capable of taking the load of the two slabs subsequently lifted, as illustrated. The steel lifting collars, which are cast into each slab around each column, provide a means of lifting the slabs and also act as shear reinforcement to the slabs around columns, and so may obviate the necessity for shear reinforcement to the slabs. Figure 7.49

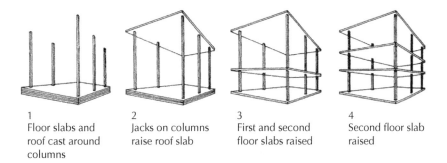

1
Floor slabs and
roof cast around
columns

2
Jacks on columns
raise roof slab

3
First and second
floor slabs raised

4
Second floor slab
raised

Figure 7.48   Sequence of lifting slabs.

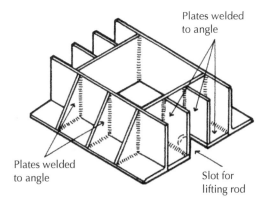

Plates welded
to angle

Plates welded
to angle

Slot for
lifting rod

Figure 7.49   Lifting collar.

is an illustration of a typical lifting collar fabricated from mild steel angle sections welded together and stiffened with plates welded in the angle of the sections.

The lifting collars are fixed to steel columns by welding shear blocks to plates welded between column flanges and to the collar after the slab has been raised into position, as illustrated in Figure 7.50. Connections to concrete columns are made by welding shear blocks to the ends of steel channels cast into the column and by welding the collar to the wedges, as illustrated in Figure 7.50. With this connection, it is necessary to cast concrete around the exposed steel wedges for fire protection. The connection of steel extension columns is made by welding, bolting or riveting splice plates to the flanges of columns at their junction. Concrete extension columns are connected either with studs protruding from column ends and bolted to a connection plate, or by means of a joggle connection.

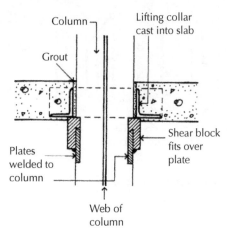

Connection of slab to steel column

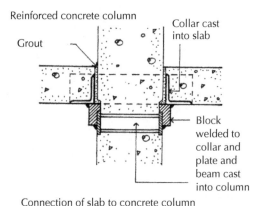

Connection of slab to concrete column

Figure 7.50   Connection of slab to columns.

Further reading

See www.thisisukconcrete.co.uk for further information about concrete and its decarbonisation. See also The concrete Society: www.concrete.org.uk.

Reflective exercises

Why would you choose a structural concrete frame over a structural steel or structural timber frame? What factors do you need to consider and how would you justify your decision to your client?

Your client is concerned about the carbon footprint of concrete and has asked you to justify its use. Your client is keen to use local materials and labour and aims to have zero-carbon building developments by 2030.

❏   How would you advise your client and how would you justify and support your reasoning?

Envelopes to Framed Buildings: Chapter 8 AT A GLANCE

What? The term 'building envelope' has come into widespread use to describe the skin that encloses structural framed buildings. The terms cladding and curtain walling are widely used to describe the materials that enclose structural framed buildings. A variety of materials may be used to form the cladding panels and curtain wall. These range from glass to concrete, stone, terracotta, metals and recycled content material.

Why? The primary function of the building envelope is to provide weather protection to the building interior. This may be achieved via a single skin or a 'double-skin' to help temper the internal environment. Cladding and curtain walling units, and particularly the joints, must be able to resist water and wind penetration to keep the building watertight and airtight, while also resisting lateral wind forces. Concomitant with wall construction, the building envelope must comply with fire regulations in terms of combustibility and surface spread of flame, while also providing thermal and acoustic insulation. The building envelope must also provide adequate natural light while minimising glare and overheating from solar radiation.

When? Once the structural frame has been erected, it is then possible to fix the cladding or curtain walling to, or between, the structural envelope to form a weatherproof envelope. It is also possible to fix the cladding and curtain walling as the frame is being erected for high-rise buildings. Over-cladding an existing building to improve thermal performance and appearance may require the installation of a suitable framework on which to fix the cladding or the replacement of existing fixings. Overcladding may have an impact on the fire resistance of the fabric, which will need careful consideration.

How? Individual units are fixed, or hung, on a framework to form a weatherproof skin. Some cladding systems include thermal insulation within the panels, but most systems rely on additional thermal insulation behind the external façade. This build up of elements needs to be designed to ensure all cavities are fire stopped in accordance with legislative requirements and manufacturers' instructions. While the majority of high-rise buildings rely on mechanical ventilation systems, it is becoming more common for designers to incorporate a degree of natural ventilation into the design of the building. A building envelope should have adequate strength to support its own weight between points of support or restraint fixing to the structural frame, and sufficient stability against lateral wind pressures. The design of the envelope must allow for differential movement to avoid damage to fixings and the materials that make up the envelope. This requires specialist design and engineering skills.

8 Envelopes to Framed Buildings

The structural frame provides the possibility of endless variation in the form and appearance of buildings that no longer need to be constrained by the limitations of loadbearing construction. Building designers may choose to express the structural frame or conceal it behind the building envelope. Concrete and steel are the most common framing materials for high-rise buildings, although timber is an alternative choice for up to around 10 storeys. Whatever the choice of frame, it is important to remember that it will form part of the building envelope and may affect the envelope's functional requirements. The use of the various materials for the external wall is, to some extent, influenced by the relative behaviour of the structural frame and the wall to accommodate differential structural, thermal and moisture movements, which affect the functional requirements of a wall. The finished appearance of the external wall is another significant consideration, and it is possible to create some highly creative buildings with the use of cladding, as illustrated in Photograph 8.1.

8.1 Terms and definitions

A variety of terms and definitions exist in relation to the external skin of framed buildings. The term used in this chapter is 'envelope', which covers a variety of construction methods ranging from cladding and curtain wall construction to adaptive façades and double skins.

The term 'cladding' came into general use as a description of the external envelope of framed buildings, which clothed or clad the building in a protective coating that was hung, supported by or secured to the structural frame. The word 'facings' has been used to describe materials used as a thin, non-structural, decorative, external finish such as the thin, natural stone facings applied to brick or concrete backing. The word 'façade' and the term 'façade engineering' are used more generally to cover the materials and technologies used to form the external envelope of a framed building. As the technologies have evolved, the amount of design freedom has also improved, as witnessed by the use of terms such as 'advanced building skins' and 'adaptive envelopes' that have come into widespread use.

The word 'wall' or 'walling' will be used to describe the use of those materials such as stone, brick, concrete and blocks that are used as the external envelope of framed buildings, where the appearance is of a continuous wall to the whole or part of several storeys or as

Barry's Advanced Construction of Buildings, Fifth Edition. Stephen Emmitt.
© 2023 John Wiley & Sons Ltd. Published 2023 by John Wiley & Sons Ltd.

Photograph 8.1   Creative use of cladding.

walling between exposed, supporting beams and columns of the frame. The word 'cladding' will be used to describe panels of concrete, glass fibre-reinforced cement (GRC), glass reinforced plastic (GRP), glass, compressed/constituted-rock-based products and metal fixed to and generally hung from the frame by supporting beams or inside light framing as a continuous outer skin to the frame. The external walls of framed buildings are broadly grouped as:

❑   Infill wall framing to a structural grid.
❑   Solid and cavity walling of stone, block and brick.
❑   Facings applied to cavity background walls.
❑   Cladding panels of precast concrete, GRC, GRP and rock-based products.
❑   Thin sheet cladding of metal.
❑   Glazed wall systems.
❑   Double wall systems.

8.2 Functional requirements

Under load, both steel and concrete structural frames suffer elastic strain and consequent deflection (bending) of beams and floors and shortening of columns. Deflection of beams and floors is generally limited to about one three-hundredth of span, to avoid damage to supported facings and finishes. Shortening of columns by elastic strain under load can be of the order of 2 mm for each storey height of about 4 m, depending on the load. Elastic short-ening of steel columns may be of the order of 1 mm per storey height. The comparatively small deflection of beams and shortening of columns under load can be accommodated by the joints in materials such as brick, stone and block and the joints between panels, without adversely affecting the function of most wall structures.

Unlike steel, concrete suffers drying shrinkage and creep in addition to elastic strain. Dry-ing shrinkage occurs as water, necessary for the placing of concrete and setting of cement, migrates to the surface of concrete members. The rate of loss of water and consequent shrinkage depend on the moisture content of the mix, the size of the concrete members and atmospheric conditions. Drying shrinkage of concrete will continue for some weeks after placing. For the small members of a structural frame, drying out of doors in summer, about half of the total shrinkage takes place in about one month and about three-quarters in six months. For larger masses of concrete, about half of the total shrinkage will take place one year after placing. The bond between the concrete and the reinforcement restrains dry-ing shrinkage of concrete. Concrete in heavily reinforced members will shrink less than that in lightly reinforced sections. Drying shrinkage of the order of 2–3 mm for each 4 m of column length may well occur. The shrinkage occurs due to temperature variation and hydration. As the concrete sets, it produces heat from the chemical reaction, which reduces over time. As the concrete hydrates, water is used in the chemical reaction and will also evaporate from the concrete – this will cause the concrete to shrink.

Creep of concrete is dependent on stress and is affected by humidity and by the cement content and the nature of the aggregate in concrete. The gradual creep of concrete may con-tinue for some time; however, shortening of columns is minimal. Depending on the nature of the concrete, shrinkage due to creep could be of the order of 2 mm for each storey height of column over the long term. Like drying, shrinkage creep is restrained by reinforcement. Creep is much more of a problem in beams than it is in columns.

The combined effect of elastic strain, drying shrinkage and creep in concrete may well amount to a total reduction of up to 6 mm for each storey height of building. Because of these effects, it is necessary to make greater allowance for shortening in the design of wall structures supported by an in situ cast concrete frame than it is for a steel frame. Solid wall structures such as brick which are built within or supported by a concrete structural frame should be built with a 12–15 mm compression joint at each floor level to avoid dam-age to the wall by shortening of the frame and expansion of the wall materials due to ther-mal and moisture movements.

Experience shows that there are generally considerably greater inaccuracies in line and level with in situ cast concrete frames than there are with steel frames. There is an engineer-ing tradition of accuracy of cutting and assembling steel that is not matched by the usual assembly of formwork for in situ cast concrete. Deflection of formwork under the load of wet concrete and some movement of formwork during the placing and compaction of concrete combine to create inaccuracies of line and level of both beams and columns in concrete frames that may be magnified up the height of multi-storey buildings. Allowances for these inaccuracies can be made where fixings for cladding are made by drilling for bolt fixings rather than relying on cast-on or cast-in supports and fixings. The advantage of the precast concrete frame is in the greater accuracy of casting of concrete under controlled factory conditions than on site.

Functional requirements

The primary functional requirements of a building envelope are:

- ❏ Strength and stability.
- ❏ Resistance to weather.
- ❏ Durability and freedom from maintenance.
- ❏ Fire safety.
- ❏ Resistance to the passage of heat.
- ❏ Resistance to the passage of sound.

Strength and stability

A building envelope should have adequate strength to support its own weight between points of support or restraint fixing to the structural frame, and sufficient stability against lateral wind pressures (see Photographs 8.2 and 8.3). The design of the envelope must allow for differential movement to avoid damage to fixings and the materials that make up the envelope. Brick and precast concrete cladding panels tend to not suffer the rapid changes of temperature between day and night that thin materials do, because they act to store heat and lose and gain heat slowly. Thin sheet wall materials, such as GRP, metal and glass, suffer rapid changes in temperature and consequent expansion and contraction, which may cause distortion and damage to fixings or the thin panel material or both.

In the design of wall structures faced with thin panel or sheet material, the ideal arrangement is to provide only one rigid support fixing to each panel or sheet, with one other flexible support fixing and two flexible restraint fixings. The need to provide support and restraint fixings with adequate flexibility to allow for thermal movement and at the same

Photograph 8.2   Terracotta cladding panel.

Photograph 8.3   Fixing of cladding panel to structural frame.

time adequately restrain the facing in place and maintain a weathertight joint has been the principal difficulty in the use of thin panel and sheet facings.

Resistance to weather

Brick and stone exclude rain from the inside of buildings by absorbing rainwater, which evaporates to outside air during dry periods. The least thickness of solid wall material necessary to prevent penetration of rainwater to the inner face depends on the degree of exposure to driving rain. Common practice is to construct a cavity wall with an outer leaf of masonry as a rain screen, a cavity and an inner leaf that provides adequate thermal resistance to the passage of heat, and an attractive finish.

Precast concrete wall panels act in much the same way as brick by absorbing rainwater. Because of the considerable size of these panels, there have to be comparatively wide joints between panels to accommodate structural, thermal and moisture movements. The joints are designed with a generous overlap to horizontal joints and an open drained joint to vertical joints to exclude rain. Non-absorbent sheet materials, such as metal and glass, cause-driven rain to flow under pressure in sheets across the face of the wall, so making the necessary joints between panels of the material highly vulnerable to penetration by rain. These joints should at the same time be sufficiently wide to accommodate structural, thermal and moisture movements and serve as an effective seal against rain penetration. The materials that are used to seal joints are mostly short-lived as they harden on exposure to atmosphere and sun, and lose resilience in accommodating movement. The 'rain screen' principle is designed to provide a separate outer skin, to screen wall panels from damage by wind and rain and deterioration by sunlight, and to improve the life and efficiency of joint seals. This can also enhance the aesthetic appeal of a building.

Durability and freedom from maintenance

The durability of the external envelope is a measure of the frequency and extent of the work necessary to maintain minimum functional requirements and acceptable appearance. Brick and stone facing require very little maintenance over the expected life of most buildings. Precast concrete wall panels, which weather gradually may become dirt stained due to slow run-off of water from open horizontal joints.

Panels of glass will maintain their finish over the expected life of buildings but will require frequent cleaning of the surface and periodic renewal of seals. Self-cleaning glass and self-cleaning panels may help to reduce the frequency and cost of cleaning. Of the sheet metal facings that can be used for wall structures, bronze and stainless steel, both expensive materials, will weather by the formation of a thin film of oxide that is impermeable and prevents further oxidation. Aluminium weathers with a light-coloured, coarse-textured, oxide film that considerably alters the appearance of the surface, although the material can be anodised to inhibit the formation of an oxide film or coated with a plastic film for the sake of appearance. Steel, which progressively corrodes to form a porous oxide, is coated with zinc, to inhibit the rust formation and a plastic film as decoration. None of the plastic film coatings are durable as they lose colour over the course of a few years on exposure to sunlight, and this irregular colour bleaching may well not be acceptable from the point of view of appearance to the extent that painting or replacement may be necessary in 10–25 years. Periodic maintenance and renewal of seals to joints between metal-faced panels will be necessary to maintain performance.

Fire safety

Buildings must be designed and constructed to limit the spread of fire within the building, over the external surface of the building, and to other buildings. The design of the building envelope will include the external finish, such as cladding panels, and the materials that lie behind the façade (i.e. insulating materials and the internal face of the wall). The primary concern is the combustibility of the materials used, the resistance of the building envelope to the external surface spread of fire and the prevention of smoke and fire spread in concealed spaces, such as cavities within the building envelope. For low- and medium-rise buildings it may be possible to design and construct the building envelope following the guidance as set out in Approved Document B and associated codes, together with fire testing. For larger buildings with floors over 18 m above ground level and high-rise buildings, it will be necessary to design the building with fire engineers and construct the building envelope in accordance with the fire safety strategy for the building. This will involve a more restricted choice of cladding materials to comply with fire legislation.

Fire resistance must be maintained in situations where the external envelope interacts with compartment floors and walls to ensure the integrity of the compartmentation is maintained. Fire stopping is required between the external envelope and the compartment wall or floor to provide the same period of fire resistance as the floor or wall. The fire stop is designed to prevent the transfer of heat and the passage of smoke. Fixings are required to maintain the integrity of the joint between the compartment wall or floor and the external envelope. The position of the fixings in relation to the fire stops and the fire resistance of the fixings need to be considered as an integral part of the fire safety design.

The need to work to the building's fire safety strategy also applies to existing buildings. Particular care is needed to ensure the fire safety strategy is not compromised when making changes to the building envelope, for example when over-cladding the building to improve the appearance and thermal performance of the building. Detailed surveys are required to establish if the building's fire safety strategy has been maintained since it was constructed, or whether it has been altered (improved or compromised) through renovations and associated work. Once this has been established it is possible to design the external fabric in accordance with the (revised) fire safety strategy for the building.

Given the bespoke nature of curtain walling and cladding systems, it is essential that designers work with manufacturers and specialist suppliers to ensure fire safety is met. Manufacturers of cladding and curtain wall systems provide notional fire resistance for their products and systems as used in a specific manner. The figures provide a useful guide to the expected fire resistance of the building envelope; however, the actual detailing of the envelope, especially at the junction with the structural frame, will affect fire safety. Testing may be required to establish the actual fire resistance of the proposed construction.

Resistance to the passage of heat

The interiors of buildings clad with large areas of glass may be prone to unwanted heat gain. Solar heat gain (and associated solar glare) can be controlled through the use of simple shading devices fixed externally and/or internally to the building fabric and via the use of double skins (double façades) that control the internal environment through a series of layers that include passive and active solar shading.

Thermal insulation is required to prevent heat from passing through the fabric, from inside to out and vice versa. Simply adding layers of insulation to the building fabric does not always provide the best solution, especially where aesthetics and space requirements are at a premium. High-performing insulation products are available, such as aerogels and vacuum-insulated panels (Figure 8.1), which offer better thermal resistance for the same thickness. Combined with more traditional insulation materials, it is possible to reduce thermal bridging and provide an effective thermal barrier within a relatively thin building envelope (Table 8.1).

Thermal bridge
Members of a structural frame may act as a thermal bridge where the wall is built up to, or built between the frame, as illustrated in Figure 8.2, where the resistance to the thermal transfer through the brick slips and beam is appreciably less than through the rest of the wall. Similarly, there is a thermal bridge across a precast wall panel and a beam and column, as illustrated in Figure 8.3. It may be difficult to provide an effective way of preventing the thermal bridge formed by the supporting structural frame. The effect of the bridge may be modified by the use of floor insulation and a suspended ceiling or by setting frame members, where possible, back from the outer face of the wall, as illustrated in Figure 8.4. Wall panels of precast concrete, GRP and GRC have been used with a sandwich or inner lining of an insulating material. This arrangement is not entirely effective because the insulating

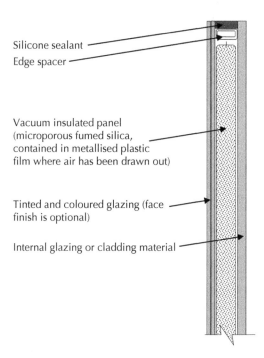

Silicone sealant

Edge spacer

Vacuum insulated panel
(microporous fumed silica,
contained in metallised plastic
film where air has been drawn out)

Tinted and coloured glazing (face
finish is optional)

Internal glazing or cladding material

Figure 8.1   Glazed vacuum-insulated cladding panel.

Table 8.1 Comparison of insulation materials used within cladding systems

Product	Properties	Approximate thickness to achieve a U-value of 0.2W/m²K (dependent on construction – for comparison only) (mm)	W/mK thermal conductivity
Cellulose fibre	Insulants used for construction where space is not at a premium	175–230	≈0.035–0.045
Glass wool fibre	In combination with rigid insulants can reduce thermal bypass – with acoustic properties – economic solution where space is not at a premium	160–220	≈0.031–0.044
Rook wool fibre	In combination with rigid insulants can reduce thermal bypass – with acoustic properties; economic solution where space is not at a premium	155–250	≈0.034–0.042
Expanded polystyrene	Rigid insulation – reduced thermal conductivity	140–170	≈0.031–0.038
Extruded polystyrene foam	Rigid insulation – reduced thermal conductivity	130–190	≈0.029–0.039
Polyurethane foam with CO_2	Mid-range insulants	160–180	≈0.034–0.036
Polyurethane foam with Pentance		110–150	≈0.022–0.030
Phenolic foam		120–130	≈0.024–0.026
Polyisocyanurate foam		110–120	≈0.022–0.024
Phenolic foam with foil face	Upper mid-range insulants	80–120	≈0.017–0.022
Polyisocyanurate foam with foil face		80–90	≈0.018
Aerogel blanket	High-performing insulants – advanced cladding; relatively high costs	60–70	≈0.010
Vacuum insulation	High-performing insulants – advanced cladding; relatively high costs	25	≈0.005

material, if open-pored as are many insulating materials, may absorb condensate water, which will reduce its thermal properties, and the edge finish to panels, necessary for rigidity and jointing, will act as a thermal bridge. Thin metal wall panel materials or composite panels, which are supported by a metal carrier system fixed across the face of the structural frame, can provide thermal insulation more effectively by a sandwich, inner lining or inner skin of insulating material with the edge jointing material acting as a thermal break in the narrow thermal bridge of the edge metal, as illustrated in Figure 8.5.

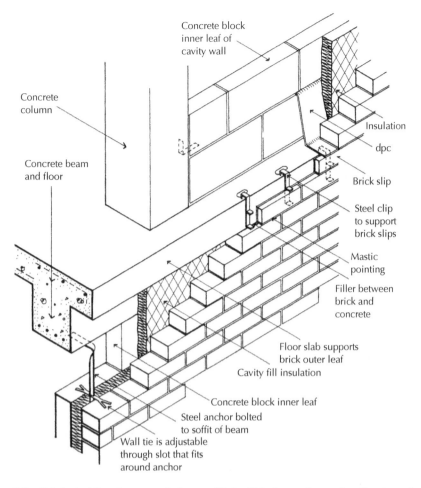

Concrete block
inner leaf of
cavity wall

Concrete
column

Concrete beam
and floor

Insulation

dpc

Brick slip

Steel clip
to support
brick slips

Mastic
pointing

Filler between
brick and
concrete

Floor slab supports
brick outer leaf

Cavity fill insulation

Concrete block inner leaf

Steel anchor bolted
to soffit of beam

Wall tie is adjustable
through slot that fits
around anchor

Figure 8.2 Brick cladding to concrete frame. (*Note:* This form of construction is no longer used due to thermal bridge across the concrete floor and brick step. It may, however, still be found in the existing building stock.)

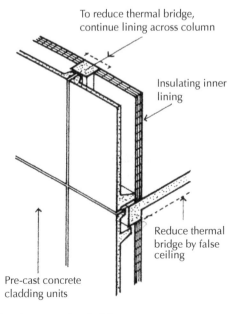

To reduce thermal bridge,
continue lining across column

Insulating inner
lining

Reduce thermal
bridge by false
ceiling

Pre-cast concrete
cladding units

Figure 8.3 Insulation lining to concrete cladding.

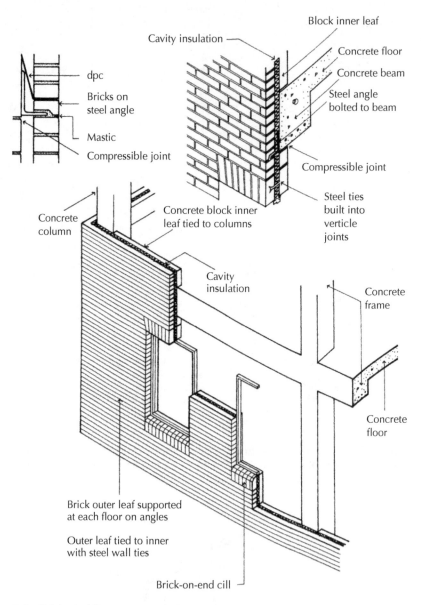

Figure 8.4 Brick cladding to structural frame.

The use of sealed glazing and effective weather seals to the joints of cladding panels and windows restrict the natural exchange of outside and inside air. Thus, it is necessary to use mechanical systems of air conditioning or natural ventilation strategies to provide sufficient air changes to ensure the comfort of the building occupants. Although the vast majority of high-rise buildings will employ mechanical air handling systems, it is possible to design buildings to benefit from natural ventilation control.

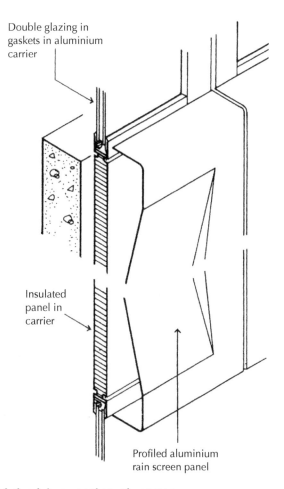

Double glazing in
gaskets in aluminium
carrier

Insulated
panel in
carrier

Profiled aluminium
rain screen panel

Figure 8.5   Profiled aluminium panel as rain screen.

Resistance to the passage of sound
Manufacturers of cladding and curtain wall systems provide notional sound-resistance figures for their products. The figures provide a useful guide to the expected noise reduction of a particular construction; however, the actual detailing of the cladding and curtain wall, especially at the junction with the structural frame, will affect the actual values. Similarly, the addition of thermal insulation behind the cladding can also help in reducing the passage of sound from outside to inside. As a general rule, the denser and thicker the material, the more effective it is as a barrier to airborne sound, as the dense mass absorbs the energy generated by the sound source. This will be influenced by the materials used for the structural frame and internal walls and floors the detailing of the junctions between the main elements. Double-skin building envelopes can also be highly effective in helping to reduce the transfer of sound from the external environment into the building.

8.3 Infill wall framing to a structural grid

Infill wall frames are fixed within the enclosing members of the structural frame or between projections of the frame, such as floors and roof slabs, which are exposed, as illustrated in Figure 8.6. The infill wall may be framed with timber or metal sections, with panels of an appropriate material secured within the frame.

The framing with its panels or sheet covering should have adequate strength and stability in itself to be self-supporting within the framing members and resist wind pressure and suction acting on it. Sufficient support and restraint fixings between the frame and the surrounding structural members are required. The framing, its panels and sheet covering must adequately resist the penetration of water to the inside face by a system of resilient mastic, drained and sealed joints. The joints between the framing and the structure should be filled with resilient filler and weather sealed with mastic to accommodate structural, moisture and thermal movement. To enhance the thermal resistance of the lightweight framing and covering materials, double glazing and/or solar control glass should be used with double skin insulated panels, insulation between framing members or behind sheet covering materials.

In the 1950s and 1960s, the infill wall frame system was much used in framed buildings, particularly for multi-storey housing in the UK. Many of the early infill wall frame systems suffered deterioration due to the use of steel framing poorly protected against corrosion, panel materials that absorbed water and poor jointing materials that gave inadequate protection against rain penetration. These failures, coupled with the introduction

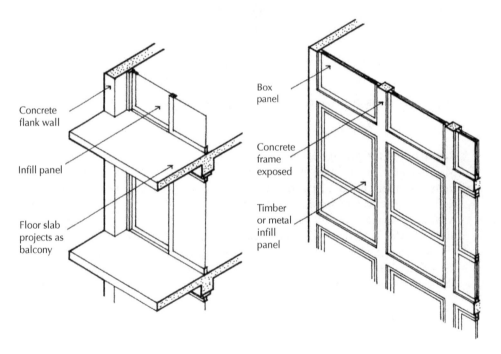

Figure 8.6   Infill panels.

of alternative walling materials such as concrete, GRC and GRP panels and glazed walls, led to loss of favour of wall infill framing. There were also problems with thermal bridging through the concrete frame, as would be the case with the building illustrated in Figure 8.6. Thermal bridging is difficult to design out of such structures, which are better suited to warm climates. In countries where summer temperatures are high and shade from the sun is a necessity, many buildings are constructed with a reinforced concrete frame with projecting floors and roof for shade and as an outdoor balcony area in summer, as illustrated in Figure 8.6. Because of the protection afforded by the projecting floor slabs and roof against wind-driven rain and the diminution of daylight penetration caused by these projections, in winter months, it is common practice to form fully glazed infill panels in this form of construction.

8.4 Cavity walling

In the early days of the multi-storey structural frame, solid masonry was used for the external walling, built as a loadbearing structure off the supporting framework. Ashlar natural stone, brickwork and terracotta blocks were used, which imposed considerable loads on the supporting frame and foundations. To improve thermal resistance and to provide a cavity as a barrier to the penetration of water to the inner face, it became usual practice to construct masonry walling as a cavity wall.

With the use of cavity walling, it was considered necessary to provide support for at least two-thirds of the thickness of the outer leaf of the wall and the whole of the inner leaf at each floor level. This posed difficulties where the external face was to have the appearance of a traditional loadbearing wall. The solution was to fix special brick slips to mask the horizontal frame members at each floor level, as illustrated in Figure 8.2. A disadvantage of these brick slips is that even though they are cut or made from the same clay as the surrounding whole bricks, they may tend to weather to a somewhat different colour from that of the whole bricks and so form a distinct horizontal band that defeats the original objective. An alternative to the use of brick slips at each floor level is to build the external leaf of the cavity walling directly off a projection of the floor slab with the floor slab exposed as a horizontal band at each floor level or to build the walling between floor beams and columns and so admit the frame as part of the façade. This technique has been largely abandoned because of the problem of thermal bridging through the exposed floor slab.

The strength and stability of solid and cavity walling constructed as cladding to framed structures depend on the support afforded by the frame and the resistance of the wall itself to lateral wind pressure and suction. As a general principle, the slenderness ratio of walling is limited to 27 : 1, where the slenderness ratio is the ratio of the effective height or length to effective thickness. The effective thickness of a cavity wall may be taken as the combined thickness of the two leaves. To provide the appearance of a loadbearing wall to framed structures, without the use of brick slips, it is usual practice to provide support for the outer leaf by stainless steel brackets or angles built into horizontal brick joints, as illustrated in Figure 8.4.

A common support for the brick outer leaf of a cavity wall is a stainless steel angle secured with expanding bolts to a concrete beam, as illustrated in Figure 8.4. Depending on the relative thickness of the supporting flange of the angle and the thickness of the mortar joints, the angle may be bedded in the mortar joint or the bricks bearing on the angle may be cut to fit over the angle. To allow for relative movement between walling and the frame, it is usual practice to form a horizontal movement joint at the level of the support angle by building in a compressible strip, which is pointed on the face with mastic to exclude water.

As an alternative to a continuous angle support, a system of support brackets may be used. These stainless steel brackets fit to a channel cast into the concrete. An adjusting bolt in each bracket allows some vertical adjustment and the slotted channel some horizontal adjustment, so that the supporting brackets may be accurately set in position to support brickwork as it is raised. The brackets are bolted to the channel to support the ends of abutting bricks, as illustrated in Figure 8.7. A horizontal movement joint is formed at the level of the bracket support. Supporting angles or brackets may be used at intervals of not more than every third-storey height of building or not more than 9 m, whichever is the less, except for four-storey buildings where the wall may be unsupported for its full height or 12 m, whichever is the less. Where support is provided at every third-storey height, the necessary depth of the compressible movement joint may well be deeper than normal brick joints and be apparent on the face of the wall.

To provide support for the wall against lateral forces, it is necessary to provide some vertical anchorage at intervals so that the slenderness ratio does not exceed 27 : 1. Fishtailed or flat anchors fitted to channels cast into columns are bedded in the face brickwork at the same intervals as wall ties, as illustrated in Figure 8.8, to provide lateral and vertical restraint. To provide horizontal, lateral restraint, anchors are fitted to slots in cast-in channels in beams or floor slabs at intervals of up to 450 mm. To provide anchorage to the top of the wall at each floor level where brick slips are used, it is usual to provide anchors that are bolted to the underside of the beam or slab and to fit stainless steel ties that are built into brickwork at 900 mm centres, as illustrated in Figure 8.2.

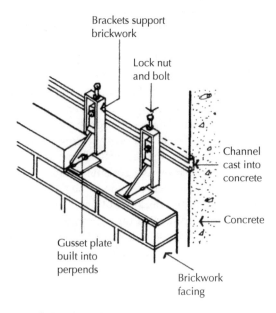

Figure 8.7   Loadbearing fixing for brickwork.

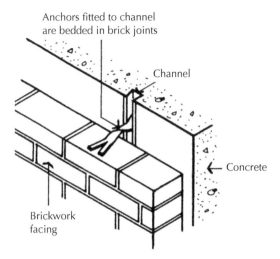

Figure 8.8 Restraint fixing for brickwork.

Where solid or cavity walling is supported on and built between the structural frame grid, some allowance should be made for movements of the frame, relative to that of the walling due to elastic shortening and creep of concrete, flexural movement of the frame, and thermal and moisture movements. Practice is to build in some form of compressible filler at the junction of the top of the walling and the frame members and the wall and columns as movement joints, with metal anchors set into cast-in channels in columns and bedded in brickwork and to both leaves of cavity walls at intervals similar to cavity wall ties.

Where cavity walling is built up to the face of columns of the structural frame and supported at every third storey, the support and restraint against lateral forces are provided by anchors. These anchors are fitted to cast-in channels and bedded in horizontal brick joints at intervals similar to cavity ties. To provide for movement along the length of walling, it is usual to form continuous vertical movement joints to coincide with vertical movement joints in the structural frame and at intervals of not more than 15 m along the length of continuous walling and at 7.5 m from bonded corners, with the joints filled with compressible strip and pointed with mastic. A wall of sound, well-burnt clay bricks should require no maintenance during the useful life of a building other than renewal of mastic pointing of movement joints at intervals of about 20–25 years.

Resistance to the penetration of wind-driven rain depends on the degree of exposure and the necessary thickness of the outer leaf of cavity walling and the cavity width. The use of cavity trays, and a damp-proof course (dpc) at all horizontal stops to cavities, is accepted practice. The purpose of these trays, illustrated in Figure 8.2, is to direct water that may collect inside the cavity away from the inner face of the wall. If the thickness of the outer leaf and the cavity is sufficient to resist penetration of water, there seems little logic in the use of these trays. To prevent water-soluble salts from the concrete of concrete frames finding their way to the face of brickwork and so causing unsightly efflorescence of salts, the face of concrete columns and beams that will be in contact with brickwork is painted with bitumen.

The requirements for resistance to the passage of heat usually necessitate the use of some material with comparatively good resistance to the transfer of heat, either in the cavity as cavity fill or partial fill with a lightweight block inner leaf, as illustrated in Figure 8.2. Where the cavity runs continuously across the face of the structural frame, as illustrated in Figure 8.4, the resistance to the transfer of heat of the wall is uninterrupted. Where a floor slab supports the outer leaf, as illustrated in Figure 8.2, there will be, to some extent, a cold bridge as the brick slips and the dense concrete of the floor slab will afford less resistance to the transfer of heat than the main cavity wall. The very small area of floor and ceiling may well be colder. Internal insulation around the floors, ceilings and columns may be used to reduce the impact of any cold bridges. Where internal insulation is used, vapour barriers should be used to prevent the warm, moisture-laden air reaching cold surfaces.

8.5 Facings applied to solid and cavity wall backings

The word 'facings' is used to describe comparatively thin, non-structural slabs of natural or reconstructed stone, faience, panels, ceramic and glass tiles or mosaic, which are fixed to the face of, and supported by, solid background walls or to structural frames as a decorative finish. Common to the use of these non-structural facings is the need for the background wall or frame to support the whole of the weight of the facing at each storey height of the building or at vertical intervals of about 3 m, by means of angles or corbel plates. In addition to the support fixings, restraint fixings are necessary to locate the facing units in true alignment and to resist wind pressure and suction forces acting on the wall. Fixing centres are calculated using wind-loading software. To allow for elastic and flexural movements of the structural frame and differential thermal and moisture movements, there must be flexible horizontal joints below support fixings and vertical movement joints at intervals along the length of the facings. Both horizontal and vertical movement joints must be sufficiently flexible to accommodate anticipated movements and must be water resistant to prevent penetration of rainwater.

Natural and reconstructed stone facings

Natural and reconstructed stone facings are applied to the face of buildings to provide a decorative finish to simulate the effect of solidity and permanence traditionally associated with solid masonry. Because of the very considerable cost of preparation and fixing, this type of facing is mostly used for prestige buildings such as banks and offices in city centres. Granite is the natural stone much favoured for use as facing slabs for the hard, durable finish provided by polished granite and the wide range of colours and textures available from both native and imported stone. Polished granite slabs are used for the fine gloss surface that is maintained throughout the useful life of a building. To provide a more rugged appearance, the surface of granite may be honed to provide a semi-polish, flame textured to provide random pitting of the surface or surface tooled to provide a more regular rough finish. Granite-facing slabs are generally 40 mm thick for work more than 3.7 m above ground and 30 mm thick for work less than 3.7 m above ground.

Limestone is used as a facing, usually to resemble solid ashlar masonry work, the slabs having a smooth finish to reveal the grain and texture of the material. These comparatively soft limestones suffer a gradual change of colour over the course of years, and this weathering is said to be an attractive feature of these stones. Limestone-facing slabs are 75 mm thick for work more than 3.7 m above ground and 50 mm thick for work less than 3.7 mm above ground. Hard limestones are used as facings for the hardness and durability of the materials. This type of stone is generally used as flat, level-finished, facing slabs in thicknesses of 40 mm for work more than 3.7 m above ground and 30 mm for work less than 3.7 m above ground.

Sandstones are used as facing slabs. Some care and experience are necessary in the selection of these native sandstones as the quality, and therefore the durability, of the stone may vary between stones taken from the same quarry. This type of stone is chosen for the colour and grain of the natural material, the colour of which will gradually change over some years of exposure. Because of the coarse grain of the material, it may stain due to irregular run-off of water down the face. Sandstone-facing slabs are usually 75 mm thick for work 3.7 m above ground and 50 mm thick for work less than 3.7 m above ground.

Marble is less used for external facings in northern European climates, as polished marble finishes soon lose their shine. Coarser surfaces, such as honed or eggshell finishes, will generally maintain their finish, provided white or travertine marble is used. Marble-facing slabs are 40 mm thick for work 3.7 m above ground and 30 mm thick for work below that level. Reconstructed stone made with an aggregate of crushed natural stone is used as facing slabs as if it were the natural material, in thicknesses the same as those for the natural stone.

Fixing natural and reconstructed stone facings

The size of stone-facing slabs is generally limited to about 1.5 m in any one or both face dimensions or to such a size as is practical to win from the quarry. Stone-facing slabs are fixed so that there is a cavity between the back of the slabs and the background wall or frame to allow room for fixings, tolerances in the sawn thickness of slabs and variations in background surfaces, and also to accommodate some little flexibility to allow for differential structural, thermal and moisture movements between the structure and the facing. The cavity or airspace between the back of the facing slabs and the background walling or structure is usually from 10 to 20 mm and free from anything other than fixings, so that the facing may suffer small movements without restraint by the background. Small differential movements are accommodated through the many joints between slabs and, more specifically, through vertical and horizontal control (movement) joints. The types of fixings used to support and secure facing slabs in position are:

❑ Loadbearing fixings.
❑ Restraint fixings.
❑ Combined loadbearing and restraint fixings.
❑ Face fixings.
❑ Soffit fixings.

These fixings are made from one of the corrosion-resistant metals such as stainless steel, aluminium bronze or phosphor bronze. Stainless steel is the general description for a group of steel alloys containing chromium and other elements. The type of stainless steel commonly used for structural fixings is austenitic stainless steel.

Loadbearing fixings

Corrosion-resistant metal angles or corbel plates are used to carry the weight of the stone facing. These fixings are bolted to, built into or cramped to slots in the background wall or structure. The loadbearing fixings provide support at each floor level at not more than 3 m. The fixings bridge the cavity to provide support at the bottom or close to the bottom of slabs, with two fixings being used to each slab. Loadbearing fixings take the form of stainless steel angles or corbel plates that fit into slots cut in the bottom edge or into slots cut in the lower part of the back of slabs at each floor level or at vertical intervals of about 3 m.

Common practice is to support each facing slab on two supports, with the angle or corbel supports fixed centrally on vertical joints between slabs so that each supports two slabs. Angle and corbel plates should be at least 75 mm wide. At vertical movement joints two supports are used, one on each side of the joint, to the lower edge or lower part of the two stone slabs on each side of the joint. These separate supports should be at least 50 mm wide. Angle loadbearing fixings are bolted to the in situ concrete or brick background, with expanding bolts. Holes are drilled in the background into which the bolts make a tight fit so that, as the bolt is tightened, its end expands to make a secure fixing. Angles may be fixed to provide support to the bottom edge of slabs, with the supporting flange of the angle fitting into slots in the bottom edge of adjacent slabs so that a narrow horizontal joint between slabs may be maintained. Angle support to the thicker sedimentary stones is often made to grooves cut in the backs of adjacent stones, some little distance above the lower edge, into which the flange of the angle fits. This fixing is chosen where the edges of these laminated stones might spall where the lower edges were cut. Figure 8.9 is an illustration of loadbearing angle support fixings.

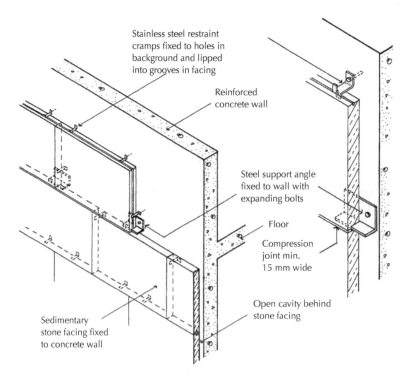

Figure 8.9   Stone facing to solid background.

Corbel plate loadbearing fixings may be used as an alternative to angle supports, particularly for the thinner stones such as granite. Flat or fishtail corbel plates are from 6 to 16 mm thick, depending on the size of slab to be supported, 75 or 50 mm wide and from 125 mm long. The purpose of the fishtail end is to provide a more secure bond to the cement grout in which the corbel plate is set. A pocket is made in the concrete or brick background by drilling holes and chiselling to form a neat pocket into which the corbel plate is set in dry, rapid-hardening cement and sand, which is hammered in around the corbel. The one-part cement to one-part sand grout is left for at least 48 hours to harden. Corbel plate supports are usually fixed to provide support by fitting into slots cut in the back of adjacent stones some little distance above the lower edge of slabs, as illustrated in Figure 8.10.

Two corbel plates are used to give support to the stone facing at each floor level or not more than 3 m. The thickness of the plates and the depth of their bed into the background have to be sufficient to provide support for all the stone slabs between floors. A common alternative to flat corbel plates is to form the protruding ends of corbels, which fit into slots in the back of stones, to slope up at an angle of 158° to the horizontal. The upward slope provides a more positive seating for stones and to some extent serves to restrain and align the stones. The setting into the background of the corbel plates and the cutting of the slots in the back of the stones require a degree of skill to achieve both an intimate fit of stones to corbels and true alignment of stone faces. For setting in brick backgrounds, fishtail-ended corbel plates are made for building in or grouting into pockets. Corbel plates may be shaped for bolting to backgrounds, as illustrated in Figure 8.11.

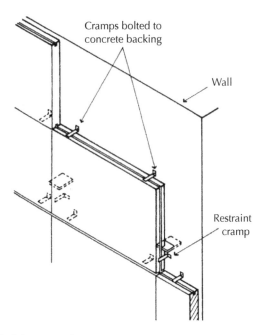

Cramps bolted to concrete backing

Wall

Restraint cramp

Figure 8.10   Corbel plate supports.

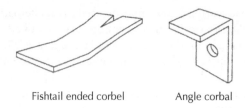

Fishtail ended corbel   Angle corbal

Figure 8.11   Corbel plates.

Restraint fixings

Restraint fixings take the form of strip metal or wire cramps shaped to hook into grooves or holes in the edges of slabs and formed for bolting to or being set into pockets or slot anchors in the background. The most straightforward form of restraint cramp consists of a narrow strip with one end bent up and holed for bolting to the background, with the other end double-lipped to fit into grooves or slots in the top and bottom edges of slabs, as illustrated in Figure 8.9. A similar cramp has a fishtail end for grouting in pockets cut in the background. Strip metal restraint cramps may be shaped to fit into slot metal anchors that have been cast into concrete walls and frames. The anchors are cast in horizontally so that the cramp may be adjusted in the slot anchor to coincide with vertical joints between stones. The cramp may provide restraint through double-lipped ends that fit to grooves in adjacent slabs or by a dowel that fits to holes in slab edges, as illustrated in Figure 8.12. Fishtail-ended strip metal cramps are made for building or grouting into the horizontal joints of brick backgrounds. Dowels fit to holes in the cramps for setting into holes or grooves cut in the horizontal joints between stones.

Stainless steel wire restraint cramps are used for the thinner granite slab facings. These dense stones lend themselves to being accurately drilled to take the hooked ends of these cramps that fit into either the side or top and bottom edges of adjacent slabs. These stainless steel wire cramps are usually screwed or bolted to solid backgrounds and shaped to fit to holes or grooves cut in the horizontal joints of adjacent stones. As an alternative, a loose wire cramp may be used to hook an upper stone to those below, as illustrated in Figure 8.13. For brick backgrounds, a looped wire cramp is grouted into brick joints, with the double-toed ends of the wire set into holes or grooves in adjacent stones. An advantage of wire cramps is that they can be bent on site to make an intimate fit to the holes or grooves cut in adjacent stones.

Combined fixings

Combined loadbearing and restraint fixings combine the two functions in one fitting. Both angle and corbel plate loadbearing fixings may be made to perform this dual function for those slabs that are supported at each floor level. Dowels welded to the angle supports may fit to holes or grooves in the stones they support and the top edges of the stones below. Similarly, double-hooked ends of cramps may perform the same function.

Stone bonder courses

To provide support for sedimentary stone-facing slabs at each floor level, a system of stone bonder courses at each floor level, supported by the structural frame, may be used.

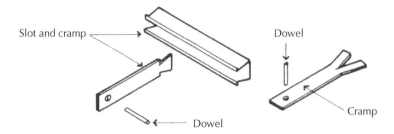

Slot and cramp

Dowel

Cramp

Dowel

Figure 8.12   Slot anchor and cramp.

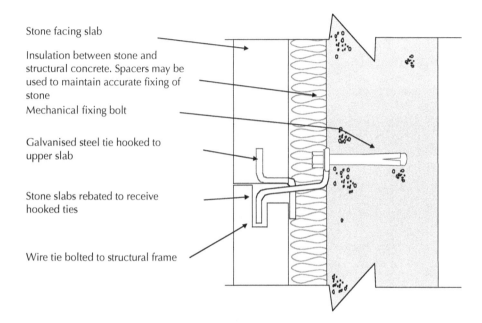

Stone facing slab

Insulation between stone and structural concrete. Spacers may be used to maintain accurate fixing of stone

Mechanical fixing bolt

Galvanised steel tie hooked to upper slab

Stone slabs rebated to receive hooked ties

Wire tie bolted to structural frame

Figure 8.13   Wire tie restraint fixing.

A course of stones is bedded on a beam to provide support for the facing slabs to the floor above, as illustrated in Figure 8.14. The bonder course of stones is of sufficient thickness and depth to both tail back and give support to the facing slabs. Restraint fixings are set into slot anchors cast in the beam and built into the brick background (Figure 8.15). At the top edge of slabs and underside of the bonder course, a movement joint is formed to accommodate relative movement of the structure and the slabs. A disadvantage of this system is that the thermal resistance of the bonder course and beam will be less than that of the walling and so act as a thermal break.

Today, it would be unusual to construct the internal leaf using a cavity wall, but cladding may be fitted to an existing cavity wall in refurbishment. Nowadays a steel cladding frame would be used, providing a structure on to which the stone can be hung and supported and in which insulation could be housed.

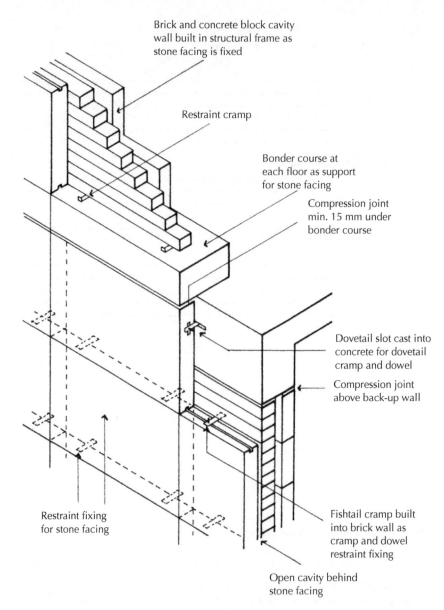

Brick and concrete block cavity
wall built in structural frame as
stone facing is fixed

Restraint cramp

Bonder course at
each floor as support
for stone facing

Compression joint
min. 15 mm under
bonder course

Dovetail slot cast into
concrete for dovetail
cramp and dowel

Compression joint
above back-up wall

Restraint fixing
for stone facing

Fishtail cramp built
into brick wall as
cramp and dowel
restraint fixing

Open cavity behind
stone facing

Figure 8.14 Stone bonder courses. Insulation would be positioned within the cavity, applied to the internal face of the stone or on the internal face of the wall.

Face fixings

As an alternative to support and restraint fixing of stone-facing slabs, face fixing may be used for thin slabs. Each stone-facing slab is drilled for and fixed to a solid background with at least four stainless steel or non-ferrous bolts. The stone slabs may be bedded in position

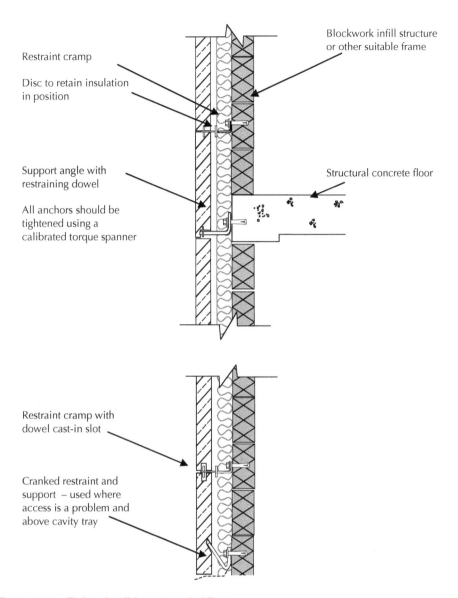

Restraint cramp

Disc to retain insulation
in position

Support angle with
restraining dowel

All anchors should be
tightened using a
calibrated torque spanner

Blockwork infill structure
or other suitable frame

Structural concrete floor

Restraint cramp with
dowel cast-in slot

Cranked restraint and
support – used where
access is a problem and
above cavity tray

Figure 8.15   Fixing detail for stone cladding.

on dabs of lime putty or weak mortar, which is spread on the back of the slabs and the slabs
are then secured with expanding bolts and washers, as illustrated in Figure 8.16.

The holes drilled in the stone for the bolts are then filled with pellets of stone to match
the stone of the slabs. Joints between stones are filled with gunned-in mastic sealant to
provide a weathertight joint and to accommodate differential thermal and structural movement. The bolts must be of sufficient size, accurately set in place and strongly secured to the
solid backing to prevent failure of the fixing.

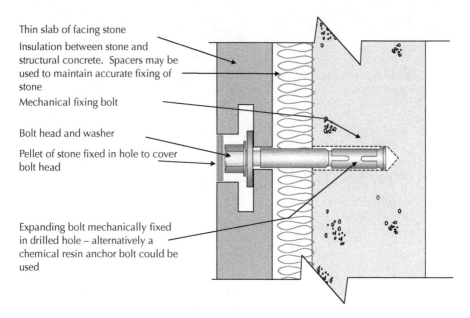

Thin slab of facing stone

Insulation between stone and structural concrete. Spacers may be used to maintain accurate fixing of stone

Mechanical fixing bolt

Bolt head and washer

Pellet of stone fixed in hole to cover bolt head

Expanding bolt mechanically fixed in drilled hole – alternatively a chemical resin anchor bolt could be used

Figure 8.16   Face fixing for stone cladding.

Soffit fixings

Support and fixing of stone facings to soffits is affected by the use of hangers and plates, cramps or dowels that are suspended in slot hangers cast in the structural soffit. Stainless steel or bronze channels are cast into the soffit of reinforced concrete beams and slabs to coincide with joints between facing slabs. Hangers fit into the lipped channel to allow for adjustment of the hanger to suit joints in stone soffit. With the system illustrated in Figure 8.17, the plates supported by the hangers may be cut to fit into grooves or slots cut in the edges between stones, or be made to fit into slots cut in the edges of four stones, depending on the thickness and weight of the stone slabs used and convenience in fixing. The number of hangers used for each soffit-facing slab depends on the size and weight of each slab and the thickness of the lipped edge of the slab bearing on and therefore carrying, the slab. At the junction of soffit slab, and wall slabs, a hanger bolted to the wall background may have a lip or stud to fit to slots or holes in the edge of the soffit slabs.

Joints between stone slabs

Joints between stone-facing slabs should be sealed as a barrier to the penetration of rainwater running off the face of the slabs. Where rainwater penetrates the joints between stone slabs it will be trapped in the cavity between the slabs and the background wall, will not evaporate to air during dry spells and may cause conditions of persistent damp. Open or butt joints between slabs should be avoided in external facework.

The joints between sedimentary stone slabs, such as limestone and sandstone, may be filled with a mortar of cement, lime and sand (or crushed natural stone) mix 1 : 1 : 6 and

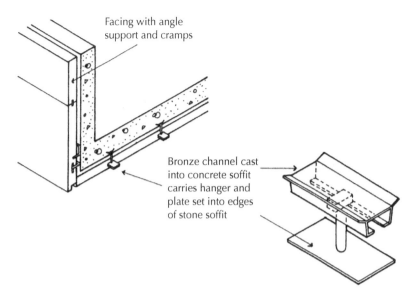

Facing with angle
support and cramps

Bronze channel cast
into concrete soffit
carries hanger and
plate set into edges
of stone soffit

Figure 8.17   Soffit fixing for stone facing (insulation omitted for clarity).

finished with either flush or slightly recessed pointing to a minimum depth of 5 mm. Joints between granite and hard limestone slabs are filled with a mortar of 1 : 2 : 8 cement, lime and sand (or stone dust) or 4 : 1 cement and sand to a minimum thickness of 3 mm. As an alternative to mortar filling the joints between stones, a sealant may be used. Sealants such as one part polysulphide, one part polyurethane/two parts polysulphide and two parts polyurethane are recommended for the majority of stones. These sealant joints should be not less than 5 mm wide. The jointing sealants will accommodate a degree of movement between stones without failing as a water seal for up to 15–20 years, when they may well need to be reformed. Mortar joints will take up some slight movement between stones but may in time not serve as an effective water seal as wind-driven rain may penetrate the fine cracks that open up. Some penetration of rainwater through joints between stones may well occur as sealants age and mortar cracks. Thus, it may be necessary to hack out and reform joints to prevent moisture penetration to the inside face of the building.

Movement joints

Much of the early elastic shortening of the columns of a structure will have taken place before a wall cladding is fixed. The long-term shortening of reinforced concrete columns, through creep, has to be allowed for in horizontal movement joints. Differential temperature and moisture movements of a wall facing relative to the supporting structure will generally dictate the need to allow some movement of joints and fixings. There will be, for example, very considerable temperature differences between facing slabs on an exposed south-facing wall and the structure behind so that differential thermal movement has to be allowed for both in joints and support and restraint fixings.

A general recommendation in the fixing of stone-facing slabs is that there should be horizontal movement joints at each storey height below loadbearing support fixings or not more than 3 m. These joints are usually 10–15 mm deep and filled with one of the elastic sealants. Where so wide a joint would not be acceptable in facework finished with narrow joints, it is usual to accommodate movement in narrower sealant-filled horizontal joints to all the facework. Vertical movement joints are formed in facework where these joints occur in the structure, to allow for longitudinal structural, thermal and moisture movements. A continuous vertical joint is formed between stone facings and filled with sealant.

Faience slabwork

Faience is the term used to describe fire-glazed stoneware in the form of slabs that are used as a facing to a solid background wall. Slabs are fixed in the same way as stone facings. The best quality slabs are made from stoneware that shrinks and deforms less on firing than does earthenware. The fired slab is glazed and then refired to produce a fire-glazed finish. The slabs are usually 300 × 200, 450 × 300 or 610 × 450 mm and 25–32 mm thick. They form a durable, decorative facing to solid walls. The glazed finish, which will retain its lustre and colour indefinitely, needs periodic cleaning, especially in polluted atmospheres. Faience slabwork was much used as a facing in the 1930s in the UK, as a facing to large buildings such as cinemas.

Terracotta

Terracotta was much used in Victorian buildings as a facing because it is less affected by polluted atmospheres than natural limestone and sandstone facings. Fired blocks of terra-cotta, with a semi-glaze self-finish, were moulded in the form of natural stone blocks to replicate the form and detail of the stonework buildings of the time. The plain and ornamental blocks were made hollow to reduce and control shrinkage of the clay during firing. In use, the hollows in the blocks were filled with concrete and the blocks were then laid as if they were natural stone. This labour-intensive system of facing is still used today.

Tiles and mosaic

'Tile' is the term used to describe comparatively thin, small slabs of burnt clay or cast concrete up to about 300 mm square and 12 mm thick. These small units of fired clay and cast concrete are used as a facing to structural frames and solid background walls. For many years, the practice has been to bond tiles directly to frames and walls with cement mortar dabs, which provides sufficient adhesion to maintain individual tiles in place. Unfortunately, this system of adhesion does not make any allowance for differential movements between the frame, background walls and tiles, other than in the joints between tiles, which can be considerable, particularly with in situ cast concrete work. To make allowance for movements in the structure and the facing, tiles should be supported and restrained by cramps that provide a degree of flexibility between the facing and the background. For economy and ease of fixing, the tiles can be cast onto a slab of plain or reinforced concrete, which is then fixed in the same way as stone-facing slabs. 'Mosaic' is the term used to describe small squares of natural stone, tile or glass set out in some decorative pattern.

The units of mosaic are usually no larger than 25 mm². A mosaic finish as an external facing should be used as a facing to a cast concrete slab in the same way as tiles.

8.6 Cladding panels

The word 'cladding' is used in the sense of clothing or covering the building with a material to provide a protective and decorative cover. Cladding panels, usually storey (floor) height, serve the function of providing protection against wind and rain, and resistance to the transfer of heat from inside to outside, without providing structural support. Available in a wide range of sizes, colours and surface finishes, they contribute to the aesthetic of the building.

Precast concrete cladding units

Precast concrete cladding units are usually storey height, as illustrated in Figure 8.18, or column spacing wide as spandrel or undercill units for support and fixing to the structural frame. Units are hung on, and attached to, frames as a self-supporting facing and wall element, which may combine all of the functional requirements of a wall element.

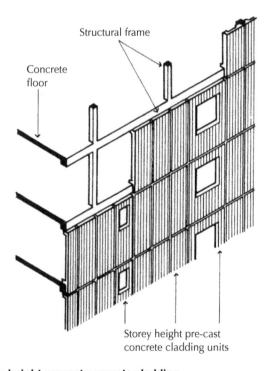

Figure 8.18   Storey height precast concrete cladding.

Precast concrete cladding units are cast with either the external face up or down in the moulds, depending on convenience in moulding and the type of finish. Where a finish of specially selected aggregate is to be exposed on the face, the face-up method of casting is generally used for the convenience and accuracy in applying the special finish to the core concrete of the panel. Cladding units that are flat or profiled are generally cast face down for convenience in compacting concrete into the face of the mould bed. Strongly constructed moulds of timber, steel or GRP are laid horizontal, the reinforcing cage and mesh are positioned in the mould, and concrete is placed and compacted. For economy in the use of the comparatively expensive moulds, it is essential that there be a limited number of sizes, shapes and finishes to cladding units to obtain the economic advantage of repetitive casting. For strength and rigidity in handling, transport, lifting and support, and fixing, and to resist lateral wind pressures, cladding units are reinforced with a mesh of reinforcement to the solid web of units, and a cage of reinforcement to vertical stiffening ribs and horizontal support ribs. Figure 8.19 is an illustration of a storey-height cladding unit.

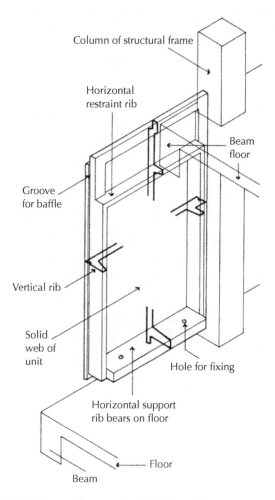

Figure 8.19   Storey height precast concrete cladding unit.

The initial wet plastic nature of concrete facilitates the casting of a wide variety of shapes and profiles, from flat solid webs enclosing panels to the comparatively slender solid sections of precast concrete frames for windows. The limitation of width of cladding units is determined by facilities for casting and size for transport and lifting. The width of the units cast face up is limited by ease of access to placing the face material in the moulds. The usual width of storey height panels is from 1200 to 1500 mm, or the width of one or two structural bays.

There is no theoretical limit to the size of precast units, provided they are sufficiently robust to be handled, lifted and fixed in place, other than limitations of the length of a unit that can be transported and lifted. In practice, cladding units are usually storey height for convenience in transport and lifting, and fixing in place. Cladding units two or more storeys in height have to be designed, hung and fixed to accommodate differential movements between the frame and the units, which are multiplied by the number of storeys they cover.

Storey-height precast concrete cladding is supported by the structural frame, either by a horizontal support rib at the bottom of the units or by hanging on a horizontal support rib at the top of the units, as illustrated in Figure 8.20. Bottom support is preferred as the concrete of the unit is in compression and is less likely to develop visible cracks and crazing than it is when top-hung. Whichever system of support is used, the horizontal support rib

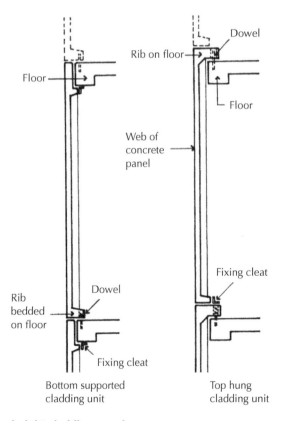

Figure 8.20   Storey height cladding panel.

must have an adequate projection for bearing on structural floor slabs or beams and for the fixings used to secure the units to the frame. At least two mechanical support and two restraint fixings are used for each unit. The usual method of fixing at supports is by the use of steel or non-ferrous dowels that are grouted into 50 mm square pockets in the floor slab. The dowel is then grouted into a 50 mm diameter hole in the support rib, as illustrated in Figure 8.21. The advantage of this dowel fixing is that it can readily be adjusted to inaccuracies in the structure and the panel. Dowel fixings serve to locate the units in position and act as restraint fixings against lateral wind pressures.

The vertical stiffening ribs are designed for strength in resisting lateral wind pressures on the units between horizontal supports and strength in supporting the weight of the units that are either hung from or supported on the horizontal support ribs. The least thickness of concrete necessary for the web and the ribs is dictated largely by the cover of concrete necessary to protect the reinforcement from corrosion, for which a minimum web thickness of 85 or 100 mm is usual. The necessary cover of concrete to reinforcement makes this system of walling heavy, cumbersome to handle and fix, and bulky looking. Restraint fixings to the upper or lower horizontal ribs of cladding units, depending on whether they are top or bottom supported, must restrain the unit in place against movements and lateral wind pressure. The restraint fixing most used is a non-ferrous or stainless steel angle cleat

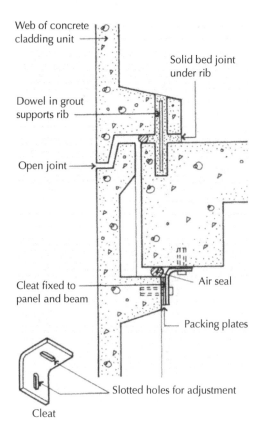

Figure 8.21   Fixing for bottom supported cladding units.

that is either fixed to a slotted channel cast in the soffit of beams or slabs or, more usually, by expanding bolts fitted to holes drilled in the concrete. The cleat is bolted to a cast-in stud protruding from the horizontal rib of the unit, as illustrated in Figure 8.21. The slotted hole in the downstand flange of the cleat allows some vertical movement between the frame and the cladding.

Another system of fixing combines support fixing by dowels with restraint fixing by non-ferrous flexible straps that are cast into the units and fit over the dowel fixing. Support and restraint fixing may be provided by casting loops or hooked ends of reinforcement, protruding from the back of cladding units, into a small part of or the whole of an in situ cast concrete member of the structural frame. The disadvantage of this method is the site labour required in making a satisfactory joint, and the rigidity of the fixing that makes no allowance for differential movements between structure and cladding. At external angles on elevations, cladding units may be joined by a mitre joint or as a wrap-around corner unit specially cast for the purpose, as illustrated in Figure 8.22.

The advantage of the wrap-around corner unit is that the open drained joint may be formed against the solid background of a column, and the disadvantage of this unit is that the small, protruding lipped edge may be damaged in lifting and handling into place. It is

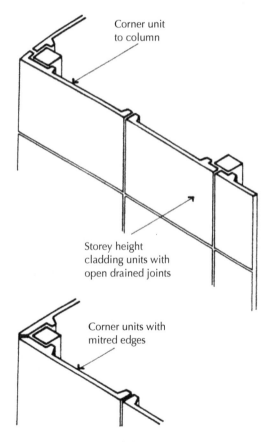

Corner unit
to column

Storey height
cladding units with
open drained joints

Corner units with
mitred edges

Figure 8.22   Corner units to concrete cladding.

not practical to make an invisible repair of a damaged edge of a unit. It is difficult to cast a neat, satisfactory drained mitre joint and to maintain the mitre junction as a gap of uniform thickness. Whichever joint is used is a matter of choice, principally for reasons of appearance. A common use for precast concrete cladding units is as undercill cladding to continuous horizontal windows or as a spandrel unit to balcony fronts. Typical undercill units, illustrated in Figure 8.23, are designed for bottom rib support and top edge restraint at columns.

Units are designed to be supported by horizontal webs that bear on a structural beam or projection of the floor slab, as illustrated in Figure 8.24, which carries the weight of the spandrel unit. Support fixing is similar to that for bottom-supported storey height panels. Restraint fixing is by non-ferrous angle cleats with slotted holes to facilitate fixing to column face and unit. The reinforced concrete cill bearing and window head ribs are adequate to stiffen the web of the units against wind pressure and suction. Where the under-window spandrel cladding is continued on a return face of the building, the regular grid of columns may be inset at corners to provide a small-span cantilever to accommodate a wraparound spandrel unit supported on the cantilever floor slab and restrained to the near to corner column.

As an alternative to bottom support, under window spandrel cladding units may be designed and cast for top, cill level support. A system of separate cill level beams is precast

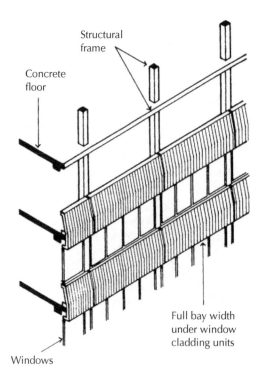

Figure 8.23   Under window (spandrel) precast concrete cladding.

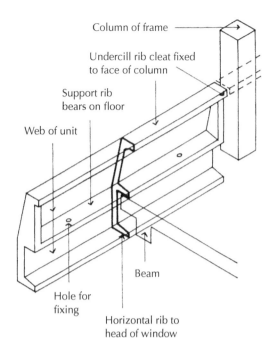

Column of frame

Undercill rib cleat fixed
to face of column

Support rib
bears on floor

Web of unit

Hole for
fixing

Beam

Horizontal rib to
head of window

Figure 8.24   Under window concrete cladding unit.

or may be in situ cast to provide support and fixing specifically for the undercill ribs of the units, with restraint fixing through a rib to the floor. The purpose of this arrangement is where structural columns are widely spaced and it is convenient for casting, transport and handling to use several units between columns. With the system of under-window cladding, the horizontal windows are framed for fixing to the underside and top of the spandrel cladding units. The junction of the window framing is usually weathered with gunned-in mastic as a weather seal. In this construction, it is probably more practical to use a sealed vertical joint between cladding units than an open drained joint, to avoid the difficulty of making a watertight seal at the junction of an open joint and window framing.

At the junction of a flat roof and a system of precast cladding units, it is necessary to form an upstand parapet either in the structural frame or in purpose-cast parapet panels. For a parapet of any appreciable depth, an upstand beam or a concrete upstand to a beam is formed and clad with purpose-cast panels, as illustrated in Figure 8.25. A non-ferrous metal capping weathers the top of the parapet. In any event, some form of upstand parapet is formed to avoid a run-off of water from the roof down the face of the cladding, which would cause irregular, unsightly staining. Where a system of parapet height cladding panels is used, the panels are cast to provide a top rib that bears on and is secured to the parapet, as illustrated in Figure 8.25, and restrained by angle cleats between the soffit of the beam and the panels. A non-ferrous capping weathers the top of the parapet and cladding panels.

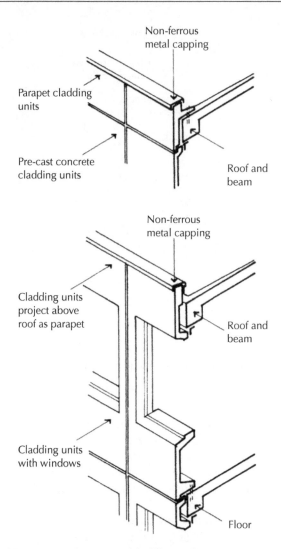

Figure 8.25   Parapets to precast concrete cladding units.

As an alternative, a system of special storey height cladding panels designed to extend over the face of, and be fixed to, the parapet may be used, as illustrated in Figure 8.26. It is possible to cast both parapet cladding panels and storey height panels that extend up to and over the parapet and down the inside face of the parapet. This may be a perfectly satisfactory finish, provided the panels can be lifted and set in place without damage. To provide protection to the sealed or open vertical joints between panels, it is necessary to fix a non-ferrous capping over the parapet and down each side over the panels. Because of the plastic nature of wet concrete, it is possible to cast cladding units in a variety of profiled and textured finishes and to include openings for windows in individual units. The cladding panels are cast with ribs for bottom support at each floor, top ribs for restraint fixing and

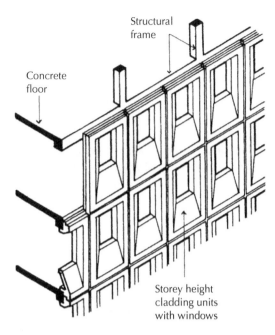

Concrete
floor

Structural
frame

Storey height
cladding units
with windows

Figure 8.26   Storey height precast cladding units.

side ribs for open drained jointing. The panels may be delivered with the window frames fixed in place ready for site glazing.

Surface finishes
To provide an acceptable finish to the exposed faces of precast concrete panels, it is usual practice to provide what is sometimes called an 'indirect finish' by abrasive blasting, surface grinding, acid washing or tooling to remove the fine surface layer and expose the aggregate and cement below. This surface treatment has the general effect of exposing a surface of reasonably uniform colour and texture. This form of surface treatment can produce a fine smooth finish by light abrasive sand or grit blasting or grinding, or a more coarse texture by heavy surface treatment.

Applied finishes
Any type of brick that is reasonably frost resistant and durable may be used as a facing material and fixed to the precast concrete panels using the mould to produce the full range of brick construction features such as corbels, string courses, piers and arches. The facing brickwork is bonded to the concrete panel either with a mechanical key or by stainless steel or nylon filament ties. A mechanical key can be provided where bricks with holes in them are used and cut along the length of the brick, so that the resulting semicircular grooves may provide a bond to the concrete. To ensure a good bond to the bricks, it is essential to saturate the bricks thoroughly before the backing concrete is placed. Where ties are used to retain the face bricks in place, the nylon filament, stainless steel wire or bars are threaded through holes in the brickwork and turned up between bricks as loops to bond with the backing concrete.

Applied finishes of brick or stone to precast concrete cladding panels are less used than they were, because of the considerable costs of casting, transporting and lifting into place heavy and cumbersome wall units. Brick walling and stone facings may be more economically applied on site to a structural frame.

Natural and reconstructed stone-facing slabs are used as a decorative finish to precast concrete panels. Any of the natural or reconstructed stones used for stone facework to solid backgrounds may be used for facings to precast concrete panels. For ease of placing the stone-facing slabs in the bed of the mould, it is usual to limit the size of the panels to not more than 1.5 m in any one dimension. Granite and hard limestone slabs not less than 30 mm thick and limestone, sandstone and reconstructed stone slabs not less than 50 mm thick are used. The facing slabs are secured to and supported by the precast panel through stainless steel corbel dowels at least 4.7 mm in diameter, which are set into holes in the back of the slabs and cast into the concrete panels at the rate of at least 11 per m² of panel and inclined at 45° or 60° to the face of the panel. Normal practice is that about half of the dowels are inclined up and half down, relative to the vertical position of the slab when in position on site. The dowels are set in epoxy resin in holes drilled in the back of the slabs. Flexible grommets are fitted around the dowels where they protrude from the back of the slab. These grommets, which are cast into the concrete of the panel, together with the epoxy resin bond of the dowel in the stone slab, provide a degree of flexibility to accommodate thermal and moisture movement of the slab relative to that of the supporting precast concrete cladding panel.

All joints between the stone-facing slabs are packed with closed-cell foam backing or dry sand, and all joints in the back of the stone slabs are sealed with plastic tape to prevent cement grout from running in. When the precast panel is taken from the mould, the jointing material is removed for mortar or sealant jointing. To prevent the concrete of the precast panel from bonding to the back of the stone slabs, either polythene sheeting or a brushed-on coating of clear silicone waterproofing liquid is applied to the whole of the back of the slabs. The purpose of this debonding layer is to allow the facing slabs free movement relative to the precast panel due to differential movements of the facing and the backing. The necessary joints between precast concrete cladding panels faced with stone-facing slabs are usually sealed with a sealant to match those between the facing slabs.

Joints between precast concrete cladding panels

The joints between cladding panels must be sufficiently wide to allow for inaccuracies in both the structural frame and the cladding units, to allow unrestrained movements due to shortening of the frame and thermal and moisture movements, and at the same time to exclude rain. The two systems of making joints between units are the face-sealed joint and the open-drained and rebated joint. Sealed joints are made watertight with a sealant that is formed inside the joint over a backing strip of closed cell polyethylene, at or close to the face of the units, as illustrated in Figure 8.27. The purpose of the backing strip is to ensure a correct depth of sealant. Too great a depth or width of sealant will cause the plastic material of the sealant to move gradually out of the joint due to its own weight.

Sealant material is applied by gun. The disadvantages of sealant joints are that there is a limitation to the width of joint in which the sealant material can successfully be retained and that the useful life of the material is from 15 to 20 years, as it oxidises and hardens with exposure to sunlight and has to be raked out and renewed. Sealed joints are used in the main for the smaller cladding units. The sealants most used for joints between precast concrete

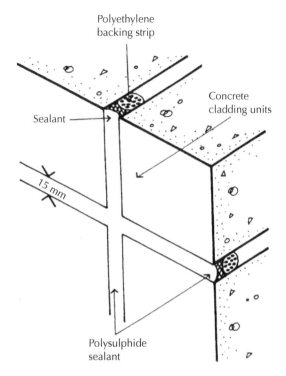

Polyethylene
backing strip

Concrete
cladding units

Sealant

15 mm

Polysulphide
sealant

Figure 8.27   Sealant joints to concrete cladding.

cladding panels are two parts polysulphide, one part polyurethane, epoxy modified two parts polyurethane and low-modulus silicone. Which of these sealants is used depends to some extent on experience in the use of a particular material and ease of application on site. The two-part sealants require more skill in mixing the two components to make a successful seal than the one-part material, which is generally reflected in the relative cost of the sealants. A closed-cell polyethylene backing strip is rammed into the joint and the sealant applied by power or hand pump gun and compacted and levelled with a jointing tool.

Open-drained joints between precast concrete cladding panels are more laborious to form than sealed joints and are mostly used for the larger precast panels where the width of the joint may be too wide to seal and where the visible open joint is used to emphasise the rugged, coarse-textured finish to the panels. Open joints are the most effective system of making allowance for inaccuracies and differential movements and serving as a bar to rain penetration without the use of joint filling material.

Horizontal joints are formed as open overlapping joints with a sufficiently deep rebate as a bar to rain penetration, as illustrated in Figure 8.28. The rebate at the joint should be of sufficient section to avoid damage in transport, lifting and fixing in place. The thickness necessary for these rebates is provided by the depth of the horizontal ribs. The air seal formed at the back of horizontal joints is continuous in both horizontal and vertical joints as a seal against outside wind pressure and driving rain.

Vertical joints are designed as open-drained joints in which a neoprene baffle is suspended inside grooves formed in the edges of adjacent units, as illustrated in Figure 8.29.

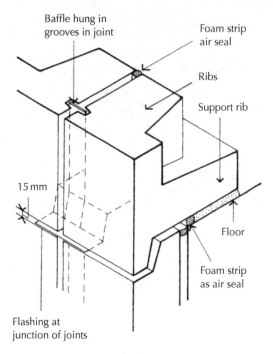

Figure 8.28   Horizontal open drained joint.

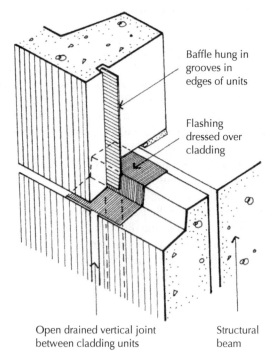

Figure 8.29   Vertical open drained joint.

The open-drained joint is designed to collect most of the rain in the outer zone of the joint in front of the baffle, which acts as a barrier to rain that may run or be forced into the joint by wind pressure. The baffle is hung in the joint so that to some extent there is a degree of air pressure equalisation on each side of the baffle due to the air seal at the back of the joint. This air pressure equalisation acts as a check to wind-driven rain that would otherwise be forced past the baffle if it were a close fit and there was no air seal at the back of the joint. At the base of each open drained joint, there is a lead flashing, illustrated in Figures 8.28 and 8.29, which serves as a barrier to rain at the most vulnerable point of the intersection of horizontal and vertical joints. As cladding panels are fixed, the baffle in the upper joints is overlapped outside the baffle of the lower units.

Where there is a cavity between the back of the cladding units and an inner system of solid block walls or framing for insulation, air seals can be fitted between the frame and the cladding units. It is accepted that the system of open joints between units is not a complete barrier to rain. The effectiveness of the joint depends on the degree of exposure to driving rain, the degree of accuracy in the manufacture and assembly of the system of walling, and the surface finish of the cladding units. Smooth-faced units will tend to encourage driven rain to sheet across and up the face of the units, and so cause a greater pressure of rain in joints than there would be with a coarse-textured finish, which will disperse driven rain and wind, and so reduce pressure on joints. The backs of cladding panels will tend to collect moisture by possible penetration of rain through joints and from condensation of moisture-laden air from outside and warm moist air from inside by vapour pressure, which will condense on the inner face of panels. Condensation can be reduced by the use of a moisture vapour check on the warm side of insulation as a protection against interstitial condensation in the insulation and as a check to warm moist air penetrating to the cold inner face of panels. Precast concrete cladding panels are sometimes cast with narrow weepholes, from the top edge of the lower horizontal ribs out to the face, in the anticipation that condensate water from the back of the units will drain down and outside. The near certainty of these small holes becoming blocked by wind-blown debris makes their use questionable.

Attempts have been made to include insulating material in the construction of precast cladding, either as a sandwich with the insulation cast between two skins of concrete or as an inner lining fixed to the back of the cladding. These methods of improving the thermal properties of concrete are rarely successful, because of the considerable section of the thermal bridge of the dense concrete horizontal and vertical ribs that are unavoidable, and the likelihood of condensate water adversely affecting some insulating materials.

It has to be accepted that there will be a thermal bridge across the horizontal support rib of each cladding panel that has to be in contact with the structural frame. The most straightforward and effective method of improving the thermal properties of a wall structure clad with precast concrete panels is to accept the precast cladding as a solid, strong, durable barrier to rain with good acoustic and fire-resistance properties and to build a separate system of inside finish with good thermal properties. Lightweight concrete blocks by themselves, or with the addition of an insulating lining, at once provide an acceptable internal finish and thermal properties. Block wall inner linings should be constructed independently of the cladding panels and structural members, as far as practical, to reduce interruptions of the inner lining, as illustrated in Figure 8.3.

Glass fibre reinforced cement (GRC) cladding panels

GRC as a wall panel material was first used in the early 1970s as a lightweight substitute for precast concrete. The principal advantage of GRC as a wall panel material is weight saving as compared to similar precast concrete panels. GRC can be formed in a wide variety of shapes, profiles and accurately finished mouldings, such as that illustrated in Figure 8.30. The material has good durability and chemical resistance, is non-combustible, not susceptible to rot and will not corrode or rust stain. The limiting factors in the use of this material arise from relatively large thermal and moisture movements and the restricted ductility of the material.

GRC is a composite of cement, sand and alkali-resistant (AR) glass fibre in proportions of 40–60% cement, 20% water, up to 25% sand and 3.5–5% glass fibre by weight. The glass

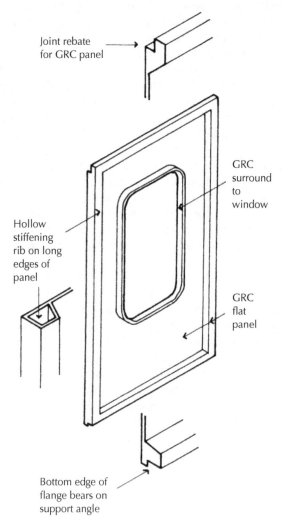

Figure 8.30 Ribbed single skin GRC panel.

fibre is chopped to lengths of about 35 mm before mixing. It is formed in moulds by spray application of the wet mix, which is built up gradually to the required thickness and compacted by roller. After the initial 3 mm thickness has been built up, it is compacted by roller to ensure a compact surface finish. For effective hand spraying, the maximum width of the panel is about 2 m. For mass production runs, a mechanised system is used with dual spray heads that spray fibre and cement, sand and water separately in the mould, which moves under the fixed spray heads. The mechanised spray results in a greater consistency of the mix and a more uniform thickness of panel than is usually possible with hand spraying. The moulds for GRC are either timber or the more durable GRP-lined, timber-framed types. Spray moulded GRC panels have developed sufficient strength 24 hours after moulding to be taken from moulds for curing. The size of GRC cladding panels is limited by the method of production as to width and to the storey height length for strength, transport and lifting purposes. It is also limited by the considerable moisture movement of the cement-rich material, which may fail if moisture movement is restrained by fixings. The usual thickness of GRC single skin panels is 10–15 mm.

As a consequence of moulding, the surface of a GRC panel is a cement-rich layer, which is liable to crazing due to drying shrinkage and to patchiness of the colour of the material due to curing. To remove the cement-rich layer on the surface and provide a more uniform surface, texture and colour, the surface can be acid etched, grit blasted or smooth ground. Alternatively, the panels can be formed in textured moulds so that the finished texture masks surface crazing and patchiness. For a uniform colour finish that can be restored by repainting on site, coloured permeable coatings are used which have microscopic pores in their surface that allow a degree of penetration and evaporation of moisture that prevents blistering or flaking of the coating. Textured permeable finishes, such as those used for external renderings, and microporous matt and glass finish paints are used. The thin single skin of GRC does not have sufficient strength or rigidity by itself to be used as a wall facing, other than as a panel material of up to about 1 m² square, supported by a metal carrier system or bonded to an insulation core for larger panels, as illustrated in Figure 8.31.

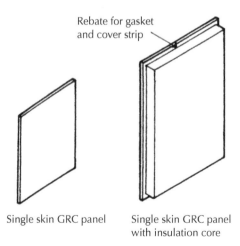

Rebate for gasket
and cover strip

Single skin GRC panel

Single skin GRC panel
with insulation core

Figure 8.31   Single skin GRC panel.

The use of GRC panels formed around a core of insulation material has by and large been abandoned. In use, the effect of the core of insulation is to cause an appreciably greater expansion of the external skin than that of the internal skin, with consequent deformation and bowing of the outer skin and possible failure. The thermal resistance of GRC is poor and it is now generally accepted that an inner skin of insulation should be provided separately from the GRC. A single-skin GRC panel larger than 1 m² does not have sufficient strength by itself and requires some form of stiffening. Stiffening is provided by solid flanges formed at the bottom and top edges of panels. The bottom flange provides support for the panel, and the top flange a means of restrain fixing. Stiffening ribs are formed in the vertical edges of panels. These stiffening ribs are usually hollow and formed around hollow or foam plastic formers to minimise weight and shrinkage of the cement-rich material as it dries. The four edges of panels are usually rebated to overlap at horizontal joints for weathering and to form weathered vertical joints. A flanged single-skin, flat GRC panel is illustrated in Figure 8.32.

Storey height, spandrel and undercill GRC panels have been extensively used as both flat and shaped panels for the advantage of the fine-grained, smooth surface of the material.

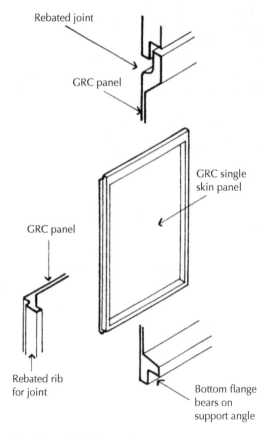

Figure 8.32   Single-skin, flanged GRC panel.

The shaping of a panel adds a degree of stiffness to the thin material in addition to the flanges and stiffening ribs. Figure 8.33 is an illustration of a shaped, window panel of GRC.

The support and restraint fixings for GRC panels are designed to allow freedom of movement of the thin panel material to accommodate differential, structural, thermal and moisture movement of the panels relative to that of the structure. To this end a minimum of fixings is used. The weight of a GRC panel is supported by the structure or structural frame at two points near the base of the panel, so that compressive stress acts on the bottom of the panel. Either one or both of the bottom supports are designed to allow some freedom of horizontal movement. To hold the panel in its correct position and allow freedom of movement, four restraint fixings are used near the four corners of the panel. These restraint fixings allow freedom of horizontal movement at the base and freedom of both horizontal and vertical movement at the two top corners. As an alternative to fixing GRC panels to the structure or structural frame, a separate stud frame is used as support for the panels, with the stud frame fixed to the structural frame with fixings designed to allow freedom of movement of the stud frame relative to the structure.

The requirement for allowances for movement in fixings and the additional work and materials involved do add considerably to the cost of the use of this material as a facing, which is one reason for the loss of favour in the use of GRC panels. The common means of

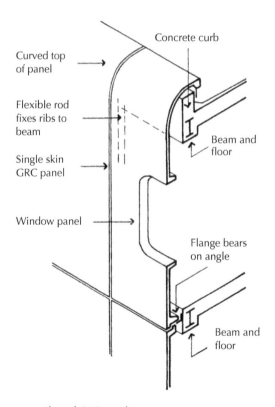

Shaped GRC panels

Figure 8.33   Shaped GRC panels.

GRC panel

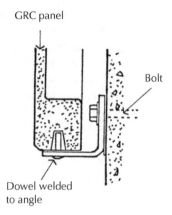

Bolt

Dowel welded
to angle

Figure 8.34 Support angle for GRC panel.

support for GRC panels is by stainless steel angles that are bolted to the solid structure or structural frame, with the horizontal flange of the angle providing support at the bottom of the panel, as illustrated in Figure 8.34.

The bottom flange of the GRC panel bears on the horizontal flange of the angle with metal packing pieces as necessary to level the panel. Either a separate restraint fixing is used or a stainless steel dowel is welded to the angle to fit into a tapered socket in the GRC flange. The socket is filled with a resilient filler to allow some freedom of movement. The dowel can serve to locate the panel in position and as a restraint fixing. This fixing allows for some horizontal movement through the resilient filler and the movement of the panel on the angle. The edge of the seating angle may be masked by the joint filler, by setting the angle into a slot in the GRC or behind a rebated joint between panels. Support angles are secured with expanding bolts to in situ concrete or brickwork, and angles fixed to structural steel frames.

Restraint fixings
Restraint fixings should tie the panel back to the structure and allow for some horizontal and vertical movement of the panel relative to the fixing. Restraint fixing is usually provided by a stainless steel socket that is cut into the back of the solid flange of a GRC panel as it is being manufactured. The socket is threaded ready for a stainless steel bolt. The bolt is fitted to an oversize hole in an angle, which is bolted to the underside of a concrete or steel beam. Rotational movement of the GRC panel is allowed by a metal tube and plastic separating sleeve that fits around the stainless steel bolt in the oversize hole in the angle. The bolt is held in place in the angle by steel washers, and some movement is provided for by washers around the angle, as illustrated in Figure 8.35. Care is required in the design to make allowance for access to make these somewhat complex fixings.

Stud frames
A large single-skin GRC panel with top and bottom flanges and edge stiffening ribs may not have sufficient stiffness to adequately resist the stresses due to moisture movement of

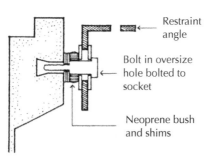

Restraint angle

Bolt in oversize hole bolted to socket

Neoprene bush and shims

Resilient bush restraint

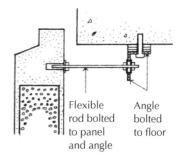

Flexible rod bolted to panel and angle

Angle bolted to floor

Flexible rod restraint

Figure 8.35   Restraint fixings.

the cement-rich material and the considerable wind forces acting on a wall. To provide support for large single skin panels and at the same time make allowance for moisture movement, a system of stud frames was developed and has been extensively used. A frame of hollow and channel steel section is prefabricated with welded joints to the top, bottom and side members, and intermediate vertical sections spaced at about 600 mm, as illustrated in Figure 8.36. L-section, 9.5 mm round steel section anchors are welded to the hollow section studs at about 600 mm centres to serve as flex anchors to the GRC panel. Near the base of the frame, T-sections are welded to the sides of the edge studs to serve as gravity anchors. Angles are also welded to the back face of the studs as seating angles to support the stud frame.

The fabricated stud frame is galvanised or powder coated to inhibit rusting. As the GRC panel is being manufactured, and the skin and flanges and ribs are formed, the stud frame is placed on the back of the compacted and still moist GRC, with the flex anchors and T bearing on the back of the panel. Moist strips of GRC are then rolled onto the back of the panel over the flex anchors and T anchors to secure them to the panel, as illustrated in Figure 8.36. The GRC strips, which are rolled over the flex anchors onto the back of the panel, provide a firm attachment to maintain the thin skin as a flat panel, yet allow sufficient rotational and lateral movement of the panel to prevent failure.

Figure 8.37 is an illustration of the fixing of storey height stud frame panels of GRC to the beams of a structural steel frame. T-section, steel beam brackets are welded to plates that are welded to the web of beams. These brackets support T-sections welded to cleats. One beam bracket is welded to beams centrally on the junction of vertical and horizontal joints between GRC panels.

The weight of the stud frame is supported by the bearing of the bottom flange of the frame on the lower flange of the beam. Top and bottom restraint fixings, cast in the side ribs of the GRC panel, are bolted to the angles of the beam brackets. Open drained or mastic or gasket joints are made to joints between panels. Stud frames are best suited to flat-storey height panels, which are supported at floor levels. Curved and profiled panels may have sufficient stiffness in their shape and not justify the additional cost of a stud frame. Similarly, small under-window and spandrel panels of GRC do not generally require a stud frame.

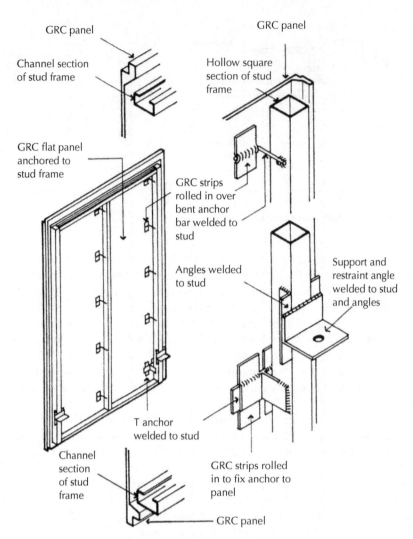

GRC panel

Channel section
of stud frame

GRC flat panel
anchored to
stud frame

GRC panel

Hollow square
section of stud
frame

GRC strips
rolled in over
bent anchor
bar welded to
stud

Angles welded
to stud

Support and
restraint angle
welded to stud
and angles

T anchor
welded to stud

Channel
section
of stud
frame

GRC strips rolled
in to fix anchor to
panel

GRC panel

Figure 8.36   Stud frame single-skin GRC panel.

Joints between GRC panels

Mastic joint

The joints between GRC panels may be square for mastic sealant, rebated for gasket joints, channelled for open-drained joints or rebated for overlap at joints. Which joint is used depends on the overall size of the panel and therefore the moisture movement that the joint will have to accommodate, the exposure of the panel to driving wind and rain, the jointing system chosen by the manufacturers and the convenience in making the joint. Appearance, as part of the overall design, may also be a consideration. For small GRC panels where the moisture movement of the cement-rich mix will be limited, a sealant joint is commonly used for the advantage of simplicity of application and renewal as necessary. The mastic

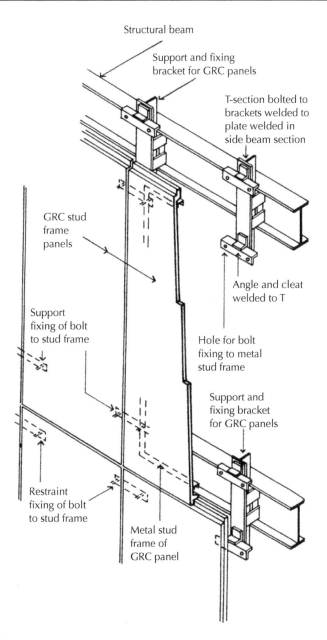

Structural beam

Support and fixing
bracket for GRC panels

T-section bolted to
brackets welded to
plate welded in
side beam section

GRC stud
frame
panels

Angle and cleat
welded to T

Support
fixing of bolt
to stud frame

Hole for bolt
fixing to metal
stud frame

Support and
fixing bracket
for GRC panels

Restraint
fixing of bolt
to stud frame

Metal stud
frame of
GRC panel

Figure 8.37   GRC stud frame fixed to structural steel frame.

sealant joint is made between the square edges of uniform-width joints by ramming a backing strip of closed-cell polyethylene into the joint as support and backing for the mastic sealant. The mastic sealant is run into the joint from a hand or power-operated gun. The sealant is finished with a tool to provide a concave surface for the sake of appearance in keeping the mastic from the panel face, as illustrated in Figure 8.38.

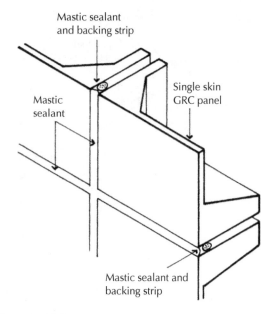

Mastic sealant
and backing strip

Mastic
sealant

Single skin
GRC panel

Mastic sealant and
backing strip

Figure 8.38   Mastic sealant joint.

A mastic sealant joint should be no wider than 15 mm as the sealant in a wider joint might sag and no longer seal the joint. The sealants most used are two-part or one-part polysulphide, one-part polyurethane, epoxy-modified polyurethane and low-modulus silicones. Which is used depends on the skill of the operative and the ease of access and application. One-part sealants are easier to use but less effective than two-part sealants. Silicones tend to be the most difficult to use and the most effective. In time, sealants may oxidise and harden and require renewal after a number of years.

Gasket joint
A gasket joint is used for larger GRC panels where accuracy of manufacture and fixing can provide a joint of uniform width. The elasticity of the gasket is capable of accommodating the moisture movements of the panel while maintaining a weather seal. A gasket joint may be used for vertical and horizontal joints or for vertical joints alone with overlapping rebated horizontal joints, as illustrated in Figure 8.39.

For gasket jointing, the edges of the panels are rebated to provide space for insertion of the gaskets. A backing strip and mastic seal is formed on the back face of the joint to exclude wind. The strip of neoprene gasket is rammed into the joint to make a weathertight seal.

Where both vertical and horizontal joints are gasket sealed, preformed cross-over gaskets are heat welded to the ends of the four straight lengths at junctions of vertical and horizontal joints and then rammed into position. Being set in position some distance behind the GRC panel faces, the gaskets are protected against the scouring effect of wind and rain and also the hardening of the material due to oxidisation caused by direct sunlight. It is a reasonable expectation that these gasket joints will be effective during the life of most buildings. The advantage of the overlapping, rebated horizontal joint illustrated in Figure 8.39

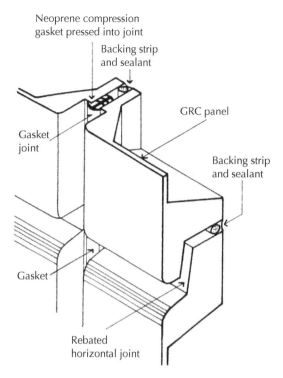

Neoprene compression
gasket pressed into joint

Backing strip
and sealant

GRC panel

Gasket
joint

Backing strip
and sealant

Gasket

Rebated
horizontal joint

Figure 8.39   Gasket joint.

is that the deep rebate will provide a degree of protection against rain running down the building face that a parallel flat face will not, where a similar vertical joint will provide protection from one direction only.

Open-drained joint
Open-drained, vertical joints are used for the larger GRC or rock-based panels, where it is not practical to form the narrow, accurately fitted joints necessary for both mastic and gasket jointing. Plastic channels are cast in a chase in the vertical edges of panels, as illustrated in Figure 8.40, inside which a plastic baffle is hung as a loose fit. The inside joint between panels is sealed with a backing strip and mastic sealant joint.

The loose baffle acts as a first line of defence against wind-driven rain, most of which will run down the baffle. Because the baffle is a loose fit, there will be a degree of wind pressure equalisation on each side of the baffle, which will appreciably reduce wind pressure on rain that may find its way past the baffle. The advantage of this joint is that it allows for movement of the panel, inaccuracies in manufacture and assembly, and acts as a reasonable barrier to wind-driven rain. This open joint acts in conjunction with a rebated, overlapping horizontal joint, which is mastic sealed on the inner face. It is accepted that this joint is not a complete barrier to rain. The effectiveness of the joint depends on the degree of exposure and the surface finish of the panels. Smooth-faced panels will tend to encourage rain to sheet across panels and so cause greater pressure of rain than would coarse textured panels.

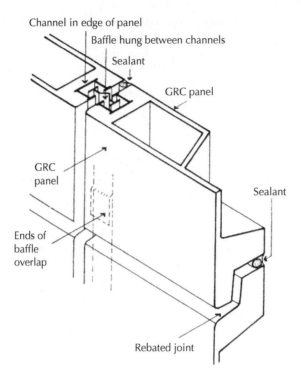

Channel in edge of panel

Baffle hung between channels

Sealant

GRC panel

GRC panel

Sealant

Ends of baffle overlap

Rebated joint

Figure 8.40   Open-drained joint.

8.7 Sheet metal wall cladding

The two metals most used for cladding panels are mild steel and aluminium in the form of hot-rolled strip. Mild steel, which has a favourable strength-to-weight ratio, suffers the considerable disadvantage of progressive, destructive rusting on exposure to atmosphere. To inhibit rusting, the mild steel is coated with zinc. Steel strip can be cold rolled or pressed to shape. Aluminium is a malleable metal, which can readily be cold rolled or pressed to shape and which on exposure to the atmosphere will form a dull, coarse-textured oxide film that prevents further corrosion. Because of this dull, unattractive coating, aluminium is usually finished with an organic coating to improve its appearance. Sheet metal cladding can be broadly grouped as:

❑   Laminated panels.
❑   Single skin panels.
❑   Box panels.
❑   Rain screens.

Laminated panels

Laminated wall panels are made from layers of metal formed around a central core of insulation with the long edges formed by welting or welding the inner and outer metal linings together. The combination of sheet metal and insulation provides a weather surface and insulation in one wall unit. The strip metal is usually of hot-rolled strip. The outer strip is usually of cold-rolled, corrugated or trapezoidal profile, and the inner of flat strip. The disadvantage of using a flat, hot-rolled outer lining of strip metal is that the outer surface will tend to distort due to the considerable expansion of the metal caused by solar energy. The solar heating of the outer lining will not be transferred to the panel due to the insulation core, and the distortion of the outer surface will show obvious unsightly rippling.

A profiled outer lining, which will also distort due to solar heating, will not show any signs of rippling due to the profiles that will take up and mask the effects of distortion. The principal advantage of profiled outer linings is that the depth of the profiles will give stiffness to the panel against bending between supports. The profiled laminated metal panel, illustrated in Figure 8.41, is made in lengths of up to 10 m for fixing with self-tapping screws to holes in horizontal sheeting rails. The vertical edges of sheets overlap as a weather seal.

The insulating core is of polyisocyanurate foam. Both inner and outer steel strip linings are galvanised, and both inner and outer linings of both steel and aluminium strip are usually coated with a coloured organic coating for protection and decoration. The difficulty of making a neat, weathertight and attractive finish to the ends of the one-direction profile panels at eaves, ground level and at corners and around openings tends to limit their use to simple, shed forms of building. Photograph 8.4 shows a detail of profiled insulated cladding and the panels being fixed to a large steel-framed building.

Single-skin panels

Single-skin panels of hot-rolled metal strip are usually stiffened by forming the strip into a shallow pan. The flanged edges of the pan provide a surface for joints to surrounding panels. The process of forming sheet metal into a shallow pan is a comparatively simple and inexpensive process. The corners of the metal are cut, the edges cold bent using a brake press and the corners welded. To provide additional stiffness to the shallow pan, and as a means of fixing to a carrier frame, it is common to weld a supporting frame of angles or channels to the back of the pan, as illustrated in Figure 8.42a. These single-skin panels are fixed to the structural frame or to a separate carrier system fixed to the structural frame with insulation fixed to the back of the panels.

Single-skin panels can be pressed to profiles, as illustrated in Figure 8.43, and fixed as a rain screen or formed as window panels, as illustrated in Figure 8.42b. The flanged, rounded edges for the panel and the window opening provide sufficient stiffness without the use of a supporting frame. The flanged edges and rounded corners are formed by drawing or pressing. This two-dimensional operation is considerably more expensive than one-way cold rolling.

Pressing of strip metal in one operation is performed by pressing around a shaped former or by deep drawing or a vacuum press where the strip is pressed or drawn to shape around a former. These one-off processes are generally limited to panels of 2×1 m for pressing and up to 5×2 m for drawing. The deep-drawn, storey height aluminium panels, illustrated in Figure 8.43, are formed as window units that are fixed to an outer carrier frame. The outer carrier is connected to the inner carrier by plastic, thermal break fixings. A gasket provides a weather seal between outer and inner carriers at open joints.

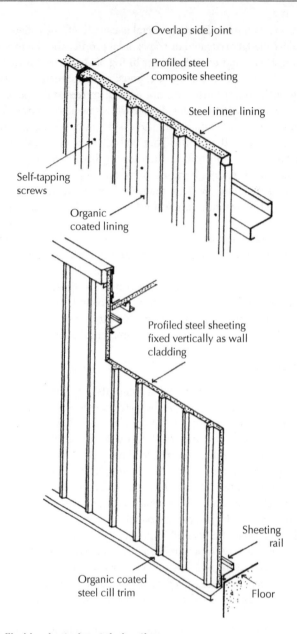

Figure 8.41   Profiled laminated metal sheeting.

The inner carrier, which is bolted to the structural frame, supports an insulated box panel system as inner lining and internal finish. The open joints and the separation of the outer carrier from the inner carrier allow some unrestrained thermal movement of the cladding panels to limit distortion due to solar heat gain. The complexities of the outer and inner carriers supported by a structural frame are necessary for the precision engineering skills required for this system of cladding.

Photograph 8.4   Insulated cladding panels (a) being fixed into position (b).

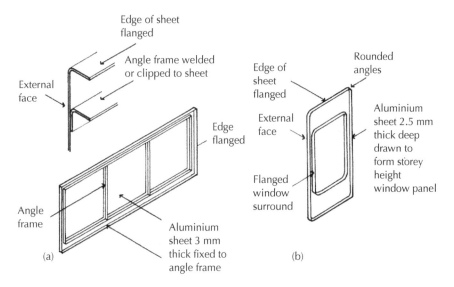

Figure 8.42   (a) Single-skin panel with frame. (b) Profiled single-skin panel.

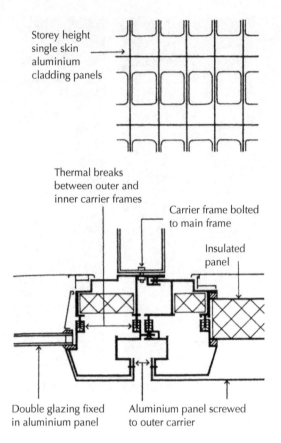

Storey height single skin aluminium cladding panels

Thermal breaks between outer and inner carrier frames

Carrier frame bolted to main frame

Insulated panel

Double glazing fixed in aluminium panel

Aluminium panel screwed to outer carrier

Figure 8.43   Single-skin aluminium cladding panels.

Box panels

Box panels are made from two single panels with flanged edges formed around a core of insulation as a box, as illustrated in Figure 8.44. The flanged edges of the inner and outer panels are pop-riveted together around a neoprene strip. The neoprene strip acts as a thermal break, which allows a degree of movement of the outer panel to that of the inner panel.

As an alternative to riveting the flanged edges of metal panels to form a box, the flanged, rebated edge of one panel may be bonded to an edging piece of wood or plastic, which is bonded to the flat edge of the other panel. Both metal panels are bonded to an insulation core. The advantage of box panels is that an inner and outer lining can incorporate an insulation core in one prefabricated unit ready for site fixing, with the metal faces ready prepared as an inner and outer finish as required. Box panels are much used as an inner lining to provide both insulation and an internal finish behind an outer sheet metal lining, and to provide the main weather and decorative finish. As an inner lining, the box panel is unlikely to suffer distortion of the faces of panels due to differential thermal movement.

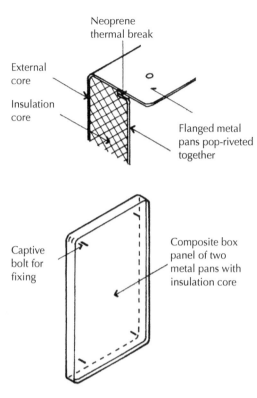

Neoprene
thermal break

External
core

Insulation
core

Flanged metal
pans pop-riveted
together

Captive
bolt for
fixing

Composite box
panel of two
metal pans with
insulation core

Figure 8.44   Metal box cladding panel.

Where box panels are used as an external cladding, there is a likelihood of the outer lining suffering surface distortion due to thermal expansion of the outer face of the panel. This likely distortion may be limited by the use of comparatively small panels and masked by profiling the outer metal skin. The aluminium strip box panels illustrated in Figure 8.45 are specifically designed for the building.

The ribbed, anodised, aluminium panels are made by vacuum forming. The outer tray is filled with Phenelux foam, and the inner tray is then fitted and pop-riveted to the outer tray through a thermal break.

Separate aluminium subframes for each panel are bolted to lugs on the structural frame. Continuous neoprene gaskets seal the open drained joints between panels, which are screwed to subframes with stainless steel screws. The horizontally ribbed face of the outer skin of the box panels serves the purpose of masking any distortion that may occur, provides some stiffening to the aluminium strip and provides a decorative finish.

Rain screens

The term 'rain screen' describes the use of an outer panel as a screen to an inner system of insulation and lining, so arranged that there is a space between the screen and the outer lining for ventilation and pressure equalisation. The open joints around the rain screen,

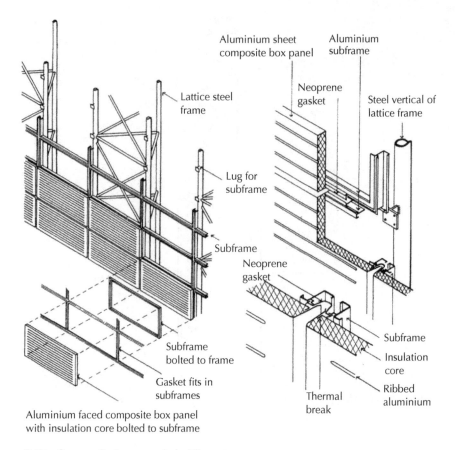

Aluminium faced composite box panel
with insulation core bolted to subframe

Figure 8.45   Composite box panel cladding.

illustrated in Figure 8.46, allow some equalisation of air pressure between the outer and inner surfaces of the rain screen, which provides relief of pressure of wind-driven rain on the joints of the outer lining behind the rain screen. The rain screen will protect the outer lining system from excessive heating by solar radiation and will protect gaskets from the hardening effect of direct sunlight.

All wall panel systems are vulnerable to penetration by rain, which is blown by the force of wind on the face of the building. The joints between smooth-faced panel materials are most vulnerable from the sheets of rainwater that are blown across impermeable surfaces. The concept of pressure equalisation is to provide some open joint or aperture that will allow wind pressure to act on each side of the joint and so make it less vulnerable to wind-driven rain. It is not possible to ensure complete pressure equalisation because of the variability of gusting winds that will cause unpredictable, irregular, rapid changes in pressure. Provided there is an adequate open joint or aperture, there will be some appreciable degree of pressure equalisation, which will reduce the pressure

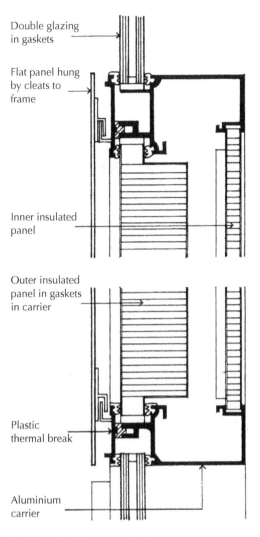

Double glazing
in gaskets

Flat panel hung
by cleats to
frame

Inner insulated
panel

Outer insulated
panel in gaskets
in carrier

Plastic
thermal break

Aluminium
carrier

Figure 8.46   Flat single-skin aluminium panel rain screen.

of wind-driven rain on the outer lining behind the rain screen. A fundamental part of the rain screen is airtight seals to the joints of the panels of the outer lining system to prevent wind pressure penetrating the lining. Because of the unpredictable nature of wind-blown rain and the effect of the shape, size and groupings of buildings on wind, it is usual practice to provide limited airspace compartments behind rain screens, with limited openings to control air movements. The profiled, single-skin rain screen panels illustrated in Figure 8.5 serve as spandrel panels between window panels to provide some protection to the insulated inner panels. The proliferation of joints between window and solid panels provides more joints vulnerable to rain penetration than there are with solid storey height panels.

Rain screens offer considerable design choice in terms of finishes, such as wood effect, light changing colours, or any RAL or NCS colour. Some panels are designed and manufactured so that they can be curved on the site rather than in a factory.

Jointing and fixing

The use of preformed gasket seals in aluminium carrier systems has developed with changes in curtain wall techniques so that the majority of composite panels are fixed and sealed in neoprene gaskets fitted to aluminium carrier systems, fixed to the structural frame as illustrated in Figures 8.43, 8.45 and 8.46, in which the carrier system supports the panels and the gaskets serve as a weather seal and accommodate differential thermal movements between the panels and the carrier system. To reduce the effect of the thermal bridges made by the metal carrier at joints, systems of plastic thermal breaks and insulated cores to carrier frames are used.

Where open horizontal joints are used to emphasise the individual panels, there is a flat or sloping horizontal surface at the top edge of each panel from which rain will drain down the face of panels. In a short time, this will cause irregular and unsightly dirt stains, particularly around the top corners of the panels, unless the panels are pre-coated to be self-cleaning. The advantage of the single-storey-height box panel is in one prefabricated panel to serve an outer and inner surface around an insulating core that may be fixed either directly to the structural frame or a carrier system with the least number of complicated joints. Some composite box panels of aluminium strip are formed with interlocking joints formed in the edges of panels. The interlocking joints in panel edges, illustrated in Figure 8.47, comprise flanged edges of metal facing, formed to interlock as a male and female locking joint to all four edges of each panel. The edge of one panel is formed by a pressed edging piece, welted to the linings and formed around a plastic insert to minimise the cold bridge effect at the joint. This protruding section fits into the space between the wings of the linings of the adjacent panel with a neoprene gasket to form a weather seal.

For this joint to be effective, a degree of precision in the fabrication and skill in assembly is required for the system to be reasonably wind- and weathertight. As with all panel systems, the junction of horizontal and vertical joints is most vulnerable to rain penetration and requires precision manufacture and care in assembly. The interlocking joint system, illustrated in Figure 8.48, is used for flat, strip steel panels, which are used principally as undercill panels between windows or as flat panels without windows.

The vertical edges of the strip metal are cold roll formed to a somewhat complicated profile, with the edges of the inner and outer linings either welted or pop-riveted together. The interlocking joint is designed to hide the fixing bolt and clamp, which connect to the edge of one panel and are bolted back to the carrier or structural frame. A protruding rib on the edge of one panel is compressed onto a neoprene gasket on the edge of the adjacent panel. Horizontal joints between panels are made with a polyethylene backing strip and silicone sealant. These strip steel panels are made in widths of 900 mm, lengths of up to 10 m and thickness of 50 mm. The steel linings are galvanised and finished with coloured inorganic coatings externally and painted internally.

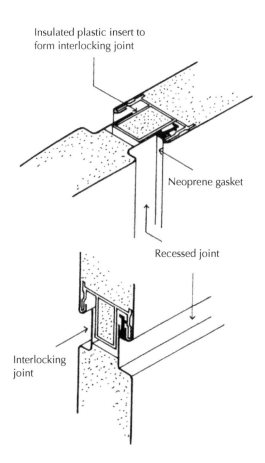

Insulated plastic insert to form interlocking joint

Neoprene gasket

Recessed joint

Interlocking joint

Figure 8.47   Interlocking panel joint with neoprene gasket.

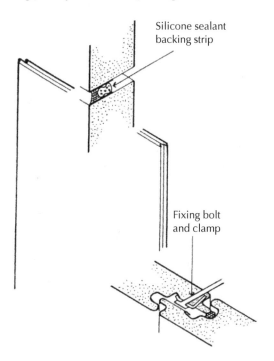

Silicone sealant backing strip

Fixing bolt and clamp

Figure 8.48   Interlocking panel joint with silicone sealant.

8.8 Glazed wall systems

With the development of a continuous process of drawing window glass in 1914 and a process of continuously rolling, grinding and polishing plate glass in the 1920s and 1930s, there was a plentiful supply of cheap window glass and rolled and polished plate glass. In the 1920s and 1930s, window glass was extensively used in large areas of windows framed in slender steel sections as continuous horizontal features between undercill panels and as large metal-framed windows. During the same period, rolled plate glass was extensively used in rooflights to factories, the glass being supported by glazing bars fixed down the slope of roofs. Many of the sections of glazing bar that were developed for use in rooflights were covered by patents so that roof glazing came to be known as 'patent glazing' or 'patent roof glazing'.

Curtain walling

The early uses of glass as a wall facing and cladding material were developed from metal window glazing techniques or by the adaptation of patent roof glazing to vertical surfaces so that the origins of what came to be known as 'curtain walling' were metal windows and patent roof glazing. The early window wall systems, based on steel window construction, lost favour principally because of the rapid and progressive rusting of the unprotected steel sections that in a few years made this system unserviceable.

With the introduction of zinc-coated steel window sections and the use of aluminium window sections, there was renewed interest in metal window glazing techniques.

Cold-formed and pressed metal box section sub frames, which were used to provide a bold frame to the slender section of metal windows, were adapted for use as mullions to glazed wall systems based on metal window glazing techniques. These hollow box sections were used either as mullions for mastic and bead glazing of glass and metal windows or as clip-on or screw-on cover sections to the metal glazing. Hollow box section mullions were either formed in one section as a continuous vertical member, to which metal window sections and glass were fixed, or as split section mullions and transoms in the form of metal windows with hollow metal sub-frames that were connected on site to form split mullions and transoms.

The complication of joining the many sections necessary for this form of window panel wall system and the attendant difficulties of making weathertight seals to the many joints have, by and large, led to the abandonment of window wall glazing systems. Glass for roof-lights fixed in the slope of roofs is to a large extent held in place by its weight on the glazing bars and secured with end stops and clips, beads or cappings against wind uplift. The bearing of glass on the glazing bars, together with the overlap of bays down the slope, acts as an adequate weather seal.

To adapt patent roof glazing systems to vertical glazed walls, it was necessary to provide a positive seal to the glass to keep it in place and against wind suction, to support the weight of the glass by means of end stops or horizontal transoms and cills, and to make a weather-tight seal at horizontal joints. The traditional metal roof glazing bar generally took the form of an inverted T-section, with the tail of the T vertical for strength in carrying loads between points of support with the two wings of the T supporting glass. For use in vertical wall glazing, it was usual practice to fix the glazing bars with the tail of the T, inside with a compression seal and on the outside holding the glass in place, as illustrated in Figure 8.49.

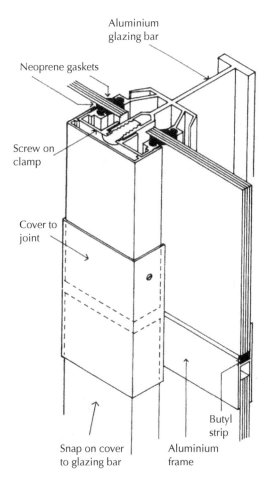

Aluminium
glazing bar

Neoprene gaskets

Screw on
clamp

Cover to
joint

Snap on cover
to glazing bar

Aluminium
frame

Butyl
strip

Figure 8.49   Aluminium glazing but used for vertical glazing.

The usual section of metal glazing bar, which is well suited to roof glazing, did not provide a simple, positive fixing for the horizontal transoms and cills necessary for vertical glazing systems. The solution was to use continuous horizontal flashings on which the upper bays of glass bore and up to which the lower bays were fitted, as illustrated in Figure 8.49. Patent roof glazing techniques, adapted for use as vertical glazing, are still in use but have by and large been superseded by extruded hollow box section mullion systems.

Hollow box section mullions were designed specifically for glass curtain walling. These mullion sections provided the strong vertical emphasis to the framing of curtain walling that was in vogue in the 1950s and 1960s, and the hollow or open section transoms with a ready means of jointing and support for glass. Hollow box section mullions, transoms and cills were generally of extruded aluminium, with the section of the mullion exposed for appearance's sake and the transom, cill and head joined to mullions with spigot and socket joints, as illustrated in Figure 8.50. A range of mullion sections was available to cater for various spans between supporting floors and various wind loads. The mullions, usually fixed at about 1–1.5 m centres, were secured to the structure at each floor level and mullion lengths joined with spigot joints, as illustrated in Figure 8.50. The spigot joints between mullions and mullions, and between mullions and transoms, head and cill, made allowance for thermal movement, and the fixing of mullion to frame made allowance for differential structural, thermal and moisture movements. Screw-on or clip-on beads with mastic or gasket sealants held the glass in place and acted as a weather seal. This form of curtain walling with exposed mullions was the fashion during the 1950s, 1960s and early 1970s.

Stick system of curtain walling

This, the earliest and for many subsequent years the traditional form of glass curtain walling, has become known as the 'stick system'. Typical of the stick system is the regular grid of continuous mullions, bolted to the structural frame, with short discontinuous transoms as a regular grid into which panes of glass are fitted and secured. The advantage of this system is

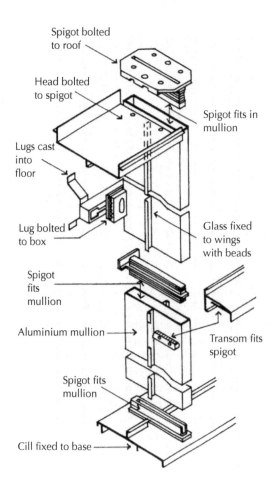

Figure 8.50   Aluminium curtain walling.

the use of a small range of standard aluminium sections that can be cut to length and joined with spigot joints as a carrier frame for glass panes. The carrier system is secured to the floors of the structural frame with bolts and plates that allow for some small relative movement between structure and carrier frame. Initially, the glass was secured in the carrier with sprung, clip-on or screw-on aluminium clips and later by gaskets that were compressed up to or both sides of the glass to secure it in place and act as a weather seal. While the many joints between the members of the carrier frame and glass make allowance for structural and thermal movements, they are also potential points for penetration of wind and rain, particularly at corners where clips and gaskets are mitred to fit.

In the early forms of glass curtain walling, the section of the hollow, extruded aluminium mullions and transoms was exposed on the face of the walling. The squares of glass were held in position against wind pressure and suction by sprung metal clips that were fitted into wings on the mullions and transoms and bore against the face of the glass. These sprung clips were adequate to hold the glass in place but did not provide a wind- and watertight seal, particularly at corners where the beads were mitre cut to fit. To show the least exposure of the aluminium carrier sections on the face of the glass walling, for appearance's sake and as a means of fixing neoprene gaskets to provide a more positive weather seal, the extruded hollow box section mullions and transoms, illustrated in Figure 8.51a, were used.

The main body of the aluminium carrier was fixed internally behind the glazing with continuous head and cill sections, and continuous mullions with transoms fitted to the side of mullions. The mullion and transom sections were fitted over cast aluminium location blocks screwed to the frame with a mastic seal or neoprene gasket to the bottom of each mullion. The advantage of the joints between the vertical and horizontal carrier sections is that there is allowance for some structural and thermal movement. To provide the least section of carrier frame externally, the section outside the glass is the least necessary for bedding glass and for neoprene gaskets that fit into the serrated faces of a groove. A slim section of aluminium and the gaskets are all that show externally. This system of extruded, hollow-section aluminium carriers is designed to support the whole of the weight of the glazing and wind pressure and suction in the position of exposure in which the building is erected. The extruded aluminium carrier system is fixed to the structure by angle cleats, which are bolted together through the mullions and bolted to the structure at each floor level, as illustrated in Figure 8.51b.

The sealed double-glazing units are fixed to the carrier frame with distance, setting and location blocks. Neoprene gaskets, mitre cut at corners, are compressed into the serrated-edged groove in the carrier frame and up against the double-glazed units. This gasket glazing system effectively secures the glass in position against wind pressure and suction and acts as a weather seal. The slender section of the aluminium carrier frame, mullions and transoms that show on the external face provide the illusion of a glass wall.

With the traditional stick system of curtain walling, the aluminium members of the carrier frame are assembled and fixed to the structure on site, and the glass panes are glazed into rebates and weather sealed with gaskets, which secure the glass in place and provide restraint against the considerable wind forces acting on the facade of multi-storey buildings. In this system of glazed wall cladding, it is generally economic to use comparatively closely spaced mullions, to support panes of glass, from 600 mm to 1 m centres to use the least section of mullion and thickness of glass to support the weight of glass and the wind forces

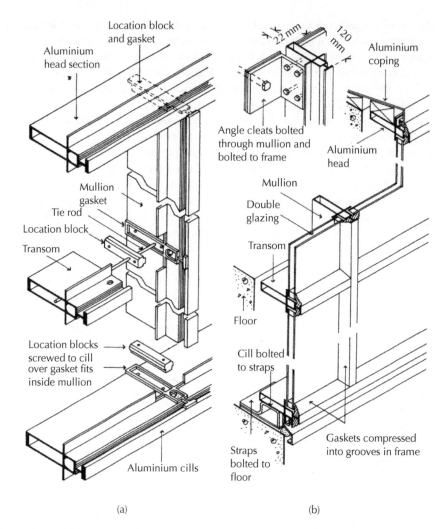

Figure 8.51 (a) Curtain wall carrier frame. (b) Double-glazed panels.

acting on the glass. The site glazing with mastic tape or gaskets used in this system may not provide long-term protection to the edge seals of double glazing units, which are vulnerable to decay due to the penetration of water, particularly at the bottom corners of glass.

Unitised or panel system of curtain walling

An alternative to the stick system is the unitised or panel system of curtain walling, in which complete panels of an aluminium frame with glazing are fabricated as units ready for hoisting into position for fixing to the structural frame or to a carrier system. The advantage of this system is the facility of precision assembly of the components and glazing undercover in the conditions most favourable for successful glazing, particularly for insulating glass (IG) units, to provide the most effective protection to edge seals. By virtue of repetitive, precision assembly, large glazed panels may be fabricated at reasonable cost for fixing to the structural frame with narrow weather-sealed joints between panels, the maximum area of glass and least exposed area of framing. In effect, this is a system of large, dead light window frames, ready glazed for fixing as a glazed wall system between floors or as a curtain wall. The members of the frame are designed to show the least area on face necessary for satisfactory support, bedding and weather sealing of glass, and adequate metal section and depth to support the weight of glass and wind forces acting on the panel where it is fixed and restrained at its top and bottom edges to structural floors. The edges of the frames may be shaped for gasket glazing compressed into rebates by metal strips bolted from behind or may be shaped so that the long edges of panels interlock and the horizontal joints overlap over gasket seals.

Large panels may be fixed to a carrier frame, which is fixed to the structural frame. The carrier frame is designed to support the dead and live loads of the panels, provide a background for gasket or sealant joints and assist in the alignment of the panels across the face of the building. Figure 8.52 is an illustration of a glazed curtain wall in which panels of

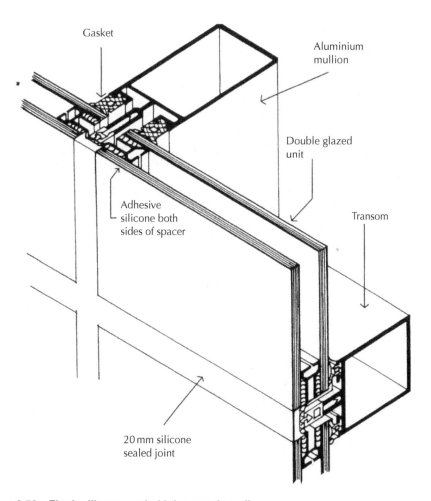

Figure 8.52   Flush silicone-sealed joint curtain wall.

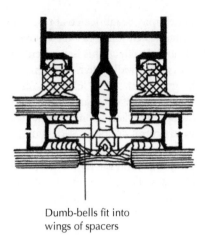

Dumb-bells fit into
wings of spacers

Figure 8.53   Fixing glazed panels.

double-glazed units are fixed to a system of aluminium mullions and transoms fixed to the structural frame at each floor level. The spacer bars at the edges of the double-glazed units are shaped so that the two panes of glass can be secured to the wings of the spacer bar with adhesive silicone. The silicone acts as a powerful, long-term adhesive.

The glass panels are secured to the carrier frame with dumb-bells that fit to a bar that is screwed to the mullion, as illustrated in Figure 8.53. The dumb-bells fit into the wings of the spacer to secure the glazed panel at intervals on all four sides to hold the glass panel in position. A gasket fixed to the bar provides a backing onto which the silicone sealant jointing is run between glass panes.

This system of curtain walling is used for the advantage of the flush silicone-sealed joints that provide a flush external face. This system is often used with coloured glass for the dramatic effect of a large expanse of reflective material. For reasons of safety, the height of this cladding is limited to about 8 m, unless a system of mechanical retention of the outer panes is used. Mechanical retention takes the form of aluminium angles around the edges of each outer pane. This glazing system is used for small areas of cladding both internally and externally.

Structural glazing sealant

The characteristics of making a powerful bond between glass and metal have been used in the system known as structural glazing. The advantage of this system is that glass may be bonded to the face of a metal frame through a narrow edge strip contact of silicone to the back of glass and the face of a metal frame. The glass is held firmly in place by the silicone, which will transfer a whole or part of the weight of glass to the frame and the whole of wind forces acting on a glass panel in the face of a building. This straightforward system of glazing, which needs no spacing, bedding or weathering, has been exploited for the facility of a flush external glass face interrupted only by very narrow open joint or silicone sealant gap-filling seals between panes of glass. Figure 8.54 illustrates the simplicity of this system of glazing with panes of glass bonded to a simple aluminium frame, with the joints between panes of glass silicone sealed. Single or double sheets of glass can be supported by an aluminium frame designed for fixing to a carrier system or curtain wall grid secured to the structural frame.

Because of the adhesion of the silicone, both the weight of the glass and wind forces acting on the glass panel are supported by the metal frame in the illustration of four-sided structural glazing shown in Figure 8.54. Because of the natural resilience of the structural silicone bond, some small thermal movement of glass relative to the frame can be accommodated. Thermal breaks wedged into the frame and bearing on the back of the glass together with the silicone bond will serve to reduce the thermal bridge at the junction of glass and metal. The silicone seal between the edges of glass can be run continuously around all joints. The seal is run onto a backer rod of polyethylene and tooled either flush with the glass or with a shallow concave finish. The joints between glass faces should ideally be no more than 20 mm wide.

Structural glazing systems are designed to give a flush facade of glass with the supporting mullions and transoms of the carrier frame just visible through a transparent glass facade. The extent to which the carrier frame is visible will depend on the intensity of light reflected off the surface of glass. For structural silicone to be most effective in bonding glass to the supporting framework, it should be applied in conditions where the cleanliness of the surfaces to be joined, temperature and humidity can be controlled.

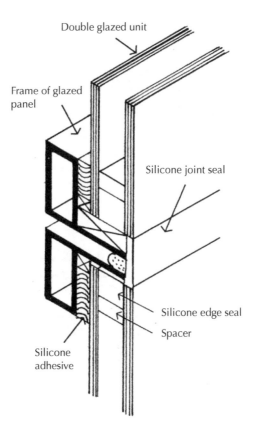

Figure 8.54   Four-sided structural sealant.

Two-sided structural silicone glazing is used for the two vertical edges of single glass panes, which are bonded to a metal frame fixed to mullions of a curtain wall system and supported at their base and restrained at their top edge by aluminium transoms. The wind forces acting on the glass are carried by the framing back to the carrier system, and the weight of the glass by the transoms of the curtain wall. Structural glazing to IG units may be applied to the inside face of the inner sheet of glass in the form of four-sided glazing bonded to an aluminium panel frame, as illustrated diagrammatically in Figure 8.54. The edge seal to the spacer bars between the glass in the IG units is made with silicone sealant to provide the most effective seal against penetration of water. As there is no means of bonding the outer glass to the frame, the outer glass is held in place by the silicone edge seal. The bond of the edge seal will be insufficient to retain the outer glass firmly in place and the glass would sink under its weight. For this reason, silicone setting blocks and the fin of the panel frame are used in the glazing rebate to provide mechanical support. As a means of forming a narrow, open or silicone sealant joint between insulating glazed units, the edge of the units is formed as a stepped edge. The outer pane of glass projects beyond the inner pane in the form of a step, as illustrated in Figure 8.55, with the panel frame behind the glass edge.

The glazed curtain wall system illustrated in Figure 8.56 is designed to utilise the advantages of precision engineering fabrication to minimise site work and gain full advantage of the application of silicone adhesive, silicone sealant and gaskets in the most favourable conditions.

Aluminium mullions are fixed to the structural frame with aluminium transoms cleat fixed to the mullions. Aluminium frame sections to each glazed panel are fixed to the edges of the outer pane with silicone adhesive, with the panel frame providing mechanical support to the bottom edge of the IG unit. Clips set into channels in the glazed panel frame and the carrier frame secure the glazed panel in position. Ethylene propylene diene monomer (EPDM) gaskets set into grooves provide a seal between the glazed panel frame and the carrier mullions and transoms and serve as thermal breaks. An 18 mm open joint is formed between the glazed panels, backed by a gasket which is set into fins on the mullion and bearing on the panel frames as a weather seal. As a precaution against the possibility of the outer panes of the IG units becoming dislodged by wind pressure, an aluminium glass retention section is fixed around the four edges of each pane and clipped back to the

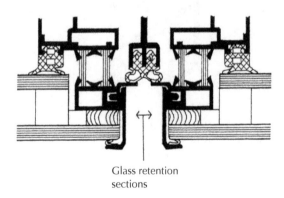

Glass retention
sections

Figure 8.55   Glass retention sections.

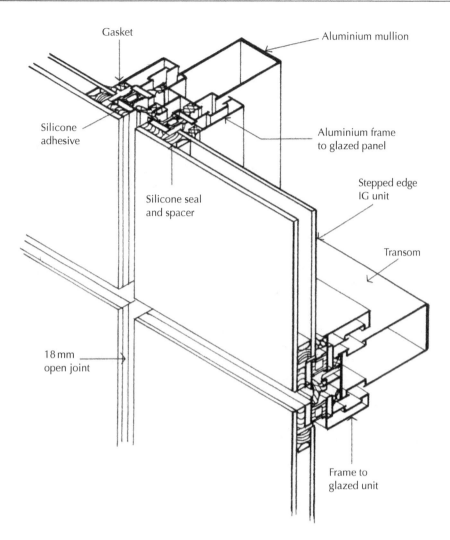

Figure 8.56   Prefabricated flush face curtain wall.

frame, as illustrated in Figure 8.55. These glass retention sections provide some mechanical restraint, particularly for the large panels of glass often used in multi-storey buildings. This sophisticated, precision system of glazed curtain walling is most used for large areas of glazed cladding, where the considerable cost of prefabrication and fixing is justified by the durability and appearance of the finished result.

High-performance structural glazing system: structural silicone bonding

High-rise buildings situated in areas prone to extreme weather, such as hurricanes and typhoons, need to be designed to resist high wind pressures over 6 kPa. Cladding systems must be capable of remaining in place when subjected to extreme wind pressure and must also be sufficiently stable when damaged by impact from flying debris, such that the unit remains largely intact and bonded to its frame. This places a large demand on the structural cladding and the silicone bond. Wind pressure tests, missile and other impact tests are used to prove the durability and the robust nature of the cladding materials. With the desire to see less of the cladding frame, greater emphasis has been placed on the properties of the silicone adhesive. The silicone is used to fix the cladding unit from the sides and from behind and must be able to withstand the structural loads of the cladding unit.

Suspended frameless glazing

Suspended frameless glazing is a system of supporting large panes of glass by bolts secured to brackets fixed to glass fins or to an independent frame, without any framing around the glass panes and with narrow joints between panes, gap filled with silicone sealant to give the appearance of a flush glass face. The principal use of this system is as a screen of glass as a weather envelope to large, enclosed spaces such as sports stadia, conference halls, exhibition centres, airports and showrooms, where a clear view of activities inside or outside of the enclosure is of advantage. The prime function of this use of glass is not the admission of daylight or as an efficient weather shield.

Suspended glazing has also been used as a glass wall screen to a variety of buildings and entrances, where little or nothing can be seen of outside or inside activities to justify the screen and the glass enclosure, and its visible supporting frame is used for the sake of appearance. Suspended glazing is usually vertical and limited to a height of about 20 m. Sloping, suspended glazing has been used as a wall screen and for roofs largely for effect.

Suspended glazing depends on the use of stainless steel bolts that pass through holes drilled in the glass to connect to stainless steel plates that are fixed to glass fins or an independent framework. Holes are drilled near each corner of a pane far enough from edges to leave sufficient glass around holes to bear the weight of the glass and resist shear stresses. Plastic washers, fitted to the accurately drilled holes, provide bearing for the bolts, prevent damage to cut glass edges and make some little allowance for thermal movement. Bolts are screwed to stainless steel plates that are fixed to glass fins or a supporting metal frame.

One, two or four back plates are used at the junction of four glass panes with fibre gaskets between the surfaces of plate and glass. Sleeves around bolts at connections to plates accommodate some small rotational movement. For appearance's sake, the external head of supporting bolts may be countersunk headed to fit to a washer in the countersinking of the glass for a near flush or flush fit. Figure 8.57 shows the connection of a bolt to glass and plate.

For the safety of those inside and outside a building, either heat toughened or heat-treated laminated glass is used for suspended glazing so that, in the event of a breakage, the glass shatters to small fragments least likely to cause injury. Toughened glass is made in maximum sizes of 1500×2600 mm and 1800×3600 mm and thicknesses of 10, 12 or 15 mm. Laminated glass is made from 4 or 6 mm glass for the thinner layer, and 10 or 12 mm for the thicker layer, in maximum sizes of 2000×3500 mm in glass thickness combinations of 4 or 6 mm with 10 or 12 mm with a 2 mm thick interlayer of polyvinylbutyral (PVB).

Insulating units of glass may be used for double-glazed suspended glazing, which is hung from bolts through both the inner and outer panes of glass with a clear plastic boss around the bolts to act as a spacer in the cavity. The usual combinations of glass are an inner pane

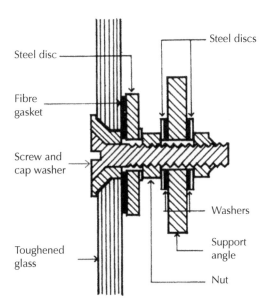

Steel discs

Steel disc

Fibre
gasket

Screw and
cap washer

Washers

Toughened
glass

Support
angle

Nut

Figure 8.57   Bolt support for suspended glazing.

of toughened glass 6 mm thick, a 16 mm airspace, and an outer pane of 10 or 12 mm thick glass. The edges of the IG units are made with spacer bars and a silicone edge sealant. Holes for bolts are normally 60 mm from edges at corners. The maximum size of glass used is 2000×3500 mm for 6 mm glass. Both body and surface-tinted glass may be used.

It is usual practice to hang the glazed screen some distance from the edges of floors and provide some separate, independent system of support and stiffening to the screen against wind forces to emphasise the effect of a flush, frameless screen of large panes of glass. Any independent system of support for suspended glazing has to provide support for each pane of glass and restraint against wind forces. Two systems of support are used. In the first, glass fins are fixed internally at right angles to the screen and bolted to angle plates bolted to the glass screen. In the second, lattice metal frames are fixed between structural floors and roofs, and bolted to the glazed screen at the vertical joints between glass panes with the frames projecting into the building. With both systems, the glass fins and the frames are designed to provide support for each glass pane of the screen and resistance to wind forces that will tend to force the screen to bow out of the vertical plane.

Glass fin support

Glass fins fixed at right angles to the screen are the least obtrusive and provide the least visual barrier to a wide, clear view. The glass fins are the least width necessary to support the weight of the glass screen and anticipated wind loads in the position of exposure. Each fin is the same height as the glass panes it supports, and is bolted through stainless steel plates to the fin above and below and the corners of the four glass panes supported, as illustrated in Figure 8.58. For light loads, separate small plates may be used to join fins to

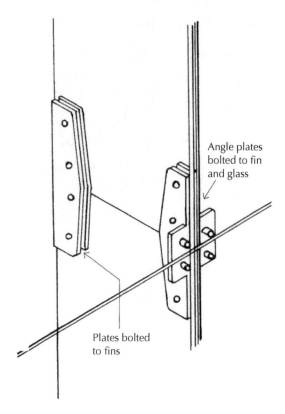

Angle plates
bolted to fin
and glass

Plates bolted
to fins

Figure 8.58   Glass fins.

the glass screen. For heavier loads, two plates are used to join fins, and two smaller plates to join the main panes of glass.

For small glazed screens, the glass fins may be used to the top and bottom panes, or the top two panes of glass with the top and bottom panes bolted to the floor and roof. For larger and heavier screens, the fins will usually extend the full height of the glazed screen, as illustrated in Figure 8.59. Whichever system is selected will be chosen as being the least visually intrusive compatible with adequate strength to give support.

Fins are usually cut from 19 mm thick toughened glass, which is holed for bolts and fixed with stainless steel plates over 1 mm thick fibre gaskets each side of the glass fin. Both single and double glazing may be used for the panes of glass to the screen for the fin system of support. The usually accepted maximum height for fin-supported glazing is 10 m.

Framed support

Glass fin support for frameless, suspended glazing, which is limited to a height of about 10 m, is used generally for sports stadia, and support to glazing hung as a screen to

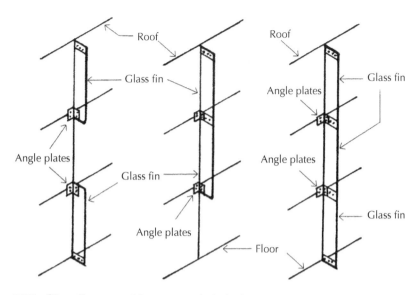

Figure 8.59   Glass fin support for suspended glazing.

conventional framed structures where a clear, unobstructed view is critical. For large enclosures, suspended glazed screens are supported by systems of lattice steel frames and tensioned cable rigging fixed between floor and roof or to an independent steel frame used to support both wall and roof glazing to single-storey enclosures where a more sturdy system of support is necessary. Various single-cell enclosures have been constructed with lattice-framed supports and tensioned cable stays to support clear glass suspended glazing for both the wall and roof, to the extent that the glass acts more as a showcase for the complicated system of frames rather than another purpose. The frame and tension cable supports may be used separately or in combination. The most straightforward system of support is by lattice steel frames anchored between floor and roof level to a structural frame or to a separate frame at the junction of walls and roof. Each frame provides support to the glazed screen at the junction of four large panes of glass. Figure 8.60 and Photograph 8.5 show typical lattice frames.

The frames are fabricated from small steel sections welded together with stays and props to support rectangular panes of glass. The panes may be hung on one long edge to provide the maximum practical width between frames for appearance sake.

As an alternative, a system of primary lattice steel frames and secondary tension cable rigging may be used in alternate vertical supports to suspended glass, as illustrated in Figure 8.60. In this secondary rigging, the cable is tensioned between floor and roof across the solid props that are bolted to the glass at the junction of four panes of glass, which act with the rigging as a tensioned truss. This tensioned rigging is used as a less obvious and unobtrusive means of support, which provides some support for the glass and more

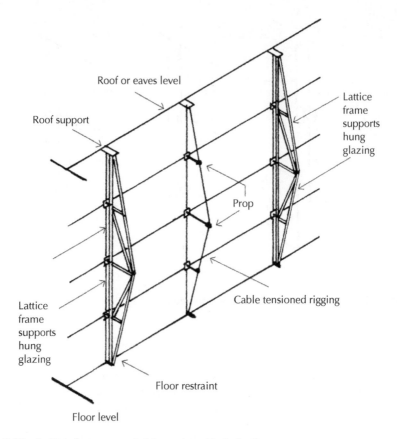

Figure 8.60   Lattice frame support for suspended glazing.

particularly as resistance to wind pressure acting on the glass screen. Lattice frame and secondary rigging systems depend on anchorages to the floor and a structural roof for support, with the top panes of glass being supported by the roof and the bottom edge of lower panes by restraint from the floor.

Where suspended glass is used as an enclosure to walls and roof, a structural steel frame is used to support the whole of the weight of glass and wind loads. Various systems of light section – vertical, horizontal and sloping lattice frames – are used together with tensioned cable rigging systems for stability and effect. A variety of plates are used for bolting glass, the most straightforward of which is one plate shaped to take the bolts at the junction of four panes of glass. Figure 8.61 is an illustration of a comparatively simple system of suspended glass supported by horizontal lattice steel frames fixed between steel portal side frames to give support to the glass. The horizontal lattice frames are braced with tension cables and braced up the face of the glass with cables. Photograph 8.6 shows a suspended glazing system.

Fixing plate with four
bolts holds glazing in
position

Latice frame supports
hung glazing

Photograph 8.5   Suspended glazing systems with lattice frame.

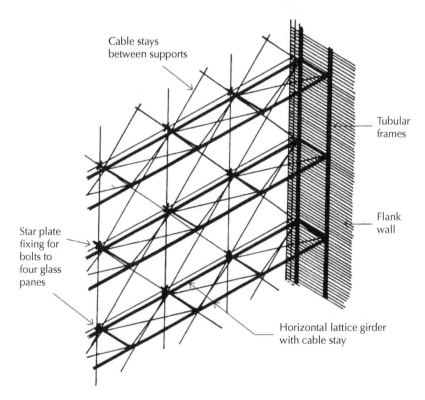

Cable stays
between supports

Tubular
frames

Flank
wall

Star plate
fixing for
bolts to
four glass
panes

Horizontal lattice girder
with cable stay

Figure 8.61   Horizontal lattice girder and cable stay supports.

Star fixing plate secures the
edge of four sheets of glass

Tubular steel frame supports
suspended glazing

Photograph 8.6   Suspended glazing system.

8.9 Double skin façades

An effective way of tempering the internal environment of buildings with large areas of
glass is to construct a double skin. The term 'double skin façade' has come into widespread
use to describe buildings with two external envelopes, essentially an outer skin protecting
an inner skin. These two walls are sometimes separated by a relatively thin intermediate
space between the walls, and in some cases by a wide corridor in which people can move
around the building. The corridor space will not be heated or cooled, but it usually provides
a sufficiently comfortable means of access that is tempered to some extent from the outside.

Thermal insulation
The calculation of the thermal insulation of a double façade needs to consider the outer
skin, the intermediate area, and the inner skin as a system. Specialist modelling software is
available to assist in this task.

Sound insulation
In simple terms, the provision of two skins to the building will help to reduce the transfer
of airborne sound from the surrounding area into the building. A figure of 20 db is sometimes used as an approximate guide to the noise reduction. Similar to the calculation of the
thermal insulation, the double envelope needs to be considered as a system when calculat-
ing the sound insulation. There is an additional challenge in that the inner skin usually has
openings into the intermediate area (i.e. windows and doors), and this leads to the transfer
of sound within the intermediate area.

Fire protection

The double skin must be designed as part of the building's fire safety strategy. The intermediate area provides an opportunity for flame and smoke to spread unless measures are put in place to compartmentalise the double skin and provide fire stops. The fire strategy will also need to consider the fire resistance of doors, windows and vents that open into the space between the two skins. Specialist advice is required to prevent unwanted sound transfer within the space between the skins.

Adaptive building envelopes

It is possible to construct buildings with an outer façade, or skin, that is capable of responding and adapting to climatic changes. This is usually known as an adaptive façade or adaptive building envelope or a smart building. The building envelope is designed to form an adaptive boundary between the inside of the building and the outside. The aim is to temper the internal environment to optimise the comfort of the building users. The underpinning philosophy is that the building skin acts in a similar way to the skin of humans. Our epidermis helps to control our temperature. The building 'skin' helps to temper the internal building environment in a similar way. Climate adaptive building skins are known by several terms, such as intelligent, kinetic, responsive and smart; terms that tend to describe the same thing.

Adaptive building skins rely on a combination of mechanical and electrical motors driven by a building management system (BMS). The BMS detects changes in the internal and external environment and sends instructions to a series of mechanical motors to open and close, typically in relation to shading devices and vents. Some of these shading devices may be made of fabric that reduces solar gain and glare, but which also let daylight into the interior. These systems tend to be expensive to run and maintain over their service life.

Alternative approaches are being tried that rely on mimicking nature, such as the process of hygromorphism. This is where materials, such as wood, undergo physical change, depending on the level of moisture to which they are exposed. As the relative humidity changes, so does the physical shape of the component.

Further reading

See the society of Façade Engineering (SFE): www.cibse.org and, for example, the Metal Cladding and Roofing Manufacturers Association homepage: www.mcrma.co.uk.

Reflective exercises

You have been appointed to construct a mixed-use development in a city centre. The proposed building is six storeys high and was approved a few years ago, with construction delayed while the client secured funding. Funding is now in place and the client has decided to progress the project. The building was designed with a structural concrete frame with timber cladding to all external walls and timber balconies. Your client is concerned about fire safety and the long-term durability of the timber cladding (specified as larch). You have been asked to look at other cladding materials.

❏ What factors do you need to consider and why?
❏ Would the situation change if the client added another two storeys to the height of the building (subject to planning approval)?

The requirements relating to the thermal performance of the envelope have become more stringent since the building was designed. You are thinking of using insulated composite cladding panels, but you are concerned about the future recovery of the panels and recycling of the materials.

❏ Why is this a challenge?
❏ What other solutions are there?

Lifts and Escalators: Chapter 9
AT A GLANCE

What? Lifts (elevators) and escalators provide the means to move people and goods from one floor to another quickly, reliably and safely. Lifts are available as traction lifts or hydraulic lifts (for medium- to low-rise developments). The traction lift relies on an electrically powered system of pulleys and wires to move the lift car. The hydraulic lift relies on a hydraulic ram to raise and lower the lift car. In both cases, the lift car and machinery are housed in a lift shaft. Escalators are used to move people from one floor, or level, to another in the form of a moving staircase. Moving walkways convey people quickly over long distances and are a feature of large buildings, such as airport terminals.

Why? With the exception of single-storey buildings, it is necessary to move people and goods from one floor to another. This is known as vertical circulation. Staircases are not a practical means for regular vertical circulation in buildings that are higher than three to four storeys in height. Lifts allow rapid and safe movement of people and goods from one floor to another. Similarly, staircases are not convenient for moving many people at the same time in comfort. Thus, escalators are popular in office and retail developments; for example, for moving customers from one sales floor to another.

When? Lift shafts are usually designed to form the structural core of high-rise buildings. Thus, the lift shaft (lift core) is usually one of the first parts of the building to be constructed. Installation of the lift car and the commissioning of the lift are usually done towards the end of the contract, usually during the fit out. This is to ensure that the lift car is installed when the building is weatherproof and when there is less chance of accidental damage during the main construction phase.

How? The lift shaft is usually constructed from in situ reinforced concrete to form a structural shaft into which the lift mechanism and car are later housed. The lift shaft must be built to precise tolerances to ensure the lift shaft is true, which allows the lift car to operate correctly. To ensure a high degree of precision is it common to use steel formwork to construct the lift shaft. Alternatively, prefabricated lift shafts may be used. The sizing of the lift shaft and car will depend on the anticipated flow of people within the building.

9 Lifts and Escalators

Quick, reliable and safe vertical circulation is an essential feature of most commercial buildings and larger residential developments. Lifts (also known as elevators) and escalators are the primary means of moving people, goods and equipment between different levels within buildings. Staircases are still required as an alternative means of escape in the event of a fire or when the lift or escalator is out of use (e.g. for routine maintenance). Lifts and escalators are prefabricated in factories by a small number of manufacturers, transported to site, installed and commissioned prior to use. These comprise 'standard' lift cars and escalators, as well as items made to specific customer requirements. Although the design and commissioning of lifts and escalators is the domain of engineers, there is a considerable amount of building work required to ensure that the mechanical equipment can be installed safely. This chapter provides a short description of mechanical transport systems.

9.1 Functional requirements

The functional requirements for staircases were set out in *Barry's Introduction to Construction of Buildings*. In buildings with a vertical change in floor level, it is necessary to provide a means of transport from one floor to another, both to improve the movement of people within the building and to allow access to all parts of the building for everyone, regardless of disability. Lifts and elevators provide quick, reliable and safe vertical movement for large volumes of people and equipment. Stair lifts and platform lifts may also be used to allow movement of wheelchairs and pushchairs from one floor or level to another. Moving walkways, or travelators, are sometimes used to accommodate relatively small differences in floor level but are mainly used to transport people over long distances within large buildings, such as airport terminals and the larger supermarkets. In all cases, there is still a requirement for the adequate provision of stairs. Stairs will need to be used in the event of an emergency and to provide an alternative route should a lift be out of order due to a mechanical breakdown or routine maintenance.

Lifts and escalators

The primary functional requirements for lifts, escalators and moving walkways are:

❏ Safety.
❏ Reliability and ease of maintenance.

Barry's Advanced Construction of Buildings, Fifth Edition. Stephen Emmitt.
© 2023 John Wiley & Sons Ltd. Published 2023 by John Wiley & Sons Ltd.

❏   Quiet and smooth operation.
❏   Movement between floors (if moving walkway, speed along floors).
❏   Aesthetics.
❏   Ease of use for all.

Safety
Safety is paramount. Lifts, escalators and moving walkways must be designed and tested to ensure the highest safety standards. EN 81-1 and EN 81-2 are the Lift Regulations standards for safety rules, construction and installation of lifts. Lifts should also comply with the Disabled Access standard EN 81-70 and Approved Document Part M. Well maintained and serviced, these mechanical transport systems should provide relatively trouble-free and safe transportation. Lifts must not be used in a fire; however, 'firefighting' lifts can be used by firefighters (see further).

Reliability and maintenance
Given that lifts and escalators are crucial to the ease of circulation of people between different floor levels, the reliability and quality of the after-sales service are an important consideration for building designers and building owners. Having mechanical transport systems out of order for a very short period of time is inconvenient, and a lengthy shutdown can be expensive in terms of loss of business and disruption to staff and customers. Many commercial buildings will require 'immediate' responses from the manufacturer's service department. Before a final choice of manufacturer is made, the proximity of the manufacturer's service branches to the building should be checked. The service level to be provided should also be checked and then clearly stated in the performance specification. It is also worth consulting with a number of previous clients to check on the quality of service provided by the manufacturer and the reliability of the lift system.

Quiet and smooth operation
The finishes of the lift car, the lift speed and the smoothness of the ride will all have an impact on the overall experience of using the lift, escalator or walkway. Modern lifts are both quiet in operation and provide smooth delivery to the required floor. Traction lifts tend to provide a gentler and faster ride than hydraulic lifts. Escalators tend to generate a small amount of noise during operation.

Speed
The comfort level of those using the apparatus limits the speed of escalators and walkways. The speed is restricted to provide a safe environment to step onto and off the escalators and moving walkways. Traction lifts are often marketed for their speed, especially in high-rise buildings. Hydraulic lifts have a maximum moving speed of 1 m/s. The Fastest traction lifts are capable of speeds of around 6 m/s, but the average speed is around 1.6 m/s. Some of the world's fastest lifts are to be found in the world's tallest buildings and are capable of ascent speeds of 16 m/s and descent speeds of 10 m/s. The elevator cars in these buildings are fitted out as airtight pressure cabins and have pressure regulation systems to ensure passenger comfort on the ascent and decent.

Aesthetics

The interior finish of the lift car and lighting, the ease of use of the call buttons, and the finish of the landing doors all contribute to the overall experience of using the lift. With escalators and walkways, the handrail/balustrade and surface finish contribute to the aesthetic of the apparatus.

Coordination and tolerances

An essential requirement when designing and planning buildings that utilise lifts and escalators is the coordination of dimensions. Lifts and escalators are manufactured to precise dimensions, and manufacturers set out specific requirements for the amount of tolerance in the structure that supports and/or encloses their equipment. Failure to coordinate dimensional drawings prior to construction, and especially failure to liaise regarding any changes that occur during construction, may result in considerable rework on site. Critical dimensions are the finished floor to finished floor dimensions, the internal size of lift shafts, the width and height of the structural opening to the lift shaft, the pit depth (ground floor level to bottom of the pit) and the headroom (height from the top of the lift car exterior to the top of the lift shaft). The lift shaft must be constructed plumb and in accordance with the vertical tolerances specified by the lift manufacturer. The formwork and finished concrete shaft must be regularly checked as work proceeds to ensure the work is within the specified tolerance.

The widespread use of laser levels for setting out and checking work as it proceeds, together with the use of steel formwork, has helped to improve the accuracy of concrete poured in situ. Similarly, the use of prefabricated units can assist in helping to achieve more accurate work. However, the quality of the work on site remains a determining factor. The work must be accurate, thus ensuring that there are no problems when the lift machinery is delivered to the site. This requires high-quality work, supervision and methodical checking of the work as it proceeds. Critical areas/dimensions are discussed further.

Electricity supply

A suitable power supply must be provided to the apparatus (lift car, escalator and walkway) as well as to the motor room, and in the case of motor roomless lifts, the lift shaft. The motor room will require lighting and emergency lighting. The lift shaft will need to be lit at the top and bottom with intermediate lights spaced at a maximum of 7 m. Thirteen-ampere switched electrical outlets will be required in the motor room and the lift shaft for power tools. Heating, ventilation equipment and thermostats will also be required to maintain an ambient temperature (as specified by the lift manufacturer). The lift car will require lighting and emergency lighting.

9.2 Lifts (elevators)

The development of the skyscraper is dependent on developments in the safety and speed of passenger lifts (elevators). Lifts are, however, not the sole domain of high-rise buildings, being a common feature in the vast majority of buildings with a change of level. There are a relatively small number of well-known manufacturers and installers of lifts. Therefore, the choice of manufacturer is rather limited, although the size of the lift car, its performance and its internal finishes are quite extensive, with lift cars designed and manufactured to suit the requirements of a particular development. There is a standard range of lift sizes available for many small- to medium-sized developments. The quality of the lift car will be determined by its function. For example, a lift car that carries people will be built to a different finish than one designed solely for transporting goods. All passenger lifts should comply with EN81-70 Accessibility to lifts for persons, including persons with disabilities. There are two types of lifts: mechanical or traction lifts and hydraulic lifts.

Traction lifts

The most common form of lift is the mechanical lift, usually described as a traction lift. The lift car is operated by a system of pulleys and steel wires, powered from a lift motor room that is usually adjacent to the lift. The motor room is usually positioned at the top of the lift shaft or sometimes within the basement. A safety system prevents the lift from falling if the steel cables break due to prolonged wear (which is extremely unlikely in a well-maintained lift). Traction lifts are the most common type of lift in use. They can be used in buildings with as little as two different ground levels (e.g. ground and first floor) to multi-storey buildings with numerous floors.

Lift motor room

The lift motor room may be positioned adjacent to the lift (Figure 9.1), or more typically at roof and/or basement level. The lift motor room must be large enough to accommodate the necessary equipment and allow clear and safe access for routine maintenance and replacement activities.

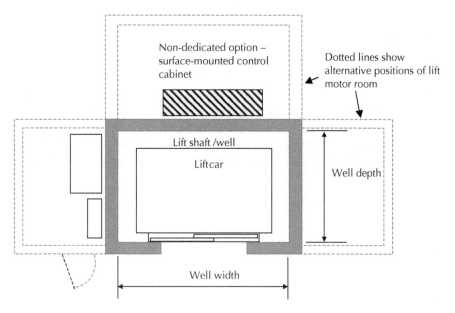

Figure 9.1   Plan of lift shaft.

Hydraulic lifts

For low- to medium-rise applications, a hydraulic passenger lift is an alternative to a traction lift. Loading on the hydraulic lift shaft is not excessive, and some manufacturers provide hydraulic lifts with their own structures, which helps keep building costs down. Hydraulic lifts require a smaller lift pit than traction lifts, since the pump can be located adjacent to the hydraulic ram, as illustrated in Figure 9.2. The hydraulic ram can be side- or rear-mounted, providing typical speeds of 0.15–1.0 m/s and a smooth operation. Where space is at a premium (e.g. in a refurbishment project), a borehole ram that extends into the ground may be used.

Hydraulic lifts offer a number of benefits over traction lifts. Life cycle maintenance costs may be lower, simply because there are less pulleys and lengths of wire ropes and hence

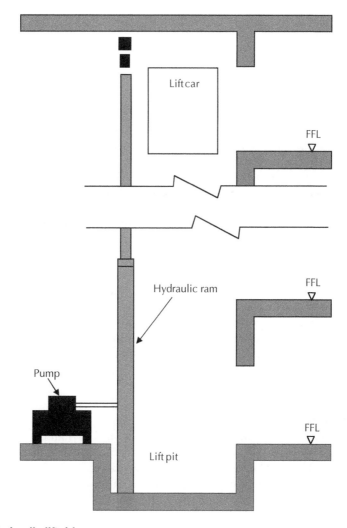

Figure 9.2   Hydraulic lift drive.

less wear. A hydraulic lift uses power only during the ascent; it uses gravity to descend and a small amount of energy to operate the valves. Energy savings may be possible compared with traction lifts. The hydraulic oil has a long lifespan and will usually last the life of the lift. If the oil needs to be replaced, it can be recycled. In an emergency (e.g. a power cut), the hydraulic lift provides safe and convenient lowering for passengers and service engineers alike. An emergency button/switch will allow the lift to lower automatically under gravity.

Non-motor room option
An alternative to the dedicated motor room is a 'motor-roomless' option for some models. This space-saving design incorporates a cabinet with control equipment that can be mounted on or recessed into a wall adjacent to the lift shaft (Figure 9.1). In addition to saving space, the surface-mounted cabinet allows maintenance work on drive and control equipment to be conducted safely (and outside the lift well).

Lift function

The function of the lift will determine the size, safe loading, speed and interior finish of the lift car. Lifts tend to be described as passenger lifts, home lifts, firefighting lifts, goods lifts, trolley lifts, service lifts, stair lifts and vertical platform lifts.

Passenger lifts
Passenger lifts are usually specified by the maximum number of people carried per lift car (e.g. a six person lift (450 kg) or an eight person lift (630 kg)). The speed of the lift (both the response time to a call and the time to travel between floors) will be a prime consideration for tall buildings. Computer software is used to calculate the number of lifts required to accommodate certain capacities and peaks in traffic. The quality of the interior finish is usually specified to match the quality of the building interior.

'Home' lifts
Small, one- and two-person domestic lifts offer an alternative to a stairlift (discussed later). These 'home' lifts are designed to have a very small footprint and be self-supporting, so that they are easy to install into existing houses with minimal building work. They are designed as a modular unit with an electric motor and drive unit contained within the lift. Some manufacturers offer units that also accommodate a wheelchair but still retain a very small footprint. The lift runs on rails and usually has a clear body to allow full visibility. The modular units can be installed quickly once the necessary opening has been made in the existing floor.

Firefighting lifts
It is possible to construct buildings with designated firefighting lifts. These lifts and the protected lift shaft have an independent electrical supply so that they can still function in a fire. They may be used by firefighters to gain access to floors if it is deemed safe to do so (depending on the circumstances of the fire or emergency). Specific safety features apply to firefighting lifts, as set out in Approved Document B and the EN-81 group of standards.

Goods lifts

Goods lifts are designed to be durable and functional. The lift car is usually constructed from mild steel sheeting with a baked enamel finish to the walls and a heavy-duty vinyl to the floor. They are usually specified by minimum size of the lift car and maximum loading. The speed of the lift car is not a prime consideration. Goods lifts may be built with a lift shaft or load-bearing wall for support. Alternatively, goods lifts with their own robust structure and motor assembly are available, which allow for greater flexibility in positioning. Self-contained lift assemblies alleviate the need for a separate motor room and are ideally suited to installation in existing buildings. Building work is required to make the necessary openings, followed by installation and commissioning. Typical loadings range from 500 to 1500 kg. The size of the lift car will depend on the type of goods being transported between floors. Goods trolleys, palletised goods, warehouse stock, furniture and other bulky goods are typical loads. If the goods lift is to be used for passengers as well, then the lift car will need to be larger.

Trolley lifts

Some goods lifts are designed to accommodate only goods trolleys. Trolley lifts provide a quick, safe and efficient way of moving heavy loads on a trolley between different floor levels. Typical loadings are 250 or 300 kg. The size of the lift car is typically around 1000 mm wide and 1000 mm deep, with a height of approximately 1400 mm. Hinged or concertina landing entrance gates are common.

Service lifts

A service lift carries relatively small loads (e.g. food, drinks, beer crates, documents and laundry) between floors of buildings. Originally termed a 'dumb waiter', the service lift allows quick, convenient and safe movement of goods between floors. Service lifts are widely used in restaurants, pubs and clubs for the transportation of bottled drinks and food; for example, where the kitchen is located above or below the dining area. Typical maximum loadings are 50 or 100 kg. The size of the car will typically be around 500–650 mm wide and 350–650 mm deep, with a service door (hatch) somewhere around 800 mm high. Service lifts are available from stock in a variety of standard sizes, although some variations in dimensions and finishes can be made on request to the manufacturers. It is usual to provide an intercom facility adjacent to the service lift to allow persons to communicate their requirements between floors.

Stair lifts

A stair lift, as the name implies, is installed on a flight of stairs to allow safe access between levels for disabled people. Stair lifts are usually installed in domestic buildings where a lift is deemed not to be necessary or cannot be economically installed in existing buildings. In public buildings, it is common to provide stair lifts (or platform lifts) in addition to lifts. Platform lifts are designed to accommodate a wheelchair and should be operational without assistance. Controls are located on the lift platform. The lift car moves up and down a rail fitted to the wall adjacent to the stairwell. For new projects, the stairs must be constructed to be wide enough to allow safe access for a stair lift and people. When installed in existing buildings (e.g. houses), the stair lift will take up the whole width of the stair when in use. The stair lift may be designed to fold away when not in use.

Vertical platform lifts

A vertical platform lift comprises an open (or closed) flat platform, with safety rails and a door. It is sometimes positioned next to small flights of stairs to allow access for wheelchairs, pushchairs and goods trolleys and comes with its own supporting structure. Some shaft sizes on the market are as small as 680×810 mm. These lifts have a very small pit depth (e.g. 300 mm), which can be replaced with a ramp if it is not possible or feasible to use a pit.

Lift specification

The design and specification of lifts involves a large number of options; the main ones are listed here:

Capacity and waiting times

The anticipated capacity of the lift car will determine its size and load-carrying capacity. Capacity will be determined by the function of the lift, be it for moving people; people equipment and goods; or goods only. The capacity, number of lifts, and waiting times will also be determined by the building's function (e.g. residential, office and hotel), the number of floors to be served, the probable population of the building and the peak demand times (e.g. up in the morning, up/down at lunchtime and down in the evening). Typical capacities of lifts are designated as low volume (e.g. 6-person lift) or high volume (e.g. 16-person lift). Waiting times are determined by traffic flow management. This should include robustness of design to allow a sufficient flow of people when a lift is out of service for routine maintenance or because it has failed.

Door type

The type of lift car door is specified by side or centre opening, and this will determine the size of the lift shaft. Lifts usually have lift car doors and landing doors, and it is common practice to use the same door opening types:

❑ Single panel (side opening).
❑ Two panel side opening.
❑ Two panel centre opening.

Entrance/exit position

There are three entrance/exit configurations, as shown in Figure 9.3:

❑ *Front only.* This is the smallest well and lowest cost.
❑ *Front and rear.* Increases shaft size and cost.
❑ *Adjacent entrances.* Available on some models. This will increase shaft size and cost.
❑ *Clear entrance width.* (e.g. 900 mm) and height (e.g. 2000 mm) will need to be specified.

Mounting

There are two main options: structure supported or wall-mounted lifts (Figure 9.4).

The structure supported lift tends to be cheaper to build, because the load-bearing wall and associated lifting beam are not required. However, this needs to be offset against a higher lift cost (10–20% more than a wall-mounted lift) and a slightly larger shaft size. In the majority of cases scaffolding is not required, which helps to save time and cost. Structure- supported lift structures are typically used for goods lifts, trolley lifts, service lifts and vertical platform lifts.

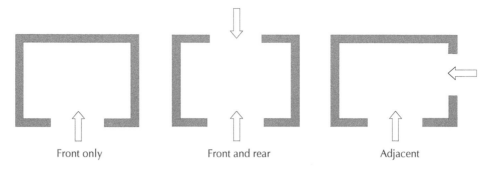

Front only   Front and rear   Adjacent

Figure 9.3   Entrance/exit positions.

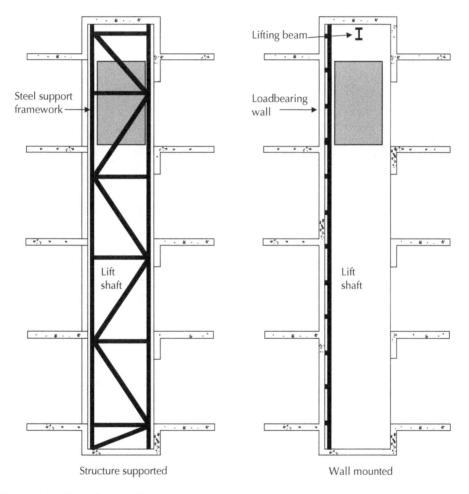

Structure supported   Wall mounted

Figure 9.4   Mounting positions.

The wall-mounted lift has a higher initial build cost due to the construction of a load-bearing wall and a lifting beam at the top. Scaffolding, or a tower scaffold, will be required to provide a safe working surface. The advantages are lower lift costs and a smaller shaft size than the structure-supported lift.

Shaft size

The interior of the lift shaft, or the shaft size, will be determined by the size of the lift car and the space required around the car (Figure 9.5). Manufacturers provide details on the minimum clear dimensions for each of their models. For example, a lift car 1000 mm wide and 1250 mm deep would require a shaft size of approximately 1500 mm deep and 1600 mm wide.

Finishes

A wide range of finishes is available for lift cars to suit different budgets and design requirements. Consideration should be given to the ceiling finish, which may be flush or suspended to suit different lighting effects. Walls may be finished with decorative panels and/or mirrors. Round-section handrails can also be added. Floors are typically finished with hardwearing vinyl, carpet or stone. A skirting helps protect wall finishes from scuffing and damage.

Consoles

Push-button consoles incorporate illuminated lights to indicate that a lift car has been called. Special pushes and raised (tactile) signage will assist people with a visual impairment. Braille may also be provided on the consoles. Recorded audio messages (voice annunciator) are usually installed as standard. Consoles also incorporate an emergency voice communication system.

Construction of the lift shaft

The lift car must be supported on a load-bearing wall or frame. Alternatively, the lift assembly could be self-supported, that is, independent of the building structure, in a robust steel framework. The usual arrangement is to build a lift shaft with reinforced concrete or load-bearing

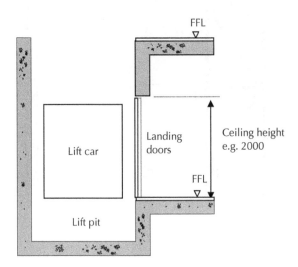

Figure 9.5   Vertical section through the lift shaft showing critical dimensions.

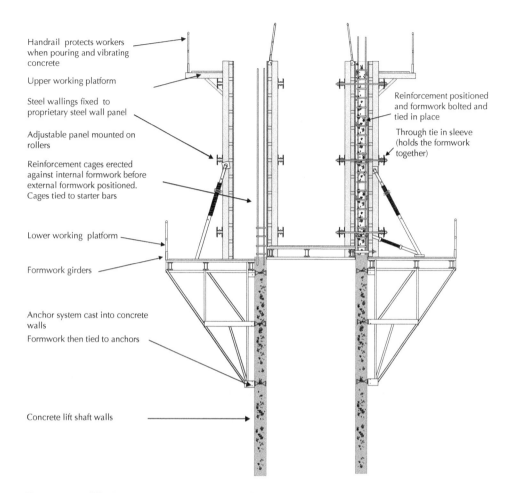

Handrail protects workers
when pouring and vibrating
concrete

Upper working platform

Steel wallings fixed to
proprietary steel wall panel

Adjustable panel mounted on
rollers

Reinforcement cages erected
against internal formwork before
external formwork positioned.
Cages tied to starter bars

Lower working platform

Formwork girders

Anchor system cast into concrete
walls
Formwork then tied to anchors

Concrete lift shaft walls

Reinforcement positioned
and formwork bolted and
tied in place

Through tie in sleeve
(holds the formwork
together)

Figure 9.6   Climbing formwork platforms. (*Source*: Adapted from http://www.peri.ltd.uk).

masonry. Concrete is now regarded as the preferred option for cores, escape routes and lift
shafts in Europe and the US. With the increased sophistication of offsite manufacturing, it is
now possible to manufacture the lift shaft and associated assemblies as prefabricated modules. This can help to improve quality and also help with dimensional coordination to ensure
that the shaft is plumb. However, currently most lift shafts are constructed on site using slip
form, climbing formwork systems (Figure 9.6). Glass lift shafts with a steel (and, in some
cases, aluminium) structure are commonly used within atria spaces, stations and airports.

Concrete lift shafts

Lift shafts are usually square or rectangular (with a corresponding square or rectangular lift
car) and constructed of reinforced concrete. The most common method of construction is
to form the lift shaft walls from in situ reinforced concrete. Circular lift shafts with corresponding circular cars are also available, although not very common.

As noted earlier, the quality of the work carried out on site is critical to ensuring that
the lift shaft is built to the specified tolerances. Tolerances are particularly important where
the lift car and guide rails are to be fixed to the walls and frame. Critical dimensions are the

Table 9.1   Well dimensions for general purpose or intensive traffic lift installations

	Internal car sizes		Recommended shaft size		
Persons carried	Width	Depth	Height	Width	Depth
8	1100	1400	2200	1800+K	2100+K
10	1350	1400	2200	1900+K	2300+K
13	1600	1400	2300	2400+K	2300+K
16	1950	1400	2300	2600+K	2300+K
21	1950	1750	2300	2600+K	2600+K
24	2300	1600	2300	2900+K	2400+K

K = wall tolerance information.
25 mm for shafts not exceeding 30 mm.
35 mm for shafts not exceeding 30 m, but not exceeding 60 m.
50 mm for wells over 60 m, but not exceeding 90 m.
Source: Adapted from Ogden (1994).

finished floor to finished floor dimensions (Figure 9.5), the internal size of lift shafts, and the width and height of the structural opening to the lift shaft. Standard internal sizes of lift shafts are shown in Table 9.1.

In order to achieve the required running clearances, door alignment and reliability of lift installation and operation, lift shafts should be constructed to high standards of verticality (within vertical tolerances). While all shafts should be constructed to the tolerances set by the lift manufacturer, a standard guide is that the shaft should not deviate from the vertical by more than 1/600 of the height or by 5 mm per storey (whichever is the greatest). However, the total deviation throughout the full height of the building must not be more than 50 mm (Figures 9.7 and 9.8 and Table 9.1).

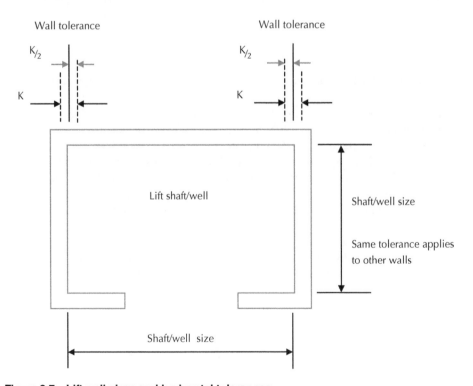

Figure 9.7   Lift well sizes and horizontal tolerances.

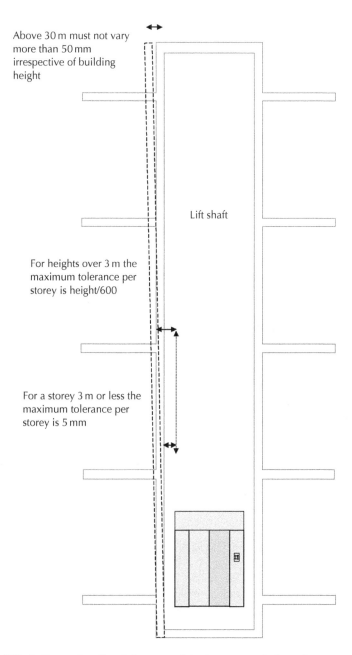

Above 30 m must not vary more than 50 mm irrespective of building height

Lift shaft

For heights over 3 m the maximum tolerance per storey is height/600

For a storey 3 m or less the maximum tolerance per storey is 5 mm

Figure 9.8   Lift shaft construction tolerances (steel or concrete frame).

The solid core of a concrete lift shaft is often used to add lateral stability to the building. The steel or concrete frame is tied into the concrete lift shaft, and the solid walls of the shaft act as bracing. Photograph 9.1 shows a central concrete lift shaft (to house three lifts) for a high-rise building. When lift and stair shafts are combined, the concrete enclosure can be designed to form the protected shaft for the purpose of escape during a fire. The concrete structure will have good heat-resistant properties if it is designed with the proper concrete

Photograph 9.1   Central concrete lift shaft under construction with access to the lift shaft covered with temporary protection.

cover to the reinforcement and aggregates that do not expand and spall when exposed to heat. If designed and placed properly, concrete can be used to form a compartment wall capable of resisting fire for up to four hours.

9.3 Escalators and moving walkways

Similar to lifts, the manufacture and installation of escalators and moving walkways or travelators are limited to a small number of manufacturers. Because of the problem of keeping the moving parts rust-free, escalators and walkways should not be used outside unless they are adequately protected from rain and snow.

Escalators

Escalators move people vertically from one level to another, and are a popular feature of commercial shopping centres and large retail stores, travel interchanges and other commercial buildings. They can be used to convey people over long distances and over more than one floor, as illustrated in Photograph 9.2. Usually installed as an architectural feature,

Photograph 9.2   Escalator adjacent to lift.

they help guide people from one level to another without the need to use staircases or lifts. The escalator comprises a moving steel mat that moulds itself to the profile of the transport system underneath. This forms a series of steps on which one stands until reaching the top or bottom of the escalator. Pitches tend to be steeper than staircases, typically around 35–45° to the horizontal.

The most important dimensions are primarily finished floor to finished floor level. A pit is required at the base of the escalator to house motors and associated equipment. This is usually covered with a steel plate, which can be removed for routine maintenance and repair.

Moving walkways

Moving walkways, or travelators, provide a flat moving surface to move people horizontally. They are common in large buildings, such as airports, where people would otherwise have to walk long distances. The idea is that people stand on the continually moving surface of the walkway and are transported from one end to the other without the need to walk. Moving walkways tend to be slightly slower than normal walking speed (remember people have to step on and off each end), and it is common for people to continue to walk along the moving surface, thus making it faster than walking on a static surface. The walkways are usually designed and constructed so that people have the option of walking alongside them.

A safety 'stop' button is positioned at the end of the walkway so that it can be stopped in an emergency. Walkways are also manufactured to be installed at a low pitch to the horizontal to accommodate small changes in floor level.

The walkway is positioned within a shallow pit in the floor slab. The depth of the pit will, to a certain extent, be determined by the manufacturer to be used. Walkways are protected with solid balustrades along their length. These are usually constructed of glass to improve visibility, and hence safety.

Further reading

See the Lift and Escalator Industry Association: www.leia.co.uk.

Reflective exercises

A large housing development has been designed as a structural timber frame with modular units (timber on a steel sub-frame). There are three lift cores within the development, all designed with concrete lift shafts and stairwells. Your client would prefer to use cross-laminated timber (CLT) for the lift shafts and stairwells.

❑ What are the implications of this potential design change in terms of the carbon footprint of the building and its structural integrity?

You are designing a large urban development, and your client wishes to use elevators in an external setting to help people of all abilities get between the different levels.

❑ How would you address concerns related to weather exposure of the elevators?

Fit Out and Second Fix: Chapter 10
AT A GLANCE

What? The terms 'fit out' and 'second fix' refer to the process of completing the building. Many commercial buildings are built to a 'shell' finish; that is, the main structure of the building is complete, but the internal fittings and decoration are not. Thus, the fit out and second fix are carried out as part of a separate contract (or contracts) to suit the client/user's requirements. Electrical trunking and light fittings, suspended floors, suspended ceilings, partition walls, shelving, display units, internal fittings, painting and decorating and signage all form part of the fit out and second fix operations.

Why? In many commercial and speculative developments, it is common to construct the building as a shell. Only when a buyer or renter has been found for the building is it completed by fitting out the shell and completing the second fix electrical, plumbing and carpentry work. This helps the developer's cash flow, allows choice for the buyer/renter of the building, and reduces the risk of theft of valuable items from the building prior to occupation. Speculative residential housing developers often take a similar approach, leaving some aspects of the fit out and second fix, such as the kitchen and bathroom fittings, light fittings, decoration and floor finishes, to be completed once they have secured a buyer.

When? Scheduling of the fit out is planned as a phased operation to allow the smooth flow of specialist trade packages. Specialist trades will complete their work in one room, or area, and move onto the next to be replaced by another trade's package. This helps to maintain an efficient and safe workflow. For example, it is common for the ceiling and light fittings to be completed first, followed by decoration and finally the floor finishes.

How? Given the nature of the work involved, it is common to sub-contract the work to specialist suppliers, tradespeople and fitters. Some manufacturers will provide a list of approved suppliers and fitters to ensure the product is fitted as intended and to the correct quality standards.

10 Fit Out and Second Fix

Many commercial buildings are often built to a 'shell' finish; that is, the main structure of the building is completed, but the internal fittings and decoration are carried out as part of a separate contract (or contracts) to suit the client/user's requirements. This allows a certain amount of flexibility and choice of finish for those occupying some or all of the building. Electrical trucking and light fittings, suspended floors, suspended ceilings, partition walls, shelving, display units, internal fittings, painting and decorating and signage all form part of the fit out and second fix operations. In this chapter, the focus is on the more common fit out systems: raised floors, suspended ceilings and internal partition walls.

10.1 Commercial fit out

The term 'commercial fit out' is normally used to cover the fit out and second fix of offices, retail units and industrial buildings. The main components of a fit out, the raised floor, suspended ceiling and internal partition walls are described further. The vast majority of raised floor, suspended ceiling and internal partition wall systems are manufactured to a patented design. The quality of the system and adaptability vary between manufacturers, from budget systems to expensive systems for the most prestigious of developments. Careful research is required to choose the most appropriate system to suit the requirements of the building users.

Offices and commercial buildings

Speculative office developments are a common feature in our towns and cities. These are usually built to a relatively standard design and completed to a shell finish under the main contract. At the most basic level, the shell finish includes the structure and external fabric of the building, staircases, lifts and service core. A more common approach is to include the raised floor and suspended ceilings, sanitary fittings and electrical outlets. Internal partition walls are usually installed as part of the fit out to suit the needs of the particular user. The office space is usually leased, and it is not uncommon for users to move to another address or lease more or less floor space at the end of the rental period. Thus, the flexibility and adaptability of office space are prime consideration.

Barry's Advanced Construction of Buildings, Fifth Edition. Stephen Emmitt.
© 2023 John Wiley & Sons Ltd. Published 2023 by John Wiley & Sons Ltd.

Retail

The term 'shop fitting' is used to describe the process of installing furniture (display shelving and units, points of sale, etc.) and associated equipment (chiller cabinets, freezer cabinets, etc.) in a retail unit. Retail units are usually termed as 'food retail' or 'non-food retail', reflecting the accepted use of the unit or building. The vast majority of retail units are designed to be relatively flexible in terms of use. This is because the units will normally be leased out to a particular business for a certain period of time. As a result, it is common practice to construct the shell of the unit, with the retailer fitting it out to match a specific house style (corporate image). This applies to shops located on the high street and also to units located in large shopping centres.

Retailers tend to change their display arrangements on a regular basis. Minor changes can usually be accommodated in adjustable shelving units, but more major changes associated with an update in corporate image usually necessitate a complete refit of the retail unit, often resulting in a lot of wasted materials. Similarly, with a change of retailer, there is usually the need for new shop fitting.

Food retail units will require additional drainage points for the condensate drain to freezer cabinets and chiller units. The store layout tends to be changed less frequently because the condensate drains determine the position of freezer and chiller units. Changes in position usually necessitate changes to drain positions, which can be disruptive to the sales area.

Shop signs

Provision for shop signage is usually provided at strategic places on the exterior face of the shop unit, and this too will usually be installed (subject to town planning consent) by the organisation leasing the shop. The shop signs are printed on to relatively thin backgrounds (e.g. acrylic sheet) or on to a translucent material so that the sign can be illuminated from behind. The shop sign is then fixed to the face of the wall, usually with screws. From a design perspective, it is important to provide a structure suitable for supporting the signage and any electrical connection required for the lighting.

Industrial

Industrial production facilities have special requirements to suit a particular manufacturing process. These may include one or a number of the following features: clean rooms, security (of staff and materials/processes), wash down facilities, special material handling areas and secure storage, specialist fire protection systems, etc. These are outside the scope of this book, but readers should be aware that these special requirements have a bearing on the choice of construction methods used and how services are integrated; for example, sealing services as they pass through compartment walls in clean room construction. Often, some compromises need to be made. For example, in fast-track projects, it would be sensible to use prefabricated, framed construction to save time on the building site; however, some production processes may require solid masonry walls.

On fast-track projects, such as new pharmaceutical production facilities, it has become common practice to manufacture and test industrial plants prior to installation in the building. The equipment is then transported to the site, moved into its final position and 'plugged-in'. It is then put through a final series of conformity and safety tests; that is, it is commissioned for use. This is usually done while the main build contract is still under way. Large access doors facilitate the delivery and subsequent maintenance and replacement of large pieces of equipment.

10.2 Raised floors

A raised floor (access floor) is used to conceal services, usually electrical cables and air-conditioning ducts, in the cavity between the raised floor and the structural floor. Raised floors are particularly useful in large open plan spaces, such as offices, where it is impractical to house services behind partition walls. Air-conditioning grilles and service outlets, such as electrical and telephone sockets, can be provided within the floor to provide a flat and even surface finish. The cables and air-conditioning ducts are hidden from view in the cavity between the raised floor and the structural floor finish, with strategically placed access panels to allow convenient access for repair, maintenance and upgrades, hence the term 'access floor'. Concealing the cables not only improves the visual appearance of the floor area but also helps to keep trip hazards to a minimum. Most manufacturers of raised floor systems also supply ancillary items such as electrical floor boxes, cable ports and cable management systems. Raised floors can also accommodate safety lighting and can be illuminated from underneath to provide additional visual interest, contributing to mood lighting and also displaying information such as advertising and location maps.

Raised floors are used in new building developments as well as in refurbishment work where there is sufficient vertical height between structural floors to add a raised floor. By adjusting the depth of the supports to suit uneven floors and changes in floor levels, it is possible to provide a flat surface and eliminate the need for ramps and short flights of stairs.

Functional requirements

The functional requirements are to:

- ❏ Accommodate and conceal services:
  - ○ Provide ease of access to services.
  - ○ Allow changes to services below the floor.
  - ○ Create a clear cavity for services.
- ❏ Be flexible:
  - ○ Provide a level, attractive and durable floor finish.
  - ○ Accommodate changes in surfaces.
  - ○ Remove changes in structural floor.
  - ○ Facilitate changes in level.
  - ○ Provide required surface level.
- ❏ Sustain and resist imposed loads:
  - ○ Be rigid and stable.
- ❏ Have good appearance and aesthetics:
  - ○ Hide unsightly structural floors.
  - ○ Present feature in its own right (e.g. lighting).

❑ Provide sound control:
  ○ Resist passage of impact and airborne sound.
  ○ Provide acoustical control (absorption and reflection).
  ○ Provide required thermal resistance and prevent formation of condensation.
❑ Provide protection against fire:
  ○ Control spread of fire and maintain structural stability in a fire.
❑ Be durable:
  ○ Resist wear and tear.
❑ Be easy to maintain:
  ○ Provide safe installation and ease of maintenance.
  ○ Provide a comfortable, easy to clean finish.

Floor assembly

There are two primary components to a raised floor: the floor panel and the support pedestal. Combined, the components provide a rigid and stable floor surface for a variety of uses. The design life of the floor panels tends to be anything up to 25 years; that of the supports is around 50 years. Loading capacity varies between systems and is typically expressed as a point load greater than 25×25 mm (e.g. 3 kN) and a uniformly distributed load (e.g. 8 kN/m^2). In situations where heavy furniture or equipment is to be installed, the loading of the system can be improved by the installation of additional support pedestals. Manufacturers will also provide details of the air leakage (at 25 Pa), the fire rating (e.g. Class 0), the combined weight of the system (e.g. 40 kg/m^2), which will vary with height and a statement on the electrical continuity of the system (e.g. that it complies with IEE Regulations). The acoustic performance of the system (which depends on the type of floor finish) should also be provided. Floor-to-wall junctions are usually sealed with an air seal.

Floor panels

Floor panels are manufactured to a nominal size of 600 × 600 mm. Depending on the construction, the depth of the panel will be around 25 to 40 mm. The majority of panels are constructed of high-density particleboard encapsulated in a galvanised steel finish, approximately 0.5 mm thick. Non-metallic systems comprise high-density particleboard or specially treated moisture-resistant chipboard, usually applied with a protective finish. It is common in UK properties to install carpet tiles on a trackifier, but timber and marble may also be used. Metallic panels are gravity-fitted and laid into the structural support grid using a hand-held suction device. Indentations in the underside of the panel allow positive location and subsequent retention within the support system. Rapid access is available through the use of a suction device to lift the panels out. The non-metallic systems tend to be screwed to the structural supports through a countersunk hole at each corner. Unscrewing the panels from the supports provides access.

Support pedestals

The pedestal is usually constructed of steel. Concrete and timber can also be used. Steel pedestals are adjustable to suit variations in the structural floor (Figures 10.1 and 10.2 and Photographs 10.1 and 10.2). They are fixed on a 600 mm^2 grid and a laser level is used to ensure a level finish. Electrical earthing is required when metallic components are used. Angled supports are also manufactured to provide a ramped floor finish. Pedestals are manufactured in a range of heights, providing clear cavity spaces from as little as 30 mm up to 1000 mm.

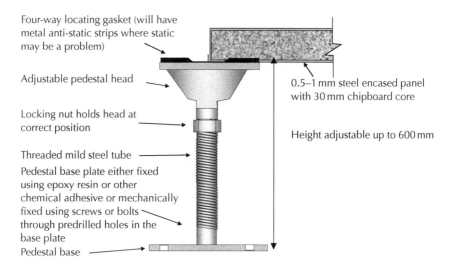

Four-way locating gasket (will have metal anti-static strips where static may be a problem)

Adjustable pedestal head

0.5–1 mm steel encased panel with 30 mm chipboard core

Locking nut holds head at correct position

Height adjustable up to 600 mm

Threaded mild steel tube

Pedestal base plate either fixed using epoxy resin or other chemical adhesive or mechanically fixed using screws or bolts through predrilled holes in the base plate

Pedestal base

Figure 10.1   Raised floor pedestal.

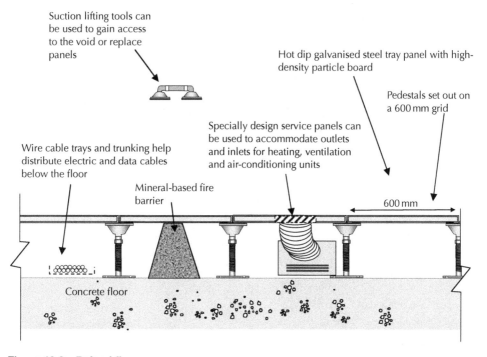

Suction lifting tools can be used to gain access to the void or replace panels

Hot dip galvanised steel tray panel with high-density particle board

Pedestals set out on a 600 mm grid

Specially design service panels can be used to accommodate outlets and inlets for heating, ventilation and air-conditioning units

Wire cable trays and trunking help distribute electric and data cables below the floor

Mineral-based fire barrier

600 mm

Concrete floor

Figure 10.2   Raised floor.

Example of pedestal and panel floor assembly

❏   31 or 32 mm overall galvanised steel/chipboard sandwich panel 600 × 600 mm, which is placed on polypropylene cruciform locating lugs and sits at the top of the pedestal. The panel can be rested on or fixed to the pedestal.

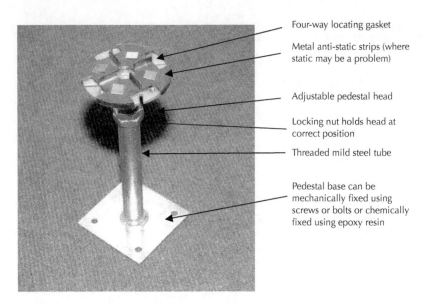

Four-way locating gasket

Metal anti-static strips (where static may be a problem)

Adjustable pedestal head

Locking nut holds head at correct position

Threaded mild steel tube

Pedestal base can be mechanically fixed using screws or bolts or chemically fixed using epoxy resin

Photograph 10.1   Raised floor pedestal.

Cable trays and service ducts are easily positioned under the raised floor

Fire barriers are installed to prevent the passage of fire through the raised floor void

Photograph 10.2   Raised floors.

❏ The pedestal is made of mild steel tube and base plate, with threaded components to allow for adjustment. Locking nuts are also used to hold the pedestals securely at the correct level.

❏ If the pedestals are over 600 mm high (deep void), they are mechanically fixed (plug and screw or gun nailed) to the sub-floor with polyurethane or epoxy adhesive at 600 mm centres. As panels are square, pedestals are fixed on a grid based on the module sizes.

❏ Panels are often constructed from a 0.5 mm steel top and bottom with 30 mm chipboard sandwiched between, or from a 1 mm steel top and bottom for heavy loading. Other types of panel construction are also available.

❏ A typical floor depth is about 600 mm, but it can be increased to over 2 m (using considerable support and framework).

❏ Fire barriers should be used under compartment walls and where necessary to prevent the cavity from acting as a conduit for fire.

Maintenance

Raised floor cavities need to be regularly cleaned to remove dust and debris. The cavity should be vacuumed using static-free tools and special air filters at least twice per year. Suction tools can be used to gain easy access into the cavity (Figure 10.2).

10.3 Suspended ceilings

Suspended ceilings are primarily used to provide an attractive finish to the ceiling while at the same time concealing (and allowing access to) services. A suspended ceiling comprises a lightweight structural grid that is supported by wire hangers attached to the underside of the structural floor, and into which lightweight tiles are placed. The structural grid is based on a 600 × 600 mm module. Suspended ceilings may also be used in refurbishment projects to provide a level ceiling finish to areas of the building that differ in height.

Suspended ceilings not only provide an attractive, level ceiling finish, with convenient access to services when required (Photographs 10.3 and 10.4), but are also used to house a wide variety of equipment, which may include ventilation grilles, light fittings, fire sprinklers, detectors (e.g. smoke and heat), security cameras, movement sensors and alarms.

Functional requirements

The functional requirements (adapted from BS 8290: Part 1 1991 Suspended Ceilings Code of Practice for Design) include:

❏ *Concealment of structure.* To hide changes in the structure, beams, floors and ties, and to provide a level and attractive ceiling finish.

❏ *Concealment of services.* To create a clear cavity for services, hide services such as ducts, equipment and cables, and allow easy access for maintenance.

❏ *Decorative appearance.* The aesthetics of finishes are always important.

❏ *Thermal insulation.* Thermal insulation may be introduced if the ceiling is adjacent to the roof. It must be designed and positioned to prevent interstitial condensation.

Services housed above the suspended ceiling grid

Suspended ceiling grid and partition wall frame installed

Ceiling tiles installed and glazed panels inserted into the frame

Photograph 10.3   Suspended ceiling. (*Source:* courtesy of D. Highfield).

- ❏ *Acoustic control.* Sound insulation and absorption. The two main aspects of acoustic control are the absorption of sound within an enclosed space and the reduction of sound passing through a material or structure. The measures necessary to control these two are different. Sound absorption is achieved by incorporating porous materials of low density in the ceiling. Sound insulation requires an impervious material of very high density. Mineral wool, with its high density, is often placed on top of the ceiling grid.
- ❏ *Fire control.* Controlling the development and spread of a fire, containing a fully developed fire, and protecting the structure against damage or collapse. A fire-resisting ceiling should not be confused with a fire-protecting ceiling; different test methods and criteria apply. A fire-protecting ceiling offers protection to the structural beams and floors. The term 'fire protection' implies that the ceiling can satisfy the stability, integrity and insulation requirements for a stated period. Sprinkler heads and systems may be part of the fire protection system.
- ❏ *Control of condensation.* Placing thermal insulation over a suspended ceiling can increase the risk of interstitial condensation. For guidance on controlling condensation, reference should be made to BS 5250 (Code of Practice for the Control of Condensation in Buildings) and BRE 143 (Thermal Insulation Avoiding Risks).

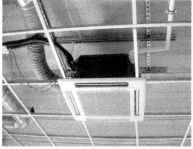

Services are suspended on their own brackets prior to the fixing of the ceiling grid

Double railed fixing bracket allows the position of the air handling unit to be accurately positioned once the ceiling grid is in place

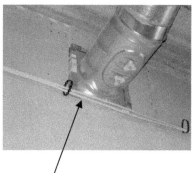

Where the ceilings pass over compartment walls fire blankets are needed to prevent the passage of fire

Services that pass through compartment walls also need to be fitted with special fire barriers, which seal the duct in the event of a fire

Photograph 10.4   Suspended ceilings: accommodation of services. (*Source*: courtesy of G. Throup).

❏ *Hygiene control.* Smooth cleanable finishes may be required to facilitate a clean room.
❏ *Ventilation.* The services may form an integral part of the suspended ceiling unit. Where services are concealed within the ceiling, they may need additional support. Each piece of plant can be individually tied to the building structure.
❏ *Heating, air conditioning and illumination.* The luminaires may be surface-mounted or supported independently from the structure. The ceiling may also be used to control the amount of light reflected and diffused into the room.

❑ *Electrical earthing and bonding.* The ceiling grid must be earthed. In the event of any electrical fault, the parts of the ceiling capable of carrying an electrical current are earthed and protected.

Ceiling assembly

A wide variety of ceiling systems are available (Figures 10.3 and 10.4 and Photographs 10.3 and 10.4). There are two primary components to a suspended ceiling: the ceiling tile and the support grid. Combined, the components provide a lightweight and level ceiling finish. The ceiling tiles have a design life of up to 25 years, and the support system has a life of around 50 years. The loading capacity of the system is sufficient to carry the weight of the system and associated equipment. Heavy fittings will require independent support from the underside of the structural floor. Manufacturers should also provide details of the air leakage (at 25 Pa), the fire rating (e.g. Class 0), and the combined weight of the system, which will vary slightly with the depth of the hangers. The acoustic performance of the system should also be provided.

Ceiling tiles

Ceiling tiles are manufactured to a nominal size of 600 × 600 mm. Tiles are made from gypsum, mineral wool, pressed lightweight steel or a lightweight material, such as particleboard, and usually have a painted finish. The majority of systems rely on a gravity fit, with

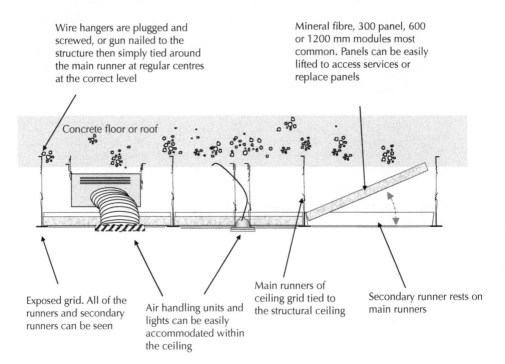

Wire hangers are plugged and screwed, or gun nailed to the structure then simply tied around the main runner at regular centres at the correct level

Mineral fibre, 300 panel, 600 or 1200 mm modules most common. Panels can be easily lifted to access services or replace panels

Concrete floor or roof

Exposed grid. All of the runners and secondary runners can be seen

Air handling units and lights can be easily accommodated within the ceiling

Main runners of ceiling grid tied to the structural ceiling

Secondary runner rests on main runners

Figure 10.3   Suspended ceiling.

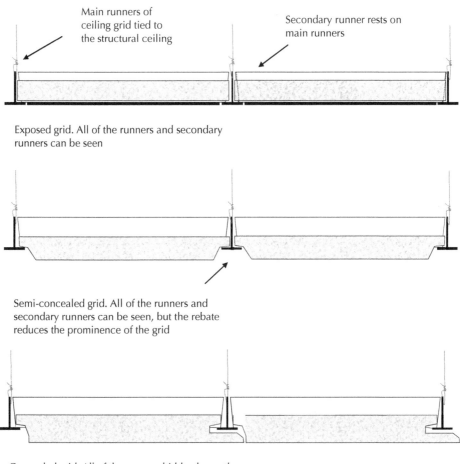

Exposed grid. All of the runners and secondary
runners can be seen

Semi-concealed grid. All of the runners and
secondary runners can be seen, but the rebate
reduces the prominence of the grid

Concealed grid. All of the runners hidden beneath
the ceiling panels

Figure 10.4   Suspended ceiling grids.

the tile placed into the grid. In areas where wind uplift may be a problem (e.g. in entrance
lobby areas), it is common to provide a more rigid fixing to avoid accidental lifting and
damage to the tiles. Most manufacturers produce plain tiles as well as tiles with a variety of
preformed apertures into which lights and sensors are placed. Installing perforated tiles can
provide additional sound insulation.

Support grid
The support grid is made from lightweight wire hangers, fixed to the underside of the structural floor. These wire hangers support the lightweight modular grid into which the tiles
are placed. Special fittings (edge details) are manufactured for intersections at structural
columns and the walls. The support grid is set out and levelled using a laser level. The grid
can be exposed, semi-concealed or fully concealed (Figure 10.4 and Photograph 10.4).

Maintenance

Careless handling and repositioning within the grid may damage tiles. To maintain an attractive finish, damaged tiles will need to be replaced. Tiles may also become dusty and marked through careless handling; thus, some cleaning may be required.

10.4 Internal partition walls

Internal partition walls are used to create discrete areas within large interior spaces. A wide variety of proprietary systems are available, which are designed for different uses and have different quality levels. Alternatively, partition walls may be constructed as a framed structure (in timber or mild steel) or as a masonry wall (brick or block). In office developments, the emphasis tends to be on flexibility and future adaptability of the workspace, while in industrial units, the emphasis tends to be more on durability and the ability to withstand minor impacts.

The term 'partition wall' is used rather loosely in the construction sector to cover both load-bearing and non-loadbearing internal walls. These have already been described in *Barry's Introduction to Construction of Buildings*. In the context of this chapter, we have limited the description of partition walls to some of the more flexible and adaptable systems that are used in commercial developments.

Functional requirements

Partition walls may be free-standing units positioned on the floor, tied to the structural floor or tied to the structural floor and the ceiling structure. In buildings with a high roof (e.g. conversion schemes), it is not uncommon for the partition wall to be supported off the floor only and a suspended ceiling to be installed to create a sensible ceiling height in relation to the room proportions. Alternatively, partition walls may be used to define space only, supported off the floor and without the need for ceilings. All partition wall systems need to be structurally stable and, especially in office space, easy to reposition without causing damage to floor or ceiling finishes.

Functional requirements are as follows:

❑ Flexibility.
❑ Adaptability.
❑ Structural stability.

Other performance requirements, which the partition wall may need to address, include:

❑ Fire resistance.
❑ Resist the spread of fire.
❑ Accommodate services.
❑ Provide required acoustic and thermal insulation.
❑ Transfer own weight and any fixtures and fittings (non-loadbearing)
❑ Transfer building loads (loadbearing).
❑ Allow the passage of light.
❑ Allow cross ventilation.
❑ Demountability.

Fire resistance

The partition may be required to prevent the passage of fire by acting as a compartment wall. Blockwork with a plastered finish provides a good resistance to fire; for example, a 100 mm block wall finished with plaster easily offers two hours' fire resistance, and a double-skinned plastered blockwork wall will achieve a fire resistance of four hours. Timber studs and proprietary walls only offer half an hour's fire resistance unless they are specifically designed as a fire-resistant structure. There are many fire-resisting plasterboards that can be easily applied to stud walls in single, double and triple thicknesses. All joints must be effectively sealed with firestops. Any gaps, service ducts, ventilation units, doors and windows provide weaknesses in fire-resisting structures and should be addressed in the detailing. It may be possible for the fire to pass around the wall under raised floors or suspended ceilings; effective fire barriers should be provided under and above the wall if it forms a compartment wall.

Fire-resisting walls should be fire-stopped at their perimeters, at junctions with other fire-resisting walls, floors and ceilings, and at openings around doors, pipes and cables. Fire- stopping materials include:

- ❏ Mineral wool.
- ❏ Cement mortar.
- ❏ Gypsum plaster.
- ❏ Intumescent mastic or tape (intumescent strip).
- ❏ Proprietary sealing systems.

Figure 10.5 provides an example of a fire-resisting partition.

Resist the spread of fire

The surface material should not allow flames to pass across it and should not fuel the fire. In public buildings, walls should be designed so that the risk of flame spreading across the surface is minimal. Compartment walls must have a low risk of spreading flame (Class 0).

Accommodate services

Allow for maintenance and repositioning of services. Some proprietary partitions are prefabricated with conduits, pipework or cables already positioned. Skirting boards and dado rails are often a good place to provide access for services. Conduits and channels run behind plastic, metal or wooden boards and entry boxes are positioned to allow services to be installed. Alternatively, the services would need to be surface-mounted.

With timber stud walls, timber floors can be easily positioned to accommodate electrical plug sockets, pipework and other fittings. Steel and timber stud walls have become increasingly popular in flats and offices. However, when installing services within these walls, care should be taken to ensure that the acoustic and fire-resisting properties of the wall are not compromised by the penetration of services through the plasterboard. Ideally, services should be surface-mounted to avoid the problem. Alternatively, sound-resisting material, such as acoustic-resisting plasterboard, should be positioned in the middle of the wall with timber or metal studs on either side of the panel, thus allowing services to be accommodated without affecting the sound-resisting material (Figure 10.5).

With masonry walls, the services can be surface-mounted or chased into the wall. Chasing should not exceed one-third of the wall's thickness vertically and one-sixth horizontally.

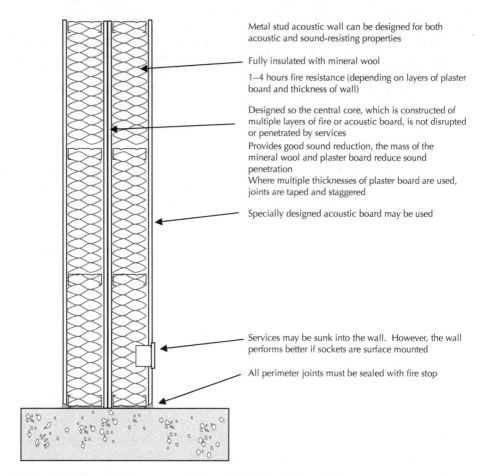

Metal stud acoustic wall can be designed for both acoustic and sound-resisting properties

Fully insulated with mineral wool

1–4 hours fire resistance (depending on layers of plaster board and thickness of wall)

Designed so the central core, which is constructed of multiple layers of fire or acoustic board, is not disrupted or penetrated by services

Provides good sound reduction, the mass of the mineral wool and plaster board reduce sound penetration

Where multiple thicknesses of plaster board are used, joints are taped and staggered

Specially designed acoustic board may be used

Services may be sunk into the wall. However, the wall performs better if sockets are surface mounted

All perimeter joints must be sealed with fire stop

Figure 10.5   Fire-resisting or acoustic metal stud partition wall.

Chases in the wall will reduce its strength, acoustic properties and fire resistance. Care should be taken to ensure that these do not compromise the specified performance.

Provide required acoustic control

The degree of sound absorption and reflection in a room is particularly important in auditoriums, lecture theatres, concert halls, etc. Lightweight porous materials will help to absorb the sound and reduce reflection, whereas heavy, hard and smooth surfaces will reflect the sound. In an auditorium or lecture theatre, reflective surfaces should be used close to the source of the sound (e.g. position of the speaker), and absorbent materials should be used towards the back of the room, where noises reflecting would cause an echo.

Transfer own weight and any fixtures mounted on the wall

The ability of a partition wall to accept fixtures is often overlooked. Overloading a wall with shelves or other fittings may cause the wall itself to break or topple over, or the load of the fittings may simply pull the fittings out of the wall fabric. The wall materials must be capable of restraining the loads applied. Studs and other reinforcing materials may need to be positioned so that the wall can accommodate the load. The head of the wall will also need to be restrained if loads are anything other than minimal.

Transfer building loads (loadbearing)

While it is common to have large, flexible open spaces divided by non-loadbearing, potentially demountable walls, it is often economical to have intermediate loadbearing supports, such as internal walls. The walls can be loadbearing, carrying loads from floors, beams and components above the wall down to the building's foundations. By using intermediate load-bearing walls, the floor beams do not need to span as far, and the section and depth of each beam can be reduced. Smaller, shallower beams are less expensive per unit length than long, deep-section beams.

Allow the passage of light

Light may be allowed to pass with or without vision through the material (usually glass). Windows and vision panels are easily introduced into stud and masonry walls. It is important that the windows are carefully selected and fitted so that they comply with the other performance criteria of the wall (e.g. fire resistance and acoustic properties).

Allow cross ventilation

Mechanical and natural ventilation ducts may be installed through the wall. If the wall is a compartment wall, these service ducts will need to be fitted with a fire stop that is capable of sealing the duct in the event of a fire.

Demountability of partition walls

To accommodate changes of use, large spaces are often divided using demountable partitions. In such situations, the ease with which a wall can be dismantled, reassembled and repositioned is of considerable importance. Any non-load-bearing wall is demountable, but it may not be possible to re-erect the structure using the same components.

Some partition walls have been designed so that they can be easily repositioned (e.g. folding concertina doors, sliding doors or walls), without the need for tools. Others are bolted, clipped or fixed into place but can be relatively easily repositioned with minimum disruption. Continuity of floor, wall and ceiling finishes may sometimes be compromised when partition walls are repositioned or removed.

Partition wall assembly

Partition walls are manufactured to a modular size. Systems are usually based on a lightweight steel stud system. Alternatively, timber studs may be used (Figure 10.6). By using multiple layers of acoustic or gypsum fire-resisting plasterboard and filling the studs with mineral wall, the fire-resisting and sound-reducing properties of the wall can be significantly increased.

Often selected purely on economic grounds, different arrangements of brick, timber, concrete, steel and patent systems can be selected to provide the required finish, flexibility, acoustics and fire-resisting properties, as illustrated in Figure 10.7.

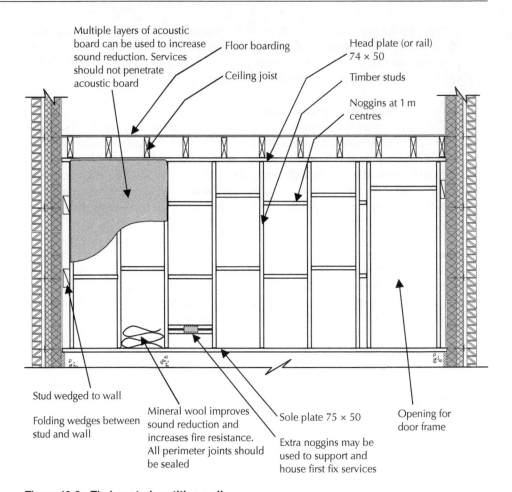

Figure 10.6   Timber stud partition wall.

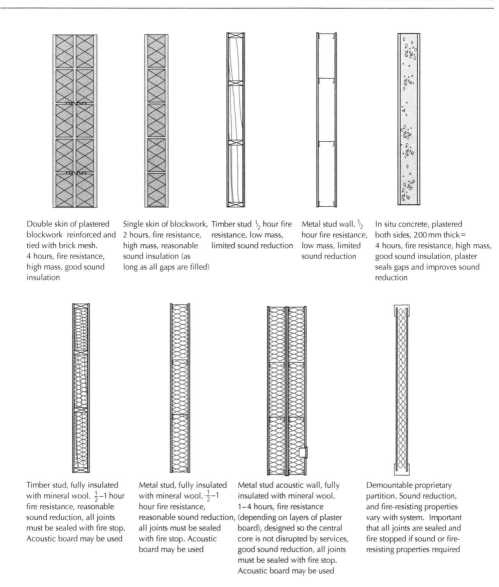

Double skin of plastered blockwork reinforced and tied with brick mesh. 4 hours, fire resistance, high mass, good sound insulation

Single skin of blockwork, 2 hours, fire resistance, high mass, reasonable sound insulation (as long as all gaps are filled)

Timber stud $\frac{1}{2}$ hour fire resistance, low mass, limited sound reduction

Metal stud wall. $\frac{1}{2}$ hour fire resistance, low mass, limited sound reduction

In situ concrete, plastered both sides, 200 mm thick = 4 hours, fire resistance, high mass, good sound insulation, plaster seals gaps and improves sound reduction

Timber stud, fully insulated with mineral wool. $\frac{1}{2}$–1 hour fire resistance, reasonable sound reduction, all joints must be sealed with fire stop. Acoustic board may be used

Metal stud, fully insulated with mineral wool. $\frac{1}{2}$–1 hour fire resistance, reasonable sound reduction, all joints must be sealed with fire stop. Acoustic board may be used

Metal stud acoustic wall, fully insulated with mineral wool. 1–4 hours, fire resistance (depending on layers of plaster board), designed so the central core is not disrupted by services, good sound reduction, all joints must be sealed with fire stop. Acoustic board may be used

Demountable proprietary partition. Sound reduction, and fire-resisting properties vary with system. Important that all joints are sealed and fire stopped if sound or fire-resisting properties required

Figure 10.7   Types of internal partition.

Further reading

There are many proprietary suspended ceiling and raised floor manufacturers and suppliers that can be found on the internet. See also: www.designingbuildings.co.uk.

Reflective exercises

You have been commissioned to upgrade the entrance area to a large multi-storey office building. The building was built in the 1960s and constructed using a concrete frame. The ground-floor entrance area has a solid floor and a suspended ceiling that was installed in the 1980s to accommodate electrical wiring and new lighting. Your initial investigations have shown that there are no fire stops in the suspended ceiling and that it does not comply with the current fire regulations.

❏   How do you proceed?
❏   What do you need to do to ensure that a new suspended ceiling complies with the current fire safety legislation?

Existing Buildings: Pathology, Upgrading and Demolition: Chapter 11 AT A GLANCE

What? Our buildings are valuable economic and cultural assets. They require regular maintenance, occasional repair and from time to time updating and upgrading to maintain their functionality. Upgrading and retrofitting can help to improve, for example, accessibility for users and the thermal performance of the fabric. Eventually, regardless of the level of care spent on a building, it will become unfit for its purpose and hence obsolete, at which time it will need to be extensively remodelled or demolished (disassembled); with components and materials recovered and recycled.

Why? Existing buildings need to be maintained, upgraded, remodelled, extended and eventually disassembled and recycled. To undertake work to existing buildings requires a thorough understanding of their context, use and construction. This is known as building pathology. Decisions about whether to demolish an obsolete building and recover/recycle materials, or bring it back to a serviceable condition are usually based on social, economic, technical, sustainable and legislative factors. Adaptive reuse of existing buildings can help to restore, regenerate and revitalise areas that have become rundown while helping to reduce the carbon emissions associated with demolition and rebuilding.

When? The timing of work to existing buildings may be in response to the identification and rectification of damage. It could also be triggered by the desire of the building owner to upgrade, remodel or demolish the property. Inspection and maintenance should be conducted on a regular basis to identify damage and decay before it becomes problematic to building owners and users. Similarly, changes in legislation relating to, for example, disabled access, can result in changes being made to the building.

How? A wide variety of techniques are brought into play when dealing with existing buildings. For historic buildings and buildings of architectural merit, many of the techniques used will rely on historic building materials and methods associated with craft to help preserve and conserve the building fabric. For the less distinguished buildings, the techniques and materials will be contemporary and are unlikely to be subject to the strict controls that apply to listed buildings. In all cases, it is essential that the building is analysed and understood prior to any interventions.

11 Existing Buildings: Pathology, Upgrading and Demolition

Buildings are a valuable economic asset. They require regular maintenance, occasional repair and from time to time updating and grading to maintain their functionality. Eventually, regardless of the level of care, a building will become unfit for its purpose and hence obsolete. Decisions about whether to demolish an obsolete building and recover/recycle materials, or bring it back to a serviceable condition are usually based on social, economic, technical and environmental (sustainability) factors. The repair, refurbishment, upgrading and/or retrofitting of buildings that have outgrown their original function may, in the majority of cases, be a more culturally sustainable option than demolition and replacement with a new artefact. However, a wide raft of factors such as cost, town planning restrictions, and technical feasibility will affect the decision-making process. In this chapter, the emphasis is on the technical factors that influence the repair and revitalisation of buildings that are deemed unfit for purpose (obsolete). Attention is also given to the retrofitting of buildings to improve their functional performance, with specific attention given to improving the thermal performance of the existing building stock and improving accessibility for all building users. This chapter concludes with demolition and recycling and reuse of materials.

11.1 The pathology of buildings

Entropy is a rule of nature that states that as soon as something reaches its desired state (i.e. maturity), it starts to decay. Buildings are no exception to this rule. Once completed, a building will start to decay as it is subjected to a variety of biological, chemical, electromagnetic, mechanical, thermal and human agents. This 'weathering' of the building as it ages can, in some circumstances, enhance its appearance, but in the majority of cases, it results in the need for regular maintenance, occasional repair and upgrading, until such a time as the building is no longer fit for purpose. The word 'pathology' is used in medicine to describe the systematic study of diseases in humans. In the built environment pathology is used to describe the 'diseases' that negatively affect the building's performance. The pathology of buildings relates specifically to the way in which a building responds to its physical environment and how it reacts to use over time. Emphasis is on understanding the symptoms, causes and treatment of problem areas. Attention could be on dealing with a specific issue, such as a leaking roof, or it could be more encompassing by addressing building obsolescence and the potential for repair, refurbishment and retrofitting, and bringing the building back to life. The implication is that we need to give our buildings a regular 'check-up' to

---

Barry's Advanced Construction of Buildings, Fifth Edition. Stephen Emmitt.
© 2023 John Wiley & Sons Ltd. Published 2023 by John Wiley & Sons Ltd.

determine their condition (health) and assess their suitability for their current, or intended use. This necessitates careful survey work to record the physical condition of the fabric and an assessment of the building's performance in use (discussed later). Informed decisions can then be taken about the level of preventative maintenance that may be required and the repair and or replacement of specific materials and components.

Building obsolescence

'Building obsolescence' is a term used to describe a building that has become outdated and unfit for purpose. An obsolete building has reached the end of its service life, which will often result in it being neglected or abandoned, resulting in further deterioration of its physical condition and it eventually becoming unsafe. Given that buildings, and the land they occupy, represent a significant economic asset for their owners, few buildings are allowed to decay and eventually collapse. Instead, the desire is for a new lease of life. Building owners are faced with a choice of; demolition and replacement with a new building; upgrading, retrofitting and repair of the building to restore the original function; or adaptive reuse, for example, to convert a redundant industrial building into residential use (which also involves upgrading the fabric to meet prevailing standards and legislation). The latter option may be associated with the desire to regenerate an area and restore or re-establish a vibrant community.

A building may be termed obsolete for one or more of the following reasons:

❏ *Physical obsolescence.* Buildings decay over time and without regular maintenance and repair, they will eventually reach a state of physical obsolescence. Given that not all building materials and components decay at the same rate, it is not unusual for the building to exhibit signs of obsolescence in a variety of areas, be it the building fabric or the internal environment. For example, the building fabric may decline at a faster rate than the structural frame, often resulting in the depreciation of the building's economic value (based on physical appearance and predicted cost of repair and refurbishment). However, the underlying structure of the building may be sound, and it may be possible to upgrade the fabric several times before the structural frame becomes unfit for purpose, at which time the entire building will need to be replaced. A similar argument can be made for the building interior.
❏ *Functional obsolescence.* Technological advances and changes in demand can often render a building obsolete in terms of its function and usability from the perspective of the users. This can be mitigated to a certain extent by designing the building to be adaptable and flexible to different demands and changing technologies, although it is not an easy task to try and predict what we might demand of our buildings 10, 20 or 50 years hence. Photograph 11.1 shows a warehouse that was derelict and subsequently upgraded to luxury waterside apartments; an example of adaptive reuse in a conservation area. The photograph helps to illustrate the sensitive introduction of new double-glazed windows to improve thermal performance of the fabric, with the new steel window frame reflecting the industrial heritage of the building.
❏ *Economic obsolescence.* As a building becomes less useful to owners and tenants, there will be a loss in economic value. At some point, the amount of investment required to maintain, repair and upgrade the building will become economically unviable based on future predictions of income. This is known as economic obsolescence. For example,

Photograph 11.1   Window detail. New windows installed as part of an adaptive reuse project: from redundant warehouse to high-quality residential apartments.

with rising energy costs it may become economically unfeasible to sufficiently upgrade the thermal insulation of a building, with the cost of the work far outweighing any future economic savings. At this point, it is likely that tenants will move to buildings that have better thermal performance and lower running costs, while owners will seek to dispose of their asset or redevelop the site.

❏ *Sustainable obsolescence.* As environmental legislation becomes ever more stringent and awareness of sustainable issues becomes more widespread, it has started to alter how we perceive our building stock. What was once a perfectly acceptable building may start to be perceived by owners and tenants as no longer sustainable, because it no longer satisfies new performance criteria (e.g. carbon reduction targets). When it is not physically possible, or economically viable, to upgrade the building to meet new environmentally sustainable guidelines and legislation, then the building will be deemed to have reached a state of sustainable obsolescence. This may result in tenants moving to buildings that better suit their organisation's environmentally sustainable values, or alternatively it may lead to a programme of upgrading and retrofitting to bring the building up to current standards and expectations.

Research and recording

A full understanding of the building's social and technical history is essential prior to carrying out any interventions. When dealing with any aspect of an existing building, there will undoubtedly be some challenges in accessing information about the building's construction and use; however, information can be collected from a wide variety of sources, helping to provide some contextual data. Measured survey drawings, as-built drawings, written descriptions, specifications and photographs will be useful. So too will local government records for planning and building regulation control and other documentary sources such as insurance records. In attempting to gather information about buildings, it is essential that the search is methodical and critical. All sources should be accurately recorded and an accurate record built up through constant cross-checking of information. A good starting point is with the original date of the building if this can be established quickly; for example, from a date stone in the fabric or through local records. Since the late nineteenth century, architects and builders have been required to submit copies of their plans and proposed construction details to the local authority building control department for approval. This body of information can provide an important source of material, the date of design, construction details, survey drawings, etc. Unlike planning records, permission to access the drawings will be required for security purposes. Information sources may comprise some or all of the following:

❏ Maps and plans.
❏ Title deeds.
❏ Newspapers and journals.
❏ Town planning records.
❏ Building control records.
❏ Records held by local builders and consultants.
❏ Local knowledge.
❏ Specialist publications and books.

Whether this exercise is conducted before, after or concurrently with an assessment of the building's condition will depend upon circumstances relating to a particular building. The important point is that it must be done before any objectives, design work or building work is carried out.

Analysis of condition

On site investigation and analysis should not be carried out until at least some of the information required has been found; this knowledge helps to focus the attention of the site survey and also aids the understanding of health and safety considerations. Designers need to be rigorous and systematic in their observation and recording of what they find. Photographs, video and thermal imaging can supplement this exercise. Photograph 11.2 provides an example of modest 'opening up' of an existing house to try and establish what was behind an existing wall.

The most common methods of data collection are:

❏ *Measured survey.* A detailed measured survey of a building and its immediate environs will enable accurate plans, elevations and sections to be produced. Undertaking this exercise also allows those conducting the survey to experience the building at close

Photograph 11.2   Opening up an existing structure to see what lies under the surface.

quarters and hence get a good feel for its character. The survey drawings may differ from historical data because of inaccuracies in original drawings, variations from the drawings during construction and/or because of unrecorded changes made to the building post-construction.

❑ *Condition survey.* Analysis of a building's physical condition is known as a condition survey and the resultant report is known as a condition report. Condition reports serve two purposes. First, they should provide an accurate and comprehensive description of the condition of the building fabric, structure and services. Second, they should act as an information source on which decisions can be made. Thus, the report must be well structured, clearly written, and contain concise conclusions and recommendations.

❑ *Post-occupancy evaluation (POE).* This term covers the monitoring of buildings to see how they are used over time, i.e. how they perform in relation to expectations (as set out in the design brief). This is an important consideration for commercial and public buildings that have to be managed to support business objectives. Common areas of concern relate to the use of energy and indoor air quality. Energy consumption can be monitored via smart meters. Indoor air quality can be measured by the installation of sensors at strategic positions within the building interior. Other performance factors, for example, thermal comfort of the users and how people use space require more extensive research and ethical approvals prior to undertaking any research. Evaluation and monitoring of an existing building are usually done for one or more of the following reasons, namely, to evaluate and monitor the:

○ Performance of the building against specified criteria (e.g. energy consumption and thermal comfort of the occupants).

○ Functionality of the building against its current (or proposed) use.

○ Building users' behaviour, with a view to improving working conditions, physical comfort and physical and mental wellbeing.

○ Maintenance and operating costs (heating costs, electricity use (lighting and equipment), water consumption, cleaning, security, etc.).

Concomitant with other data collection exercises, the purpose should be clearly defined and necessary approvals sought and granted before data collection begins. Similarly, methodologies for evaluation should be kept simple, have measurable outcomes, be properly resourced and have a realistic timeframe. Whatever method is used, the data recorded should be used to aid decision-making. The outcome is likely to be one of the following:

❑   Do nothing.
❑   Conserve (and or preserve) the building.
❑   Retrofit to upgrade the functionality of the building.
❑   Remodel the building to suit a new use.
❑   Demolish and recycle.

The 'do nothing' option may be taken due to financial and practical limitations; however, putting off a decision to, for example, repair a part of a building usually results in more expensive repairs in the future. The options of conserving, retrofitting, remodelling and demolition are discussed after we understand the underlying reasons.

11.2 Decay and defects

All building materials, products and services are finite in their life span. Just as materials have unique coefficients of expansion, they also have unique coefficients of decay; thus, elements of the building will be decaying at different rates. The effect of weathering is to erode, dissolve and discolour the building fabric, often resulting in staining and eventually in the need for specialist cleaning and repair. Some materials are enhanced by weathering; for example, stone, seasoned timber such as oak and moss-covered roof tiles. Other materials may fare less well when exposed to the elements. Even the best designs may look drab because of the wrong choice of materials, poor detailing and insensitivity to a building's micro-climate.

Agents of decay

Over time, buildings are subjected to attack from a number of different sources. Sometimes these agents of decay act independently, although it is more common that they act in conjunction with one another. The rate of decay can be reduced through sensitive detailing and materials selection, competent construction and proactive management of the building during its life. Agents of decay may be classified as being biological, chemical, electromagnetic, human, mechanical and thermal.

Biological agents

Biological agents include animals and microbiological agents. Animals include birds, insects and rodents; unwelcome guests that can cause damage to the building fabric. Domestic pets may also cause damage to property, especially surface finishes. Microbiological agents include bacteria, fungi and moulds. These are usually a sign of poorly ventilated and damp buildings. Seeds can penetrate roofs and cracks in buildings, germinate and grow into large plants if left unchecked, causing damage to the fabric as they do so. Plant roots may cause structural damage to foundations and ground floors, which will depend on the soil, the foundations and the type of plant.

Chemical agents

The range of chemical agents of decay is extensive, ranging from water and water-based solvents to acids, sugars, salts, and oxidising agents, base chemicals such as lime and dust, and products brought into buildings such as detergents and bleaches. Water ingress, be it via a leaking roof or more dramatically via flooding, will cause extensive damage to building materials. It is important that the fabric is allowed to dry out fully following any leaks or flood damage. Once dry the building work can commence to rectify the damage.

Electromagnetic agents

Electromagnetic agents include radiation and magnetic fields. Radiation mainly comes from solar radiation or radioactive radiation from the ground. Solar radiation has the effect of degrading many materials, leading to them becoming brittle and more prone to damage. Radiation will primarily come from radon in the ground and measures can be put into place to prevent the gases getting into buildings. Magnetic fields are usually generated from devices within buildings and tend to affect the building users rather than the fabric. This category also includes lightning, and lightning conductors can be placed onto tall and isolated buildings (parts of buildings) to mitigate damage.

Human agents

Buildings are designed to be used for a specific function. Over time it is inevitable that the building fabric will deteriorate through general use and unintentional damage. Buildings may also be subjected to deliberate damage, vandalism and arson. Malicious damage to property, such as theft and vandalism, can have a long-term effect on building performance if not rectified promptly, often leaving a building vulnerable to damage from the environment. For example, the theft of lead flashings from the roof of existing buildings can lead to a rapid deterioration of the property through water ingress. Left unchecked, damage will occur directly from the water penetration and indirectly through the possible development of wet and/or dry rot given the right conditions. Arson is potentially the most dangerous act of malicious damage, which leads to serious damage through the fire and in its containment and extinguishment.

Mechanical agents

Mechanical agents are related to gravity and forces imposed upon, or retained within, the building fabric. Loading will come from snow and water (both rainfall and floodwater), ice pressure and wind loading. This category also includes expansion and contraction as a result of thermal and moisture changes. Changes to live and dead loads also occur due to occupancy and changes of use over time. Seismic activity (earthquakes) affects the building through the release of kinetic energy and associated ground heave and vibration. Fortunately, the magnitude and frequency of seismic activity in the UK is quite low, resulting in minimal damage to buildings and infrastructure. Ground movement includes landslips, subsidence and changes in ground water pressure. Vibration from traffic (vehicular and aircraft), machinery and explosions may also damage buildings.

Thermal agents

Thermal agents include damage caused by changes in temperature, with rapid change in temperatures known as thermal shock. The design of buildings must allow adjacent materials and components space to expand at different rates without generating stress on their neighbours. This is usually achieved with control joints and careful selection of materials.

Construction defects

Despite everyone's best intentions, it is possible that some faults and defects will be found in the completed building. Some of these will be evident at the completion of the construction contract, but some may not reveal themselves until sometime in the future and are known as 'latent defects'. Many years may pass before the defect becomes apparent, especially where it is hidden within the building fabric. Performance monitoring during and post-construction can help to identify some of the defects before they pose a threat to the building fabric. Defects can usually be traced back to one or more of the following:

❑ An inability to apply technical knowledge.
❑ Inappropriate detailing and specification.
❑ Non-compliance with regulations and codes.
❑ Incomplete information.
❑ Late information.
❑ Late design changes.
❑ Poor work.
❑ Inadequate site supervision.
❑ Inappropriate alterations rendering an otherwise good detail ineffective.
❑ Insufficient maintenance.

We can, for simplicity, divide defects into two categories: those concerning products and those associated with the process of design and construction.

Product defects

With the constant drive to improve the quality of materials and building components from the manufacturing sector, it is unlikely that there will be a problem with building products, assuming that they have been carefully selected, specified correctly and assembled in accordance with the manufacturers' instructions. Reputable manufacturers have adopted stringent quality control and quality management tools to ensure that their products are consistently of a specified quality, are delivered to site to schedule and technical support is available as required. A well-written performance specification or a carefully selected proprietary specification, combined with careful implementation on site, should help to reduce or even eliminate product-related defects. Problems can usually be traced back to hastily prepared specifications, cost-cutting and specification of lesser quality products, and/or poor management and practice on the building site. Products recently launched onto the market and to a lesser extent products new to the specifier/user carry an increased degree of uncertainty over their performance, and hence a perceived increase in risk.

Process defects

Problems with the process of design and construction are the most likely cause of defects. The design and construction process, regardless of the degree of automation, relies on people to make decisions and to implement the result of those decisions. Designers record and communicate their decisions primarily through drawings and the written specification. Thus, the quality of the information and the timing of the delivery of the completed information (i.e. communication) will influence the likelihood of defects occurring. Quality of work on site will depend on the interpretation of the information provided, control and monitoring of the work, and the influence of the weather and physical working conditions. Design changes, especially during construction, may cause problems with constructability and subsequent maintenance and may have a detrimental effect on neighbouring products and assemblies. If a fault or defect is discovered then it needs to be recorded, reported and appropriate action agreed to correct the defect without undue delay.

The performance gap

The term 'performance gap' has come into widespread use in recent years. It is used to explain the difference between the designed (anticipated via modelling/simulation) performance of a building and the actual performance of a building in use. It is most commonly related to the thermal properties of buildings, but it can also be used to describe differences between expectations and reality in a variety of areas, such as acoustics and lighting. The difference in performance is usually related to how the building was constructed and how it is operated, although some of the modelling and simulation techniques and the assumptions made at the design process also have a part to play in explaining some of the differences. As a general rule, we should take our calculations and computer models as a guide to how the building will perform and not be too surprised if it varies, for better or worse, a little from what was predicted.

Maintenance and repair

Deterioration cannot be prevented, but it can be retarded through a combination of good detailing, good building and regular inspections and maintenance. Recurrent maintenance costs are a financial drain on building owners, and the act of maintenance may also be disruptive to the building users. This sometimes leads to maintenance work being postponed, often with consequences for the building. Efforts to reduce the frequency and extent of maintenance through sensitive selection of good-quality building products and sensitive detailing are likely to result in reduced life cycle costs for the building owner. However, there is still a requirement for regular inspection, cleaning and routine maintenance, which must be factored into the whole life costs of a building at the design stage.

Maintenance and repair should benefit the building, not hinder its aesthetic appeal of technical performance. The repair of buildings is often undertaken in an *ad hoc* manner, in stark contrast to the time and effort spent on the original building project. Inconsistency will usually devalue a property and may lead to unforeseen problems with the performance of the building fabric. It is essential that those carrying out maintenance and repairs understand the way in which the building was designed and assembled so that maintenance does not compromise performance. This means that those responsible for maintenance must have access to the as-built drawings and associated documentation.

11.3 Conservation of buildings

Views on the importance of preserving, restoring and conserving our built environment vary, although most would agree that some degree of preservation and conservation is important to protect and enhance our built and cultural heritage. Legislation relating to listing and conservation areas imposes restraints on the owner's rights to do what he or she likes with the property, without first obtaining consent from the local authority town planning department. The terms in use are:

❏ *Preservation*. This is concerned with the retention (or reinstatement to its original form) of a structure deemed to be of cultural importance to society and future generations.
❏ *Restoration*. This is concerned with returning a building, or part of a building, to the condition in which it would have been at some point in the past. Restoration has a role to play in the preservation and conservation of historic buildings and buildings deemed to be of architectural and cultural importance.
❏ *Conservation*. This is concerned with retaining (and enhancing) the cultural significance of a building. Conservation enshrines the idea that buildings are used by people and thus make up part of the living tapestry of the built environment. Therefore, alterations, improvements and change of use are to be expected to help keep the building alive and fulfilling a function. The techniques used to conserve a building will be influenced by its architectural character and the degree to which it is protected by legislation, or not.

Listing

Listing aims to protect a building from demolition or insensitive alterations and repairs, helping to retain the architectural character and cultural importance of certain buildings. Buildings may be listed because of their age, architectural merit, rarity and their method of construction. Buildings may also be listed because of their cultural significance, for example, being the birthplace of an important person.

Buildings, ranging from industrial buildings to pubs and post-war schools, may be surveyed and considered for listing once they are 30 years old. There is an additional rule that allows exceptional buildings between 10 and 30 years old to be considered for listing if they are threatened with demolition or alteration. The listing grades for England are explained further,

❏ *Grade I. This* covers buildings that are of exceptional interest, for example of national importance and some of international importance. This listing category covers approximately 2.5% of all listed buildings.
❏ *Grade II **. Of significant regional importance and some of national importance, being more important than those in the special interest category. This covers approximately 6% of listings.
❏ *Grade II*. This includes the majority of listed buildings that are of special interest. These are primarily buildings of significant local importance, warranting effort to protect them.

Listings and further information can be obtained from the local authority responsible for a particular geographical area. Once buildings are listed, alterations or demolition cannot be undertaken without first applying for and receiving listed building consent from the local authority planning department. Listing does not mean that buildings cannot be altered, but any proposed alterations will receive rigorous scrutiny to make sure they are sympathetic to the existing character of the building. Listings provide greater protection to buildings than a local authority-declared conservation area does. In the majority of cases, listing will improve the financial value of a property.

Work to existing (historic) buildings

When working on an existing building, the design solutions will be influenced by the building's existing character and context, each providing limitations and opportunities. For a listed building, the main objective will be to conserve the building through stabilisation of the fabric and structure, and sensitive repair work will help to extend the serviceable life of the building. How this is achieved will depend upon the importance of the building and its intended use. Photograph 11.3 shows a window in a listed building, being repaired using traditional materials and techniques to retain the existing character of the building.

Photograph 11.3   Window repair. Rotten timber replaced with new timber in a conservation area.

Replacing only the rotten timber with new timber also helps to reduce material waste by not replacing the entire window.

New uses for redundant buildings require a complete understanding of the building's construction, structural system, material content and services provision, as well as an appreciation of the cultural and historical context in which the building is set. A checklist would need to cover the following issues:

- ❑ Access limitations.
- ❑ Acoustic and thermal properties of the fabric (and potential for upgrading).
- ❑ Assessment of embodied energy.
- ❑ Condition of the fabric.
- ❑ Condition of the services.
- ❑ Contaminants (e.g. presence of asbestos, lead paint, etc.).
- ❑ Economic factors and life cycle analysis.
- ❑ Fire protection and escape (to current legislation).
- ❑ Health and safety factors.
- ❑ Historical context of the building and its immediate surroundings.
- ❑ Legislation, including site-specific constraints.
- ❑ Potential for reuse and recovery of materials (partial demolition).
- ❑ Reuse or demolish.
- ❑ Scope for new use (and future reuse).
- ❑ Social context.
- ❑ Stability of the structure and foundations (capacity for increased loading).

These factors need to be considered before the brief is finalised or design work commences. They should form an essential part of the critical condition survey and feasibility study.

Architectural character

Alterations and extensions, no matter how minor, will affect the building's character. The application of new construction techniques to regional traditions of building, using locally available materials and labour, may be one (sustainable) approach to enhancing the character of a building. Responding to the existing building fabric and the spirit of the place is a good starting point for many designers and is often the preferred approach of town planning officers and the immediate neighbours. However, it is possible to introduce modern materials and methods to existing buildings and hence enhance their architectural character. Successful remodelling of buildings is usually achieved by employing one of two design strategies:

- ❑ *Match existing.* Use of materials and building techniques to match those used previously. A continuation of tradition through colour, texture, application, scale and design philosophy. Specialist publications and design guides are essential reference tools. The successful completion of the building work will rely on specialist craft skills and the availability of craftspeople to undertake the work.
- ❑ *Contrast existing.* Use of materials and building techniques to contrast with those used previously. A break with tradition, through the use of new materials, contrasting textures, new techniques, different scale and new design philosophy. This allows users of

the building to see the new interventions, and in doing so appreciate the existing fabric that remains adjacent to the new. This may represent a relatively small part of the building, or it could be more significant in scale and impact.

Both are sympathetic approaches that are usually successful, the philosophy adopted depending upon the wishes of the client, town planners, designers, constructors, context and the resources available. If done well, the building will outlive its custodians and will probably be remodelled again in the future. If done badly, the value of the structure can be affected negatively, and future alterations and maintenance are likely to be more expensive than they should have been.

11.4 Façade retention methods

Existing buildings, both on the site to be developed and also those on neighbouring sites, affect the development of many sites in urban and semi-urban areas. Abutting buildings may need to be supported and protected for the duration of the project, during which time structures are removed and the new structure assembled. Temporary supporting works may need to be provided to ensure that work can be undertaken safely while restoring and renovating properties, demolishing structures, retaining façades and refurbishing buildings.

Façade retention

Not all existing buildings have sufficient structural properties for the proposed new use, and a considerable amount of structural work may need to be undertaken to ensure that the structure is made good. In many cases, the foundations may need to be strengthened and underpinned and the structure reinforced. In some cases, the structural work is so extensive that the only part of the original structure retained is the façade. Façade retention involves retaining only the external building envelope or specific aspects of the external fabric. This may be all of the existing walls, or in some cases, it may be as little as one elevation of the building only. The internal structure and majority of the building fabric is demolished to make way for a new structure behind the retained (historic) façade. Removing the main structural and lateral support (walls and floors) from the façade will render it unstable. A temporary support system must be put in place to hold the façade firmly in place while the existing structure is removed and the new structure installed. The temporary support system must be able to provide the necessary lateral stability and resist wind loads. The support systems may be located:

- ❏   Outside the curtilage of the existing building – external support.
- ❏   Inside the curtilage of the existing building (behind the façade) – internal support.
- ❏   Both external and internal to the existing building – part internal and part external.

Figures 11.1 and 11.2 provide examples of external, internal and part internal-part external façade retention systems. Each support system is designed specifically to suit the façade that is being supported and the process used to construct the new building. As well as supporting the walls of the façade, it may also be necessary to support adjacent buildings that previously relied on the support from the original building.

Rigid frame designed to resist lateral loads

Upper scaffold designed as a fully braced frame to transfer lateral loads to rigid portal framed pavement gantry

Fully braced frame acts as vertical cantilever

Trussed to resist lateral load

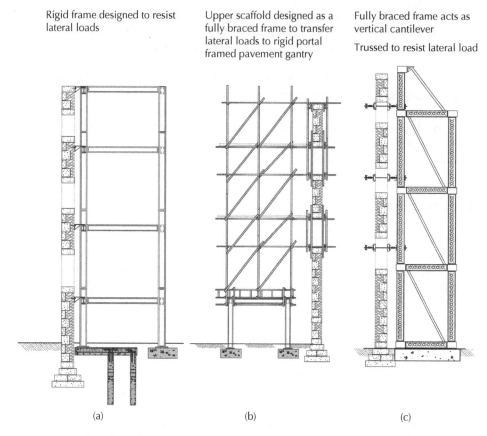

(a)   (b)   (c)

Figure 11.1 Façade retention internal and external façade support. (a) Internal support provided by new frame. (b) External support: steel tube scaffold. (c) External support: proprietary support system.

Various methods of retaining the façade and constructing the new works are used (see Photograph 11.4). This is a specialist field, so a summary of the principal issues that must be addressed is provided as follows:

❑ Temporary support to the façade – throughout the works.
❑ Must retain the façade, prevent unwanted movement, allow for differential movement and resist wind loads.
❑ Permanently tying back the façade to the new structure.
❑ Façade ties must restrain the façade and prevent outward movement away from the new structure.
❑ The ties must not transmit any vertical loads from the existing structure to the façade.
❑ Allowance for differential settlement between the new structure and the retained façade.
❑ Ties to the new structure must be capable of accommodating such movement.
❑ Ensure the new foundations do not impair the stability of the retained façade.
❑ Underpinning may be necessary to ensure that settlement is controlled.

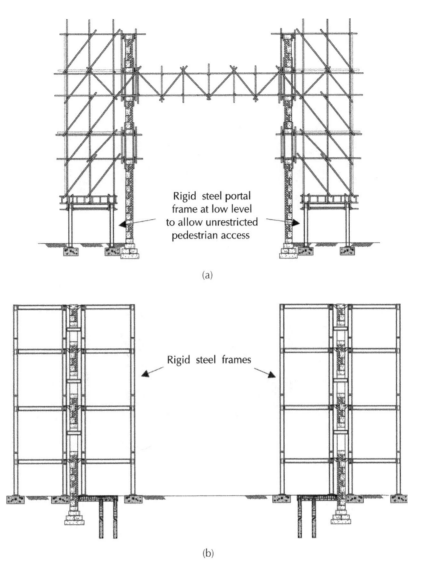

(a)

(b)

Figure 11.2   Part internal-part external façade retention. (a) Tabular steel scaffold with flying truss. (b) Temporary external frame used to provide support.

Temporary support

Temporary support to the façade can be provided by steel tubular scaffolds constructed to hold the fabric firmly in place until the new structural frame is built (Figure 11.3). When the temporary support is in position, the demolition operations to the main structure can start. As the new structural frame is constructed, the façade can be tied to it. Where possible, the scaffolding façade ties are taken through the window openings to avoid the need for breaking through the façade or drilling into the masonry to fix resin or mechanical anchors. Drilling and other potentially damaging operations should be avoided where feasible, especially if the façade is of architectural merit and/or town planning restrictions apply.

Photograph 11.4   External bespoke façade retention scheme – fabricated from rolled steel beams and columns.

Where the wall is clamped with a through tie, timber packing either side of the tie is used to provide a good contact with the surface. Surfaces of a façade are often irregular and the thickness of the remaining structure may vary. Timber packing felt and other slightly resilient and compressible materials should be used to secure the façade surface and protect it by preventing direct contact with metal supports.

The lateral forces applied by the wind may mean that the scaffolding needs to be trussed out, with kentledge applied (load to hold the support down), or flying shores may be needed to transfer the loads (Figure 11.4). The design of the shoring system is dependent upon the position and number of walls and floors retained and the position of the new structure and its floors. The installation of shoring should be coordinated with the demolition and installation of the new frame. During the demolition operations, it is essential that the integrity of the remaining structure be maintained. Only when the final structure is erected and the façade fully tied to it, can the shoring be removed completely.

Flying shores may be used to brace the structure against wind loads. These can only be used where there is an adequate return (e.g. the opposite face of the building). Depending on the direction of the force, the loads are transferred across the truss, down the opposing scaffold and to the ground. Foundations to the scaffold are used to ensure that the loads are adequately transferred. The scaffolding and the shores may be designed to resist both

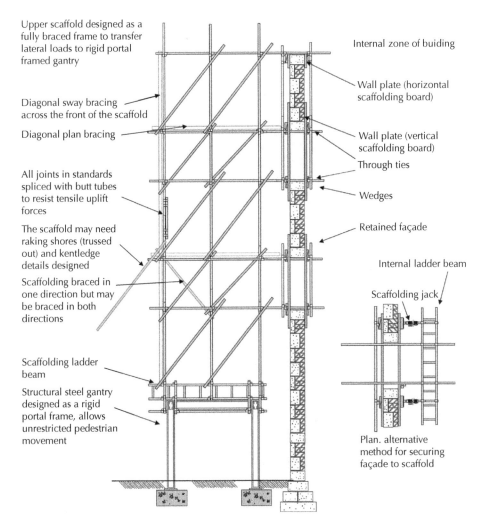

Upper scaffold designed as a fully braced frame to transfer lateral loads to rigid portal framed gantry

Diagonal sway bracing across the front of the scaffold

Diagonal plan bracing

All joints in standards spliced with butt tubes to resist tensile uplift forces

The scaffold may need raking shores (trussed out) and kentledge details designed

Scaffolding braced in one direction but may be braced in both directions

Scaffolding ladder beam

Structural steel gantry designed as a rigid portal frame, allows unrestricted pedestrian movement

Internal zone of buiding

Wall plate (horizontal scaffolding board)

Wall plate (vertical scaffolding board)

Through ties

Wedges

Retained façade

Internal ladder beam

Scaffolding jack

Plan. alternative method for securing façade to scaffold

Figure 11.3 Steel scaffold tube: temporary façade support system.

compression and tension, so the foundations must be capable of resisting uplift and compression, acting as kentledge and a thrust block.

Deflection must be limited to preserve the integrity of the retained structure. Flying trusses are constructed with a camber to reduce the impact of sagging. Intermediate scaffold towers can be used to reduce sagging, although the supports may impede the work. Where there is sufficient room outside the structure, raking shores will be used to provide the main structural support to the façade. The construction of reinforced concrete lift shafts and service towers can be used to transfer the loads to permanent structures at an early phase in the construction process.

Hybrid and proprietary façade support systems

A variety of structural formwork or falsework systems can be used, in conjunction with steel tubular scaffolding, as a hybrid system. Alternatively, the tubular, manufactured or patented systems can be used on their own to provide the required support (Photographs 11.5 and 11.6). The system illustrated in Figure 11.5 provides a schematic of the RMD support

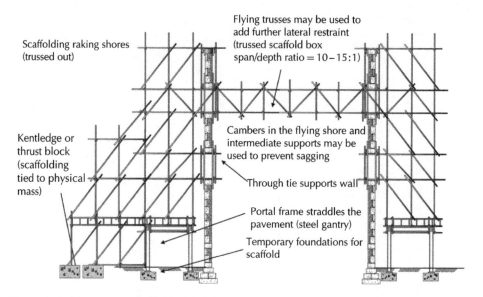

Scaffolding raking shores (trussed out)

Flying trusses may be used to add further lateral restraint (trussed scaffold box span/depth ratio = 10–15:1)

Kentledge or thrust block (scaffolding tied to physical mass)

Cambers in the flying shore and intermediate supports may be used to prevent sagging

Through tie supports wall

Portal frame straddles the pavement (steel gantry)

Temporary foundations for scaffold

Figure 11.4   Temporary scaffold with flying shores (truss), kentledge and basing.

Photograph 11.5   Proprietary façade retention systems.

Photograph 11.6   (a) Proprietary façade retention system, external support. (b) Proprietary façade retentions system providing external support allowing total demolition of internal structure. (c) Support system clamps the façade through wind opening. Adjustable shores ensure exact positioning and adjustment of the support. (d) Façade retention and support system need to ensure that abutting properties are adequately supported and works do not affect neighbouring properties.

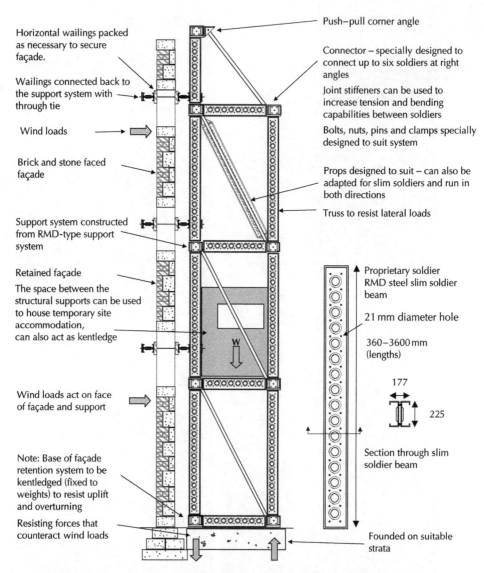

Horizontal wailings packed as necessary to secure façade.

Wailings connected back to the support system with through tie

Wind loads

Brick and stone faced façade

Support system constructed from RMD-type support system

Retained façade

The space between the structural supports can be used to house temporary site accommodation, can also act as kentledge

Wind loads act on face of façade and support

Note: Base of façade retention system to be kentledged (fixed to weights) to resist uplift and overturning

Resisting forces that counteract wind loads

Push–pull corner angle

Connector – specially designed to connect up to six soldiers at right angles

Joint stiffeners can be used to increase tension and bending capabilities between soldiers

Bolts, nuts, pins and clamps specially designed to suit system

Props designed to suit – can also be adapted for slim soldiers and run in both directions

Truss to resist lateral loads

Proprietary soldier RMD steel slim soldier beam

21 mm diameter hole

360–3600 mm (lengths)

177

225

Section through slim soldier beam

Founded on suitable strata

Figure 11.5   Façade retention using proprietary RMD support system: fully braced frame. (*Source*: http://www.rmdkwikform.net; adapted from Highfield, 2002).

system; this can be used externally, as shown in the diagram, internally or as a part internal-part external frame. The components fix together to provide a strong rigid frame. Often the areas between each lift of the supporting structure are used to house temporary site accommodation. In larger structures, multiple bays are used, which are fully braced to provide the required support (Photograph 11.5).

Fabricated support systems and use of new structure
It is also possible to limit the amount of temporary scaffolding and support systems by making use of the new structural frame (Figures 11.6 and 11.7). Although logistically complicated, in some buildings, it is possible to bore through ground floors to construct new foundations, puncture holes through upper floors and walls, and erect part of the new

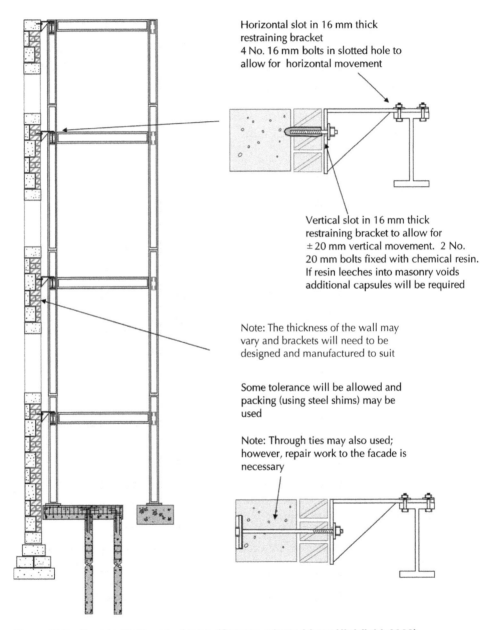

Horizontal slot in 16 mm thick restraining bracket
4 No. 16 mm bolts in slotted hole to allow for horizontal movement

Vertical slot in 16 mm thick restraining bracket to allow for ± 20 mm vertical movement. 2 No. 20 mm bolts fixed with chemical resin. If resin leeches into masonry voids additional capsules will be required

Note: The thickness of the wall may vary and brackets will need to be designed and manufactured to suit

Some tolerance will be allowed and packing (using steel shims) may be used

Note: Through ties may also used; however, repair work to the facade is necessary

Figure 11.6   Façade tied to new frame. (*Source*: adapted from Highfield, 2002).

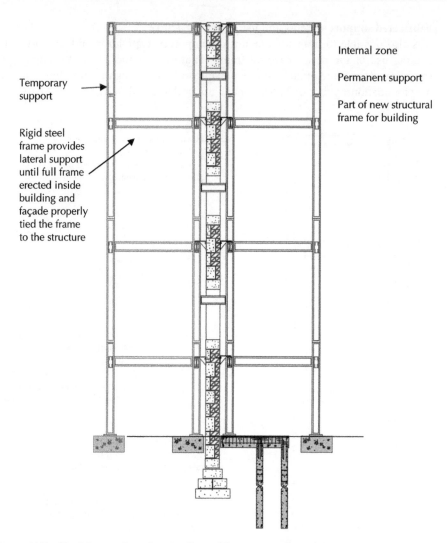

Temporary support

Rigid steel frame provides lateral support until full frame erected inside building and façade properly tied the frame to the structure

Internal zone

Permanent support

Part of new structural frame for building

Figure 11.7   Steel frame: façade retention with temporary steel frame.

structural frame before removing the main supports of the existing structure. In order for such operations to be undertaken, a thorough structural survey is required, and careful planning of the demolition sequence is necessary. Figures 11.6 and 11.7 show the façade tied to part of the new structure.

All façade retention schemes are expensive, so rather than using a scaffolding system, some retention schemes make use of specifically designed and fabricated rolled steel beams and columns to provide support to the external face of the façade. Because the structure surrounding the exterior of the building is only temporary, each of the beams and columns can be recycled and reused, and therefore the system may not be as expensive as it first appears.

Inspections and maintenance

Appropriate safety inspections must be carried out, similar to those outlined for scaffolding systems. In situations where the works are prolonged, metal support systems are susceptible to rusting and may lose their loadbearing properties.

Slight movement of the façade is to be expected; thus, the retained façade must be monitored for the duration of the works. If movement of the wall is detected and/or cracks develop in the fabric, investigation should be carried out to ensure that the wall is still structurally stable. Where cracking patterns suggest that de-lamination of the wall is occurring or the façade is losing its structural integrity, remedial works will be necessary. It is essential that monitoring and maintenance is continuous throughout the entire process and any operations necessary to ensure the façade and temporary works remain structurally sound are undertaken.

11.5 Retrofitting

Environmental legislation and attention to the environmental impact of buildings has helped to emphasise the importance of building durability and adaptability. This has been given additional importance with concerns over climate change and a warming planet. Reuse of our existing building stock is often desirable for environmental, cultural, economic and social reasons. Only around 1% of the existing building stock turns over each year, so our attention should be directed to improving the performance, durability and adaptability of our existing buildings. This is particularly relevant to reducing the carbon footprint of existing buildings. Retaining and upgrading existing buildings will also retain the embodied energy of the building and will reduce the need for additional energy embodied in new materials, as well as transportation of materials and material waste. Photograph 11.7 shows a domestic house that was upgraded and extended to better suit the needs of a growing family and also to improve the thermal performance of the fabric. With the exception of new double-glazed windows, the thermal upgrading (cavity and roof thermal insulation) is not visible to the occupants. However, the improvement in the thermal insulation values was reflected in significant reduction in heating costs and improved thermal comfort for the residents. This retrofit also improved the economic value of the property.

Retrofitting is a term used to describe the addition of new technologies and products to an existing building to improve its performance and/or functionality. Some common examples are the addition of photovoltaic (PV) panels to existing roofs, upgrading the energy performance of the building fabric, flood protection, anti-theft and anti-terrorist measures, and addition of ramps and chairlifts to improve access to and within existing buildings. The focus in this section is on upgrading thermal performance and upgrading of accessibility, usability and (thermal) comfort, together with some common measures to mitigate damage from flooding.

Upgrading thermal performance

The UK Government is committed to reducing carbon emissions by 100% by 2050 compared to 1990 levels. Approximately 20% of the building stock is more than 100 years old, the majority of which has solid walls, making the transition particularly challenging.

Photograph 11.7   Retrofitting and revitalising a domestic property.

Addressing the global climate change emergency has resulted in towns and cities implementing policies and housing retrofit plans to address this challenge and achieve net zero by 2050 (to meet the Paris Agreement). The building envelope is critical in ensuring that the thermal performance of buildings complies with modern standards, and energy retrofitting to existing buildings is required to achieve the decarbonisation of existing buildings.

Homes are the third highest emitter of carbon, and few of the existing 22 million homes in the UK operate close to the current energy standards expected and legislated for. Many have poor energy ratings. Typically EPC ratings of E, F or G. This means that a large proportion of the domestic building stock requires a thermal upgrade to bring them into line with low carbon targets for homes. This is also the case in the non-domestic sector, where the majority of existing buildings fail to meet current standards for thermal insulation. Housing is particularly challenging, given the wide range of dwelling types that exist, often requiring bespoke solutions for thermal retrofits. This is further compounded by the fact that people often have to move out of their homes for the retrofitting to be undertaken safely and efficiently. Approximately half a million homes are listed, and this presents challenges with upgrading the thermal fabric, especially windows and doors. Many conservation officers will not permit single-glazed windows to be replaced with double or triple-glazed units, or to allow doors to be replaced with an alternative that may change the appearance of the building. There is Catch-22 situation here, with legislation to protect architectural and cultural heritage impeding the desire to improve the energy efficiency of the building stock. Insulating solid walls is equally challenging, resulting in a change of external appearance or the loss of internal space.

Improving the thermal insulation of buildings requires a thorough understanding of the existing building fabric. Failure to appreciate that many interventions will change how the building 'breathes' and reacts to changes in temperature and use may lead to a deterioration of internal air quality, surface condensation and interstitial condensation, leading to deterioration of the fabric. Interventions must be considered in relation to the whole building and the detailing adjusted to suit the physical personality of the building. Sensitivity is required to ensure that the efforts to reduce energy consumption do not make the building's environmental credentials worse. We also need to take a long-term view, not a short-term 'quick fix' and consider the environmental impact of potential upgrades. Typical interventions that can be carried out to existing buildings may be to:

❏   Improve the thermal insulation of ground floors, walls and roofs.
❏   Replace single-glazed windows with double or triple-glazed units, or install secondary glazing.
❏   Replace existing doors with thermally insulated, air-tight, external door sets.
❏   Improve the air tightness of buildings, e.g. by sealing all service penetrations.
❏   Install heat recovery ventilation systems (once the airtightness has been addressed).
❏   Install solar collectors (solar thermal) and PV cells (electricity) on roofs.
❏   Replace gas boilers with heat pumps (once the thermal upgrades have been done).
❏   Replace lighting with low-energy, LED, fittings and motion sensors.

It may also be possible to reduce the thermal bridging in some buildings, although this can be technically challenging, highly disruptive and expensive, unless it is done as part of an extensive retrofitting exercise, with building users relocated during the work. The challenge for building owners, especially homeowners, is that the payback period on investing in retrofitting can be lengthy, therefore a more holistic vision is necessary. Retrofitting can, however, have a positive effect on user comfort and physical and mental wellbeing, which is not easy to allocate a cost to. There is, for example, intrinsic value in working in a thermally comfortable office compared to one that is not, often leading to better performance and happier employees.

Ground floors – retrofitting to improve thermal insulation values

Retrofitting of existing timber floors is usually carried out to improve thermal insulation and to reduce air leakage. This is practical in refurbishment and renovation projects where there are no residents in the building. It is, however, unlikely to be a realistic option for many householders due to the level of disruption and the cost involved in adding thermal insulation to the existing floor. In situations where there is a problem with the structural integrity of the floor, such as rotten floor joists, there is an opportunity to upgrade the thermal performance of the floor when undertaking the repair. As well as adding thermal insulation it is common to seal the floor, especially the junction between floor and wall, to prevent uncontrolled air filtration and hence help to further reduce heat loss.

It is not uncommon to replace suspended timber floors with solid concrete floors and rigid insulation, especially in situations where there may be problems with damp and/or difficulties with effectively ventilating the under-floor area. Replacing timber with a rein-forced concrete slab is often a quicker and more practical solution compared to replacing the timber floor. There may be, however, situations where cost or conservation concerns require replacement with a new timber floor and the addition of thermal insulation to the floor construction. The type of floor finish will affect actual insulation values and air permeability.

Many existing houses with solid concrete and stone floors do not have any thermal insulation within the floor assembly, other than the insulation provided by the earth. Improving the thermal insulation is challenging because adding insulation to an existing floor will result in raising the height of the floor in relation to existing walls, ceilings and door openings. This is often impractical. The alternative is to dig up and floor and dig down to a depth that will allow space for a new concrete floor, the thermal insulation and a floor finish. This can be challenging to do and usually, the work has to be conducted using hand tools because of physical space restrictions.

Walls

Solid masonry, predominantly built of loadbearing brick, stone, concrete block, or flint have proved to be a durable form of construction. In the majority of cases, the buildings tend to be expensive to heat due to poor thermal insulation values of the walls and air infiltration. The wall construction may be inconsistent in thickness and contain air gaps and voids due to the way it was constructed. Most stone loadbearing walls contain rubble between the external leaves. Thus, the term 'solid' wall may be a little misleading and a thorough survey is required to better understand the properties of the wall. Typical U values have been estimated and measured to be somewhere around 1.2 W/m² k for a 225 mm thick brick loadbearing wall and around 1.5 W/m² k for a 450 mm thick stone loadbearing wall. Some of the thicker masonry walls, typically built of local stone, can offer a significant level of thermal inertia, and these may not require any additional thermal insulation. Attention should also be given to the windows, doors and floors to ensure that the join between the component and the wall is airtight.

Walls will have existing hygrothermal properties. 'Hygrothermal' is a term used to describe the thermal and moisture transfer through the building fabric. Improving the thermal insulation values by adding insulation to the wall will reduce heat transfer, and in doing so this will affect moisture transfer within the wall. Applying thermal insulation to the wall will also reduce uncontrolled ventilation through the wall and at intersections with doors and windows. This will change the hygrothermal properties of the wall, and one unintended consequence may by the creation of interstitial and/or surface condensation. It is, therefore, important to consider the transfer of moisture within the proposed wall assembly so that appropriate materials can be used. Typically, a vapour barrier or a vapour control material will be required for internal and external insulation, although it is possible to have a 'vapour open' system (see below). Software modelling tools are available to assist designers in making the most appropriate interventions and advice should also be sought from manufacturers and suppliers.

In addition to considering the hygrothermal balance of the wall, it is necessary to reconsider such factors as the fire resistance of the wall assembly and the environmental impact of the insulation materials and associated materials, such as adhesives and fixings. Non-fossil fuel-derived insulation includes materials such as mineral wool, sheep's wool. Innovative materials such as aerogels are also available where space is at a premium. Thermal insulation can be added to the exterior face of the wall, the internal face of the wall, or to the air gap in cavity walls. Typical solid and cavity wall assemblies are illustrated in *Barry's Introduction to Construction of Buildings*. In every case, it is important to ensure that the wall is structurally sound and that there is no damp or water penetration through the existing fabric.

External insulation
External insulation can be applied to buildings that have little architectural merit, helping to improve the thermal insulation values and the visual appearance of the building. The work can be carried out with minimal disturbance to the building users and is often done as part of a larger renovation of the building fabric to include the retrofitting of windows and doors. The main challenges relate to buildings in conservation areas and listed buildings, where it is more likely that internal wall insulation will be more appropriate. It will be necessary to seek planning permission as this will change the appearance of the building. It is common to apply insulation to all walls, although it may not always be possible, and in some situations only some of the walls will be retrofitted with thermal insulation.

External insulation is usually achieved by applying insulation slabs to the wall, finished with an external render or cladding to protect the insulation and to provide a durable exterior wall finish. This will provide additional weather protection and additional acoustic insulation to the wall. It will also provide additional thermal mass to the walls. The depth of the insulation, ranging from approximately 150 to 300 mm, will result in windows being set a little further back in the wall. This can have benefits in terms of providing some shading from solar radiation in the summer. It may have some disadvantages in slightly reducing the amount of light passing through the window. Alternatively, if it is not possible to apply a thick insulating material, an alternative approach is to use a thermally insulating external render, sometimes in combination with internal insulation to achieve the required thermal performance. In both cases, the existing wall will need to be prepared prior to any work being undertaken. Adding insulation to the wall will also highlight any existing thermal bypassing at the window and door junctions. This is difficult to resolve in existing buildings without making more significant structural changes to the openings. Care will also be required at eaves level, as what was once a sufficient roof overhang may be compromised with the addition of external insulation, making this junction more vulnerable to water penetration.

Insulating materials that allow the wall to breathe include mineral wool, wood fibre, hemp, wool and cork. Supplied as panels, these are mechanically fixed to the face of the external wall, finished with two coats of lime or hemp lime render and a lime wash. The breathable solutions are best suited to stone walls. Rigid foam insulation finished with a cement render or cladding will create a non-breathable finish, and these systems are best applied to solid brick loadbearing walls. A variety of proprietary products are available that are designed to achieve the U-values set out in the Approved Documents.

Internal insulation

Internal insulation may be a more practical solution if the building has existing architectural and or aesthetic merit that would be lost by installing external wall insulation. There will be a small loss of internal floor area corresponding to the thickness of the insulation and the wall finish. The wall finish will need to be decorated and it will be necessary to adjust skirting boards, electrical socket outlets and radiators. It is possible to insulate some rooms and not others if finances are limited. For example, some retrofit schemes have concentrated on the ground floor rooms, which tend to house the living areas, and which benefit from being warmer than, for example, bedrooms on the upper floor(s). When insulating buildings that are listed it may be necessary to install the insulation in such a way that it can be removed at a future date without damaging the existing fabric of the interior wall. As with externally applied insulation, the challenge is to minimise the amount of thermal bypass at the window and door junctions. There are a number of approaches:

Insulated plasters and renders. These are applied by trowel or sprayed onto the existing wall surface to create a plumb surface finish and are suited to irregular wall surfaces. It will be necessary to remove non-porous finishes prior to application. Materials used include lime, hemp, cork perlite and cements to create a breathable wall finish. These products also help to improve fire resistance and acoustic insulation. A typical coat of insulating render would be 20–25 mm thick and usually two coats would be required to achieve the thermal performance required. A 3 mm finishing coat of mortar or lime putty provides a smooth internal finish, ready for decorating.

Sprayed polyurethane foams. These are typically used in conjunction with metal studwork, which is mechanically fixed to create a 50 mm gap between the wall and the studwork. The polyurethane foam is then sprayed onto the wall and behind the metal studwork, helping to mitigate thermal bypassing, and creating an airtight finish. Plasterboard is then fixed to the framework to form a plumb finish. Hybrid foams are also available that employ a similar principle.

Thermal insulation fixed to the existing wall using adhesive glue or by a mechanical method such as wall brackets or anchors so that it continuously adheres to the existing wall surface. Using a material such as wood fibre board finished with a lime plaster provides a vapour open wall that allows vapour movement through the wall. In this approach, there is no vapour check layer. It is important to make sure that there are no air pockets within the new assembly, thus the existing wall will need to be prepared to ensure a smooth and stable surface prior to the fixing of the insulation.

Thermal insulation boards adhered to the existing wall using plaster dabs, which forms a non-continuous fixing that contains air pockets between the insulation boards and the existing wall. The plaster dabs help to deal with the unevenness of existing walls but also create varying depths of air pockets. The boards need to be sealed at the junctions with floors and ceilings to prevent air and moisture movement through the wall, which could lead to condensation and mould growth within the void.

Thermal insulation fixed between a stud framework. A timber or metal frame is mechanically fixed to the wall to create a plumb surface, and fibrous insulation batts fixed between the framing. Plasterboard is then fixed to the framework to form an internal finish, with boards taped at joins to help prevent air infiltration. Surface finishes are then applied to the plasterboard. It is common to use a vapour barrier in this form of construction. It is also necessary to add a vertical DPC between timber battens and the existing wall to prevent the battens being in contact with a damp wall. This form of construction will create thermal bypasses through the framing material, which can be mitigated to a certain degree by adding a further insulating layer over the framework. This adds additional thickness to the wall assembly.

Thermal insulation fixed between and to a timber or metal frame that is not fixed to the existing wall. This may be required for listed building interiors where it is not permitted to 'damage' or permanently alter the existing wall finish. The framework of timber or metal studs is fixed to the floor and ceiling to create a continuous air gap of around 100 mm between the existing wall and the new wall; essentially an internal partition wall placed close to the existing wall finish.

Pre-fabricated systems aim to reduce the amount of work on the site, by using dry modular systems or by moving the cutting of materials to a factory environment. A laser scan will be used to inform the manufacture of the panels and cutting of boards to size. It will be necessary to prepare the wall by fixing a vapour-proof membrane to the wall. Modular systems are based on a metal framework, to which preformed insulated wall panels are fixed and bonded to one another using adhesives. Joints are sealed using a sealant to create a smooth finish ready for decorating. There is no need for a finishing coat of plaster. Other systems rely on cutting insulated boards in a factory, which are then delivered to site as a 'kit' and fixed in position by specialist contractors. This helps to minimise waste and increases the speed of installation, leading to minimal disruption to the use of the building. The boards will be finished with a finishing coat of plaster ready to be decorated.

Cavity wall insulation
Cavity walls can also be upgraded by adding thermal insulation to the exterior or interior face of the wall. This may be a practical approach where there are concerns about the integrity of the cavity. Alternatively, the cavity can be filled by injecting a thermal insulating material into the cavity. Specialist companies will drill holes in the outside walls, usually in the mortar course, then use specialist equipment to blow mineral wool, polystyrene beads, or polyurethane foam into the cavity. Once complete the holes are sealed with cement. Holes are typically 22 mm in diameter and positioned at 1 m intervals in the wall. The sealed drill holes will be visually prominent until the new cement weathers down to match the existing mortar course. It is estimated that over 75% of cavity walls in the UK have some form of insulation. Specialist surveys will be required to determine the effectiveness of the insulation and the existing U value of the wall. In situations where there is a need for additional levels of thermal insulation e.g. to the earlier insulated cavity walls that no longer meet current thermal standards, it will be necessary to install insulation on the inside or outside of the wall. This will create a complex 'sandwich' construction, which may influence the hygrothermal performance of the wall assembly. Specialist software modelling is available that aims to simulate moisture movement through the wall to allow the most appropriate solution to be chosen.

Windows

The thermal performance of single-glazed windows can be improved by draught-proofing the junction between the frame and the wall, which will help to reduce heat loss by reducing uncontrolled ventilation, and hence save energy. Proprietary products are available, usually comprising foam draft-proofing strips of material. Silicone sealant may also be used to help seal the gap between the frame and the wall. This is a relatively cost-effective way of helping to reduce loss of heat through uncontrolled ventilation. Secondary glazing can improve the thermal performance of windows (and also improve acoustic performance). Care needs to be taken to ensure the building is adequately ventilated to prevent unintended consequences of inadequate ventilation, discussed later. Specially developed products, such as ultra-thin double-glazing units, are available for historic conservation/upgrading projects. Typically, these offer a similar thermal performance to 6 mm wide units but are deemed to have less of an impact on the appearance of the window compared to many double-glazed window frames. Retrofitting of external shutters and awnings can help to reduce solar gain and hence reduce the likelihood of overheating on the hotter days.

Doors

The thermal performance of many external doors is considerably poorer than the walls in which they are installed. Traditional timber doors offer a limited amount of thermal insulation and the door sets are usually not particularly airtight. Draft-proofing can be an effective solution to reduce some heat loss, especially in situations where it is not possible or feasible to replace the door. Upgrading the door and frame to a composite door (a door that has a solid core of foam insulation) and an airtight seal will improve the thermal performance of existing buildings. It may also help to improve acoustic insulation. The new door and door set should also be manufactured in accordance with 'secure by design' to maximise security. Upgrading historic buildings can be challenging in a similar way to upgrading the thermal performance of windows as the new door and frame is likely to differ in appearance, and materials, to that it is replacing.

Roofs

Pitched roofs usually provide space for thermal insulation to be installed relatively easily, typically placed between and over the ceiling rafters. Rolls of mineral wool insulation are a relatively cost-effective upgrade. Flat roofs are more challenging, often necessitating the removal of the roof at the end of its service life and replacement with a thermally efficient alternative. See *Barry's Introduction to Construction of Buildings* for details of typical roof constructions.

Upgrading accessibility, usability and comfort

Alterations to facilitate disabled access are a major challenge for many building owners, especially some of the older and historically valuable buildings. Changes in floor level and various widths of access may contribute to the character of a building, but these features can, and often do, create barriers to access for people. Providing equal access for all often requires structural alterations and careful detailing, which must be done sensitively if the character of the building is not to be unduly affected. The most common retrofits relate to the adjustment of floor levels to create a floor without small changes in level and the addition of lifts and chairlifts. Small steps can be substituted for ramps and in some cases, a section of floor can be raised to create a flat floor finish. Installing lifts has structural implications and may well necessitate the creation of fire lobbies and safe refuge areas on each floor to comply with fire safety legislation. Equally, the implementation of (non-intrusive) fire detection and security equipment requires sensitivity to the building's visual character.

Upgrading the usability of interior space and the overall comfort of the building users is another concern. Buildings must be seen in the context of the society and the people who interact with them; thus, user feedback is crucial in formulating the design brief. Asking users how much control they wish to have over their internal environment can be instrumental in formulating design solutions. For example, the ability of users to have local control over light levels, heating and airflow may influence their perception and comfort of their internal space. These are important issues for the usability and comfort of building users as well as for the operation of the building.

Indoor air quality and condensation

With the drive to save energy has come the need to thermally insulate buildings to a high standard and to restrict the loss of heat caused by air leakage. The result is highly insulated airtight buildings. Unfortunately, in meeting one set of requirements, in this case improving thermal performance, it is also possible to unwittingly create problems, such as interstitial condensation in the fabric. This should have been considered and designed out at the design stage for new buildings, but the issues are not so straightforward when upgrading and retrofitting existing buildings, which may have been perfectly balanced for years.

Problems tend to relate to insufficient airflow within the building, which may cause problems with the health of the occupants and also result in condensation on, and within, the building fabric. A common problem is related to the replacement of windows and doors in existing buildings to improve the thermal performance of the building. The poorly fitting windows and doors would have been allowing air infiltration, which reduces the thermal performance of the building through unwanted airflow, but also allows excessive moisture to leave the building through air changes. In replacing the windows and doors with new airtight units, the flow of air into and out of the building is significantly reduced. Similarly, the relatively innocent act of blocking up or removing a chimney can have a dramatic effect on airflow. The result is that the internal climate may become stale through insufficient airflow and excessive moisture in the air is unable to escape the building, resulting in condensation on cold surfaces and within the fabric.

Indoor air quality

Sick building syndrome (SBS) is a term used to describe an unhealthy internal environment within a building. This may lead to allergic reactions, asthma and a general feeling of lethargy. Potential contaminants may be present in the building materials used and also in the fittings and furnishings introduced into the building after completion. Sealing a building to prevent air infiltration (and unnecessary heat loss) makes it necessary to introduce controlled airflow – either by natural or mechanical means – to create air changes and hence remove stale air from within the building. Air changes allow the removal of gases and moisture, particulates and other airborne contaminants, such as dust, mineral fibres and allergenic substances. This helps to prevent surface and interstitial condensation and contributes to a healthy internal environment.

Surface condensation

Surface condensation occurs when air becomes saturated (100% relative humidity), resulting in water droplets forming on cold impermeable surfaces, such as glass, ceramic tiles and metal. Left unchecked, this will lead to mould growth, the risk of corrosion, and damage to textiles and other materials. Improving ventilation when cooking, drying clothes and bathing – such as opening a window and/or switching on an extraction fan – can help to reduce the relative humidity of the air and hence reduce the risk of surface condensation.

Interstitial condensation

Interstitial condensation forms within the building fabric, for example, within a wall or a roof. As the water-laden air passes through the permeable fabric (e.g. plaster, blocks and bricks), it will move from warm air to cooler air. As the air cools, its capacity to hold moisture is reduced and 100% relative humidity is reached at the dew point. This is where condensation forms. Interstitial condensation can occur if the building fabric has not been designed correctly or constructed precisely. Over time the condensation will cause timber to rot and metal to corrode, resulting in structural damage. Unlike surface condensation, interstitial condensation cannot be seen without opening up the building fabric; thus, it is a hidden problem until such time as the damage becomes evident in some form of visible damage.

Retrofitting to mitigate damage from flooding

Many buildings are prone to flooding, positioned close to rivers that may overspill their banks in periods of prolonged wet weather or positioned in comparatively low-lying ground that may be prone to flash flooding. In locations where there is a history of flooding, it is prudent to retrofit buildings to try to minimise the possibility of flood water entering the property. If flood water does get into the building it is important to limit the damage caused to try to mitigate the disruption to residents and business premises. Flood prevention and mitigation measures for existing buildings range from relatively cheap interventions to more costly and disruptive retrofitting work. Some of the most common approaches include:

❑ Air brick covers. These can be fitted to prevent water from entering the void under suspended timber floors. They should be removed after the water has subsided. Water-resistant airbricks are also available to retrofit.
❑ Fit non-return valves to drains. This prevents flood water pushing foul water back up pipes into toilets and sinks.
❑ Fit external flood barriers to doors. These are boards that fit into a framework to protect door openings during a flood.
❑ Position electrical sockets at least 1.5 m above finished floor level. Cables should run in the ceiling/upper floor construction to protect them from the flood water, thus any electrical cables under the suspended ground floor need to be repositioned.
❑ Raise the external threshold of external doors if there is enough headroom to do so.
❑ Replace suspended timber floors with solid concrete and ceramic tile finish and raise the height of the floor if there is headroom to do so.
❑ Replace materials that are prone to water damage, such as MDF skirtings and architraves with varnished or painted timber, uPVC or ceramic tiles.

- ❏ Remove plaster to internal partition walls to expose the brick or stone, if desirable.
- ❏ Replace plaster with specialist plaster that is resilient to moisture and water damage.
- ❏ Fit quick-release internal doors. These have hinges that allow the door to be lifted off if there is a flood warning, to be moved somewhere dry to prevent water damage.
- ❏ 'Flood resistant' doors and windows are available to retrofit.
- ❏ Landscaping. Use permeable materials and design external areas to divert water away from the building.

11.6 Demolition, disassembly and recycling

There are a number of reasons why a building may need to be demolished or disassembled. Some buildings may simply have outlived their functional use, and it may not be economical to alter and upgrade them to suit current standards. Some buildings may have become derelict through a prolonged period of non-use and hence uneconomical to repair. Others may still be perfectly functional but need to be removed to make way for a new development. Buildings that have become structurally unsound through neglect or damage (e.g. by fire) may need to be demolished so that they do not pose a threat to the safety of those passing in proximity to the building. Local authorities have the power to issue a dangerous structures notice on the building owners, requiring immediate action. In all cases, the appropriate town planning office should be contacted to discuss the proposed demolition and then the appropriate consents applied for prior to any demolition work commencing.

Once a decision has been taken to demolish or disassemble a building, emphasis turns to the most economical and safe method of removing the structure. It is during these deliberations that aspects of material recovery and recycling are also addressed. Demolition or disassembly of any structure carries many risks and all demolition activities should be carried out in accordance with current legislation and guidance.

Planning

It is essential that a full structural and condition survey is undertaken so that a detailed method statement can be prepared and the appropriate demolition techniques determined. The survey may be supplemented with details of as-built drawings and structural calculations (if available). Special measures are required for the controlled removal of hazardous materials, such as asbestos and the segregation of materials to ensure maximum potential for recycling/reuse. Demolition operations must be carefully planned and each stage monitored so that the structure can be taken down without any risk to those working on the site and those in the local vicinity of the building.

Information that should be collected on a demolition survey includes:

- ❏ Existing services – live and unused services.
- ❏ Natural and man-made water courses.
- ❏ Presence of asbestos and other hazardous materials.
- ❏ Distribution of loads.
- ❏ Building structure, form and condition.
- ❏ Evidence of movement and weaknesses in the structure.

❑ Identification of hazards.
❑ Distribution and position of reinforcement – especially post-tensioned beams.
❑ Allowable loading of each floor (for demolition plant).
❑ Stability of the structure.
❑ Survey of adjacent and adjoining structures.
❑ Loads transferred through adjoining structures.
❑ Loads transferred from adjoining structures.
❑ Access to structure – allowable bearing strength of access routes.

Prior to demolition, the following tasks should be undertaken:

❑ Conduct a site survey.
❑ Contact neighbours and relevant authorities (local authority, police) to discuss the options.
❑ Identify any structural hazards and reduce or eliminate associated risks.
❑ Select an appropriate demolition or disassembly technique.
❑ Identify demolition phases and operations.
❑ Identify communication and supervision procedures.
❑ Organise logistics and identify safe working areas and exclusion zones.
❑ Erect hoardings, screen covers, nets and covered walkways to provide protection to the public.
❑ Identify need for temporary structures and controlled operations to avoid unplanned structural collapse.
❑ Select material handling method.
❑ Identify procedure for decommission services and plant.
❑ Identify recycling/reuse of materials and components and disposal methods and processes.
❑ Ensure health and safety processes are not compromised during the process.

After demolition, the following tasks should be undertaken:

❑ Conduct a survey to determine the extent of any damage to neighbouring properties, and agree on measures to repair any damage.
❑ Clean up all dust and debris from surrounding areas on a regular basis.

It is often necessary to provide restraint to walls that are to remain after buildings have been demolished. Photograph 11.8 shows flying shores bridging the gap created when a terraced house was demolished. When party walls need support as existing buildings are removed, lateral restraint can be provided in the form of flying trusses made out of tubular scaffolding (Photograph 11.9). Raking shores are often used to provide lateral restraint to walls adjacent to demolition works.

Demolition activities are usually conducted in stages. First is the 'soft strip' (or 'stripping out') process, in which the most valuable materials, components and equipment are removed. This can be a lengthy and labour-intensive process to allow the careful and safe extraction of items so that they are not unnecessarily damaged, thus maximising their

Photograph 11.8   Flying shores used to support existing structures during demolition.

Photograph 11.9   Flying trusses providing support to a party wall.

reuse potential and hence their value. The type of materials, components and equipment extracted at this stage will be heavily influenced by their commercial value. The soft strip is followed by the main demolition stage, which usually starts at the top (roof covering) and finishes with the main structural elements and foundations.

Demolition methods

A wide range of approaches may be taken to the demolition of a building, ranging from controlled explosions to reduce an entire structure (i.e. a high-rise block of flats) to a pile of debris within a few seconds, through to a sensitive and time-consuming piece-by-piece process of disassembly by hand. The methods chosen are determined through evaluating the risks and value inherent with each method in relation to the specific context of the building and its immediate environment. Timber and steel-framed buildings tend to lend themselves to dismantling. Many masonry buildings lend themselves to demolition by mechanical pushing by machine using a pusher arm and/or demolition using grapples and shears. Some of the more common methods for demolishing concrete elements include:

❏ *Ball and crane.* One of the oldest and most common methods for demolishing masonry and concrete buildings is a crane and a wrecking ball. The heavy ball is either dropped or swung into the structure, causing significant damage and gradual collapse of the building. Some additional work may be needed to cut reinforcing in concrete elements to facilitate demolition. Limitations with this method relate to the size of the building and the capacity of the crane and wrecking ball as well as safe working room. Constraints on working may relate to surrounding structures and overhead power lines. This method creates a significant amount of noise, dust and vibration, and there is always the risk of flying debris.

❏ *Bursting by chemical and mechanical pressure.* In situations where noise, vibration and dust need to be kept to a minimum it will be necessary to use bursting methods. Pressure is induced in the concrete by chemical reaction (insertion of expansive slurry) or mechanical means (application of hydraulic pressure). Holes are drilled into the concrete and force applied to the hole; the lateral forces build up over time, resulting in the concrete to split (crack). The smaller units are then removed by crane or by hand.

❏ *Cutting by thermal and water lance, drills and saws.* Thermal and water lances may be used to cut through steel and concrete. Diamond-tipped saws and drills may also be utilised.

❏ *Explosives.* Often used for removing large quantities of concrete. Explosives are inserted into a series of boreholes and remotely detonated. The explosive charges are timed in a sequence to ensure the building collapses (implodes) in the desired manner with the minimum of damage to surrounding buildings from debris. Surrounding buildings may need to be protected from damage by vibration and air blast pressure. Roads will need to be closed around the site and inhabitants removed from nearby buildings prior to the controlled explosion(s). After the explosion, there will be a need to clean up dust from surrounding areas and make good any damage.

❑ *Pneumatic and hydraulic breakers*. Machine-mounted hammers are used to break up concrete floor decks, bridges and foundations. The hammer size will be determined by the strength of the concrete and the amount of steel reinforcement contained within the structure. Telescopic arms (booms) and remote control allows access to otherwise difficult-to-reach areas. Disadvantages include noise, vibration and dust generation.

Recycling demolition waste

A high proportion of demolition waste can be recycled and/or reused. Material recovery from the demolition makes environmental and economic sense, and the amount of material being recovered and reused is steadily increasing due to environmental concerns and the cost of taking materials to licensed waste sites. Examples of materials that can be recovered and recycled include:

❑ Aggregates (sub-bases to roads and foundations).
❑ Concrete (including products extracted in their original form (e.g. blocks and slabs).
❑ Glass.
❑ Gypsum.
❑ Masonry (bricks and blocks).
❑ Metals (aluminium, copper, lead, steel, tin, zinc).
❑ Mineral waste (tarmacadam and road planings).
❑ Paper-based products.
❑ Paving slabs and flags.
❑ Plastics.
❑ Soil (top soil and excavation spoil).
❑ Stone and granite sets.
❑ Timber.

The design and construction of buildings should consider the whole life cycle of the building, which includes demolition (disassembly) and materials recovery. This requires clear decisions to be taken at the design and detailing phases about the materials to be used, the manner in which they are assembled and how they are fixed to neighbouring components. Method statements should clearly describe the assembly and disassembly strategy to aid future materials recovery and reuse without unnecessary damage to components.

11.7 Reuse and recycled materials

The careful dismantling (disassembly) of buildings provides an opportunity to use reclaimed components and materials in new construction projects. With a little thought, it is possible to divert materials and components from landfill to reuse and recycling. This can help to reduce the amount of new material extracted/used and also help to reduce the amount of material sent to landfill, thus helping to reduce the impact of construction activities on the environment. Materials and components can be reconditioned and reused (termed 'architectural salvage') or they can be recycled and incorporated into new building products. Photographs 11.10 and 11.11 show concrete and brick crushing and grading machines. The plant crushes and grades the concrete from roads, concrete blocks and bricks, so that it can be used as hardcore on the same site.

Photograph 11.10   Concrete crushing and grading plant.

Photograph 11.11   Plant crushes the brick and concrete for use as hardcore.

Salvaged materials

Materials recovery from redundant buildings has occurred throughout history, with materials being reclaimed and reused in a new structure. Stone and timber were reused in vernacular architecture, while more recently steel and concrete have been recovered and reused.

Architectural salvage, taking materials such as roof slates, bricks and internal fittings from redundant buildings for use on new projects, such as repair and conservation work, is a well-established business. The cost of the material might be higher than that for an equivalent new product, because of the cost of recovery, cleaning/reconditioning, transport and storage associated with the salvage operations. Reuse of materials and components in situ may be possible for some projects, which may help to reduce the cost and associated transportation. There is also a price premium for buying a scarce resource that will have a weathered quality that is difficult, if not impossible, to replicate with new products. However, the use of weathered materials may be instrumental in obtaining planning permission for some projects located in or adjacent to conservation areas, and so these materials can provide considerable value to building projects.

Quality of reclaimed materials is difficult to assess without visiting the salvage yard and making a thorough visual inspection of the materials for sale, and even then there is likely to be some waste of material on the site. For example, the reuse of roof slates will be dependent upon the integrity of the nail hole, and many slates will need additional work before they are suitable for reuse. In some cases, the slates will be unsuitable for reuse because of their poor quality. For work on refurbishment and conservation projects, the use of reclaimed materials is a desirable option. However, the increased cost premium for using weathered materials with a reduced service life may not be a realistic option for some projects. Another option is to use building products that have been made entirely from, or mostly from, recycled materials.

New products from recycled materials

A relatively recent development is for manufacturers to use materials recovered from redundant buildings, as well as household and industrial waste. Over recent years, there has been a steady increase in the number of manufacturers offering new materials and building products that are manufactured partly or wholly from recycled materials. These are known as recycled content building products (RCBPs), innovative products that offer greater choice to designers and builders keen to explore a more environmental friendly approach to construction. Many of these products are also capable of being recycled at a future date, thus further helping to reduce waste. A few examples are listed here:

❏ *Glass.* Recycled and used in the manufacture of some mineral thermal insulation products and as expanded glass granules in fibre-free thermal insulation.
❏ *Rubber car tyres.* Used in the manufacture of artificial stone and masonry products (see Photograph 1.2).
❏ *Salvaged paper.* Used in the manufacture of plasterboards and thermal insulation.
❏ *Plastics (including PET plastic drinks bottles).* Used for cable channels and sorted plastics recycled and used for foil materials and boards.

The issue of material choice and specification was discussed in Chapter 1, where the perception of risk associated with new products and techniques was discussed. The perception of risk associated with the use of new products is likely to be higher than that for the established and familiar products, which have a track record. Many of the recycled content products have different properties to the existing products they aim to replace. For example, inspection chamber covers and road kerbs made of recycled content plastics have different structural and thermal properties to the more familiar iron and concrete products, and this will need to be considered in the design and specification stage. The majority of products also being manufactured from recycled materials are produced by relatively new manufacturers to the market, offering products that may have little in the way of a track record in use. Thus, the perception of risk is likely to be high until the products have been used (by others) and are known to perform as expected over many years exposure to weather and use. This should not, however, stop designers, specifiers and builders from doing their own research and making informed decisions about how best to realise a more sustainable built environment.

Further reading

As noted in this chapter, work to historic buildings requires specialist advice and expertise. Additional guidance can be found at, for example, Historic England: www.historicengland .org.uk and Society for the Protection of Ancient Buildings (SPAB): www.spab.org.uk.

For advice on thermal retrofitting see Energy Saving Trust: www.energysavingtrust.org .uk and Low Carbon Homes: www.lowcarbonhomes.uk. See also the Green Alliance; www .green-alliance.org.uk

For guidance on demolition see *BS 6187:2000, Code of Practice for Demolition*

Several of the Approved Documents that are relevant to the issues discussed in this chapter, the main ones being:

Approved Document B: Fire safety
Approved Document C: Site preparation and resistance to contaminants and moisture
Approved Document E: Resistance to sound
Approved Document F: Ventilation
Approved Document L: Conservation of fuel and power
Approved Document M: Access to and use of buildings
Approved Document O: Overheating

Reflective exercises

You have been asked to retrofit a historic building located in a conservation area within a major city in England. The building was built in the Georgian period and has not been upgraded for many years. The building has solid walls, single-glazed windows, timber front and rear doors, 100 mm of insulation in the roof (which was installed 20 years ago), with an open fire in each room. There is no central heating.

❏ What measures are required to bring the building up to the current standards for thermal insulation as set out in the Approved Documents?
❏ How would you ensure that the interior has a comfortable internal environment without relying on burning fossil fuels in the open fireplaces?
❏ What are the potential challenges to discuss with the conservation officer?

Index

Barry's Advanced Construction of Buildings, Fifth Edition. Stephen Emmitt.
© 2023 John Wiley & Sons Ltd. Published 2023 by John Wiley & Sons Ltd.